ANALYTICAL ROUTES TO CHAOS IN NONLINEAR ENGINEERING

ANALYTICAL ROUTES TO CHAOS IN NONLINEAR ENGINEERING

Albert C. J. Luo

Southern Illinois University, USA

This edition first published 2014

Registered office
John Wiley & Sons Ltd, The Atrium, Southern Gate, Chichester, West Sussex, PO19 8SQ, United Kingdom

For details of our global editorial offices, for customer services and for information about how to apply for permission to reuse the copyright material in this book please see our website at www.wiley.com.

Library of Congress Cataloging-in-Publication Data

Luo, Albert C. J.
Analytical routes to chaos in nonlinear engineering / Albert C.J. Luo.
pages cm
Includes bibliographical references and index.
ISBN 978-1-118-88394-5 (cloth)
1. Systems engineering. 2. Chaotic behavior in systems. 3. Nonlinear systems. 4. Nonlinear control theory.
I. Title.
TA168.L86 2014
629.8′36 – dc23

2014001974

A catalogue record for this book is available from the British Library.

Typeset in 10/12pt TimesLTStd by Laserwords Private Limited, Chennai, India

1 2014

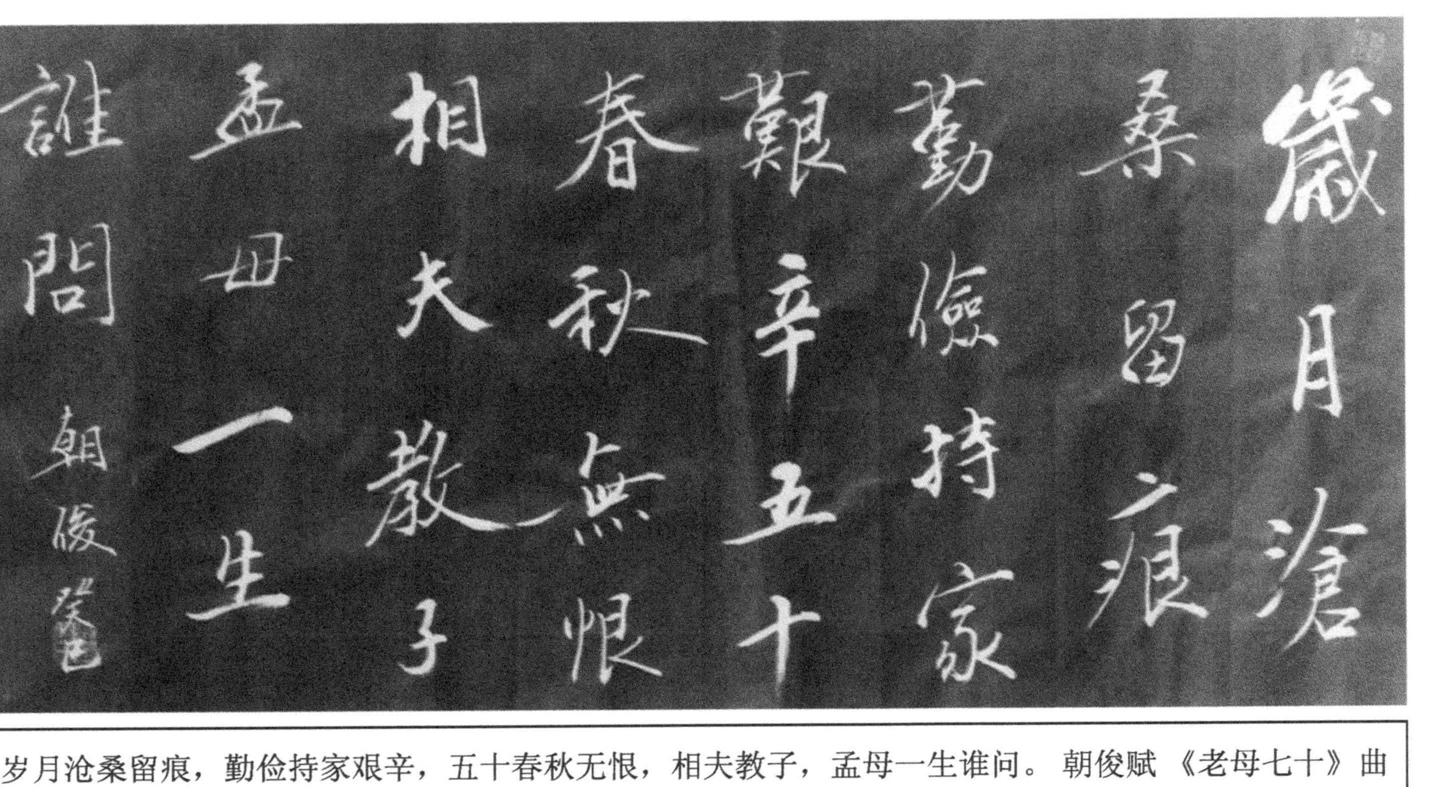

岁月沧桑留痕，勤俭持家艰辛，五十春秋无恨，相夫教子，孟母一生谁问。朝俊赋《老母七十》曲一首献给母亲七十生日。Dedicated to mother for her 70th birthday with a poem.

Contents

Preface

Periodic motions in nonlinear dynamical systems extensively exist in engineering and such periodic motions are paramount in engineering application. Since 1788, Lagrange used the method of averaging to investigate the gravitational three-body problem through a two-body problem with a perturbation. In the nineteenth century, Poincare developed the perturbation method to investigate the periodic motion of the three-body problem. In 1920, van der Pol used the averaging method to determine the periodic motions of self-excited systems in circuits. In 1945, Cartwright and Littlewood discussed the periodic motions of the van der Pol equation and proved the existence of periodic motions. In 1948, Levinson used a piecewise linear model to describe the van der Pol equation and determined the existence of periodic motions. In 1949, Levinson further developed the structures of periodic solutions in such a second order differential equation through the piecewise linear model, and discovered that infinite periodic solutions exist in such a piecewise linear model. On the other hand, in 1928, Fatou provided the first proof of asymptotic validity of the method of averaging through the existence of solutions of differential equations. In 1935, Krylov and Bogolyubov developed systematically the method of averaging. Thus, the perturbation method becomes a main analytical tool to investigate periodic motions of nonlinear oscillators in engineering. For example, in 1973 Nayfeh used the multiple-scale method to investigate Duffing oscillators for nonlinear structural dynamics. In the 1980s, since chaotic motions are observed in nonlinear vibrations, one tried to use the perturbation theory to describe chaotic motions. From the idea of the Lagrange standard form, the normal forms of nonlinear dynamical systems at equilibrium cannot be used for periodic motions and chaos in the original nonlinear dynamical systems. In 2012, the author systematically developed an analytical method to determine period-m flows in nonlinear dynamical systems. Thus this book will employ the analytical method to determine the analytical routes of periodic motions to chaos in nonlinear engineering.

This book presents analytical routes to chaos in a few typical engineering nonlinear dynamical systems through the recently developed analytical method. This book consists of five chapters. Chapter 1 gives a literature survey of analytical methods in nonlinear dynamical systems, including, the Lagrange standard form, the method of averaging, the Poincare perturbation method, and the generalized harmonic balance method. These analytical methods will be presented through theorems. In Chapter 2, the analytical bifurcation trees of period-m motion to chaos for the Duffing oscillator will be presented since the Duffing oscillator is extensively applied in structural dynamics. In Chapter 3, the period-m motion in the periodically forced, van der Pol oscillator will be presented analytically, and the analytical bifurcation trees of periodic motions to chaos in the van del Pol-Duffing oscillator will be discussed.

In Chapter 4, the analytical solutions of period-m motions in parametric nonlinear oscillators will be presented through both a parametric, quadratic nonlinear oscillator and a parametric Duffing oscillator as two sampled problems. In Chapter 5, the bifurcation tree of periodic motions to chaos in a nonlinear Jeffcott rotor dynamical system will be presented, and the periodic motions to quasi-periodic motion will be discussed. All materials presented in this book will help one better understand nonlinear phenomena in nonlinear engineering.

Finally, I would like to appreciate my students (Jianzhe Huang, Arash Bagaei Laken, Bo Yu, and Dennis O'Connor) for applying the recently developed analytical method to nonlinear engineering systems and completing numerical computations. Herein, I would like to thank my wife (Sherry X. Huang) and my children (Yanyi Luo, Robin Ruo-Bing Luo, and Robert Zong-Yuan Luo) again for tolerance, patience, understanding, and continuous support.

Albert C.J. Luo
Edwardsville, Illinois, USA

1

Introduction

In this chapter, analytical methods for approximate solutions of periodic motions to chaos in nonlinear dynamical systems will be presented briefly. The Lagrange stand form, perturbation method, method of averaging, harmonic balance, generalized harmonic balance will be discussed. A brief literature survey will be completed to present a main development skeleton of analytical methods for periodic motions in nonlinear dynamical systems. The weakness of current approximate, analytical methods will also be discussed in this chapter, and the significance of analytical methods in nonlinear engineering will be presented.

1.1 Analytical Methods

Since the appearance of Newton's mechanics, one has been interested in periodic motion. From the Fourier series theory, any periodic function can be expressed by a Fourier series expansion with different harmonics. The periodic motion in dynamical systems is a closed curve in state space in the prescribed period. In addition to simple oscillations in mechanical systems, one has been interested in motions of moon, earth, and sun in the three-body problem. The earliest approximation method is the method of averaging, and the idea of averaging originates from Lagrange (1788).

1.1.1 Lagrange Standard Form

Consider an initial value problem for $\mathbf{x} \in D \subset \mathbf{R}^n$ and $t \geq 0$,

$$\dot{\mathbf{x}} = \mathbf{A}(t)\mathbf{x} + \varepsilon\mathbf{f}(\mathbf{x}, t), \mathbf{x}(0) = \mathbf{x}_0 \tag{1.1}$$

where $\mathbf{A}(t)$ is an $n \times n$ matrix and continuous with time t. $\mathbf{f}(\mathbf{x}, t)$ is a C^r – continuous vector function of t and $\mathbf{x}$. The unperturbed system is linear ($\varepsilon = 0$) and such a linear system has n independent basic solution to form a fundamental matrix $\mathbf{\Phi}(t)$. That is,

$$\dot{\mathbf{x}}^{(0)} = \mathbf{A}(t)\mathbf{x}^{(0)} \Rightarrow \mathbf{x}^{(0)} = \mathbf{\Phi}(t)\mathbf{c} \tag{1.2}$$

Analytical Routes to Chaos in Nonlinear Engineering, First Edition. Albert C. J. Luo.

where $\mathbf{c}$ is constant, determined by initial conditions. As in Luo (2012a,b), a linear transformation is introduced as

$$\mathbf{x} = \mathbf{\Phi}(t)\mathbf{y}. \tag{1.3}$$

Substitution of Equation (1.3) into Equation (1.1) gives

$$\dot{\mathbf{\Phi}}(t)\mathbf{y} + \mathbf{\Phi}(t)\dot{\mathbf{y}} = \mathbf{A}(t)\mathbf{\Phi}(t)\mathbf{y} + \varepsilon\mathbf{f}(\mathbf{\Phi}(t)\mathbf{y}, t). \tag{1.4}$$

With $\dot{\mathbf{\Phi}}(t) = \mathbf{A}(t)\mathbf{\Phi}(t)$, we obtain

$$\begin{aligned}
\dot{\mathbf{y}} &= \varepsilon\mathbf{\Phi}^{-1}(t)\mathbf{f}(\mathbf{\Phi}(t)\mathbf{y}, t) \equiv \mathbf{g}(\mathbf{y}, t),\\
\mathbf{y}_0 &= \mathbf{\Phi}^{-1}(0)\mathbf{x}_0
\end{aligned} \tag{1.5}$$

The foregoing form is called the Lagrange standard form.

Consider a vibration problem as

$$\ddot{x} + \omega^2 x = \varepsilon f(x, \dot{x}, t). \tag{1.6}$$

From the basic solution of the unperturbed system, we have a transformation as

$$\begin{Bmatrix} x \\ \dot{x} \end{Bmatrix} = \begin{bmatrix} \cos\omega t & \frac{1}{\omega}\sin\omega t \\ -\omega\sin\omega t & \cos\omega t \end{bmatrix} \begin{Bmatrix} y_1 \\ y_2 \end{Bmatrix}. \tag{1.7}$$

Using this transformation, Equation (1.6) becomes

$$\begin{aligned}
\dot{y}_1 &= \varepsilon g_1(y_1, y_2, t),\\
\dot{y}_2 &= \varepsilon g_2(y_1, y_2, t).
\end{aligned} \tag{1.8}$$

where

$$\begin{aligned}
g_1(y_1, y_2, t) &= -\frac{1}{\omega} f(x, \dot{x}, t)\sin\omega t,\\
g_2(y_1, y_2, t) &= f(x, \dot{x}, t)\cos\omega t.
\end{aligned} \tag{1.9}$$

If the function $g_1(y_1, y_2, t)$ and $g_2(y_1, y_2, t)$ is T-periodic with $T = 2\pi/\omega$,

$$\begin{aligned}
\dot{y}_1 &= \varepsilon g_1^0(y_1, y_2),\\
\dot{y}_2 &= \varepsilon g_1^0(y_1, y_2).
\end{aligned} \tag{1.10}$$

where

$$\begin{aligned}
g_1^0(y_1, y_2) &= \frac{1}{T}\int_0^T g_1(y_1, y_2, t)dt = -\frac{1}{T}\int_0^T \frac{1}{\omega} f(x, \dot{x}, t)\sin(\omega t)dt,\\
g_2^0(y_1, y_2) &= \frac{1}{T}\int_0^T g_2(y_1, y_2, t)dt = \frac{1}{T}\int_0^T f(x, \dot{x}, t)\cos(\omega t)dt.
\end{aligned} \tag{1.11}$$

1.1.2 Perturbation Methods

In the end of the nineteenth century, Poincare (1890) provided the qualitative analysis of dynamical systems to determine periodic solutions and stability, and developed the perturbation theory for periodic solution. In addition, Poincare (1899) discovered that the motion of a nonlinear coupled oscillator is sensitive to the initial condition, and qualitatively stated that the motion in the vicinity of unstable fixed points of nonlinear oscillation systems may be stochastic under regular applied forces. In the twentieth century, one followed Poincare's ideas to develop and apply the qualitative theory to investigate the complexity of motions in dynamical systems. With Poincare's influence, Birkhoff (1913) continued Poincare's work, and a proof of Poincare's geometric theorem was given. Birkhoff (1927) showed that both stable and unstable fixed points of nonlinear oscillation systems with two degrees of freedom must exist whenever their frequency ratio (also called resonance) is rational. The sub-resonances in periodic motions of such systems change the topological structures of phase trajectories, and the island chains are obtained when the dynamical systems are renormalized with fine scales. In such qualitative and quantitative analysis, the Taylor series expansion and the perturbation analysis play an important role. However, the Taylor series expansion analysis is valid in the small finite domain under certain convergent conditions, and the perturbation analysis based on the small parameters, as an approximate estimate, is only acceptable for a very small domain with a short time period. From Verhulst (1991), the perturbation solution of dynamical system can be stated as follows.

Theorem 1.1 *Consider a dynamical system*

$$\dot{\mathbf{x}} = \mathbf{f}(\mathbf{x}, t, \varepsilon), \mathbf{x}(t_0) = \mathbf{a} \tag{1.12}$$

with $\mathbf{x} \in D \subset \mathscr{R}^n$, $|t - t_0| < h$, *and* $0 \le \varepsilon \le \varepsilon_0 < 1$. $\mathbf{f}(\mathbf{x}, t, \varepsilon)$ *is a* C^r *– continuous vector function of* t, $\mathbf{x}$, *and* ε. *Assume* $\mathbf{f}(\mathbf{x}, t, \varepsilon)$ *can be expanded in a Taylor series with respect to* ε *as*

$$\mathbf{f}(\mathbf{x}, t, \varepsilon) = \sum_{k=0}^{m} \varepsilon^k \mathbf{f}_k(\mathbf{x}, t) + \varepsilon^{m+1}\mathbf{R}(\mathbf{x}, t, \varepsilon) \tag{1.13}$$

with $|t - t_0| \le h$ *and* $0 \le \varepsilon \le \varepsilon_0$. $\mathbf{f}_k(\mathbf{x}, t)$ $(k = 0, 1, 2, \ldots, m)$ *is continuous in* t *and* $\mathbf{x}$ *with* $(m + 1 - k)$ *times continuously differentiable with* $\mathbf{x}$, *and* $\mathbf{R}(\mathbf{x}, t, \varepsilon)$ *is continuous in* t, $\mathbf{x}$, *and* ε, *and satisfies Lipschitz – continuous in* $\mathbf{x}$. *Suppose there is a* ε*-series of* $\mathbf{x}$ *as*

$$\mathbf{x}(t) = \sum_{k=0}^{m} \varepsilon^k \mathbf{x}_k(t). \tag{1.14}$$

Application of Equation (1.14) to Equation (1.12), using the Taylor series expansion of $\mathbf{f}_k(\mathbf{x}, t)$ *with respect to power of* ε, *and equating coefficients with the initial condition*

$$\mathbf{x}_0(t_0) = \mathbf{a} \text{ and } \mathbf{x}_k(t_0) = \mathbf{0}\ (k = 1, 2, \ldots, m) \tag{1.15}$$

generates an approximate solution of $\mathbf{x}(t)$ *with*

$$\left\| \mathbf{x}(t) - \sum_{k=0}^{m} \varepsilon^k \mathbf{x}_k(t) \right\| = O(\varepsilon^{m+1}) \tag{1.16}$$

on the time-scale 1.

Proof. The proof can be referred to Verhulst (1991). ■

Assume that $\mathbf{f}(\mathbf{x}, t, \varepsilon)$ in Equation (1.12) can be expanded in a convergent Taylor series with respect to ε and $\mathbf{x}$ in a finite domain. Consider an unperturbed system in Equation (1.12) as

$$\dot{\mathbf{x}}_0 = \mathbf{f}(\mathbf{x}_0, t, 0), \mathbf{x}(t_0) = \mathbf{b} \tag{1.17}$$

Using a transform

$$\mathbf{x}(t) = y(t) + \mathbf{x}_0(t) \tag{1.18}$$

Equation (1.12) becomes

$$\dot{\mathbf{y}} = \mathbf{F}(\mathbf{y}, t, \varepsilon), \mathbf{x}(t_0) = \mathbf{c} \tag{1.19}$$

where $\mathbf{c} = \mathbf{a} - \mathbf{b}$ and

$$\mathbf{F}(\mathbf{y}, t, \varepsilon) = \mathbf{f}(\mathbf{y} + \mathbf{x}_0, t, \varepsilon) - \mathbf{f}(\mathbf{x}_0, t, 0) \tag{1.20}$$

Thus, the Poincare perturbation theory for nonlinear dynamical systems can also be stated as follows:

Theorem 1.2 *(Poincare) Consider a dynamical system*

$$\dot{\mathbf{y}} = \mathbf{F}(\mathbf{y}, t, \varepsilon), \mathbf{y}(t_0) = \mathbf{c} \tag{1.21}$$

with $\mathbf{y} \in D \subset \mathscr{R}^n$, $|t - t_0| < h$, *and* $0 \leq \varepsilon < 1$. $\mathbf{F}(\mathbf{y}, t, \varepsilon)$ *is a* C^r *– continuous vector function of* t, $\mathbf{y}$, *and* ε. *If such a vector function can be expanded in a convergent power series with respect to* $\mathbf{y}$ *and* ε *for* $\|\mathbf{y}\| < \rho$ *and* $0 \leq \varepsilon < 1$, *then* $\mathbf{y}(t)$ *can be expanded in a convergent power series with respect to* $\mathbf{c}$ *and* ε *in the vicinity of* $\mathbf{c} = \mathbf{0}$ *and* $\varepsilon = 0$ *on time scale 1.*

Proof. The proof can be referred to Verhulst (1991). ■

In the perturbation theory, the Poincare-Lindstedt method is discussed herein. Consider a vibration problem as

$$\ddot{x} + x = \varepsilon f(x, \dot{x}, \varepsilon), \ (x, \dot{x}) \in D \subset \mathscr{R}^2. \tag{1.22}$$

For $\varepsilon = 0$, with initial condition $(x, \dot{x})|_{t=0} = (a, 0)$

$$x = a \cos t. \tag{1.23}$$

For variation of a foregoing solution with ε, the following transformation is introduced as

$$\omega t = \theta, \ \omega^{-2} = 1 - \varepsilon\eta(\varepsilon). \tag{1.24}$$

Application of Equation (1.24) to Equation (1.22) gives

$$\begin{aligned} x'' + x &= \varepsilon[\eta x + (1 - \varepsilon\pi)]f(x, (1 - \varepsilon\eta)^{-\frac{1}{2}}x', \varepsilon) \\ &\equiv \varepsilon g(x, x', \varepsilon, \eta) \end{aligned} \tag{1.25}$$

with initial conditions

$$x(0) = a(\varepsilon), x'(0) = 0. \tag{1.26}$$

From the solution of $\varepsilon = 0$, by the variation of constant, Equation (1.25) gives

$$x(\theta) = a \cos\theta + \varepsilon \int_0^\theta \sin(\theta - \tau) g(x(\tau), x'(\tau), \varepsilon, \eta) d\tau \tag{1.27}$$

From the periodicity, $x(\theta) = x(\theta + 2\pi)$ in the foregoing equation yields

$$\int_{\theta}^{\theta+2\pi} \sin(\theta - \tau) g(x(\tau), x'(\tau), \varepsilon, \eta) d\tau = 0. \tag{1.28}$$

Thus,

$$\begin{aligned} &\int_{0}^{2\pi} \sin \tau g(x(\tau), x'(\tau), \varepsilon, \eta) d\tau = 0, \\ &\int_{0}^{2\pi} \cos \tau g(x(\tau), x'(\tau), \varepsilon, \eta) d\tau = 0; \end{aligned} \tag{1.29}$$

from which we obtain

$$F_1(a, \eta) = 0 \text{ and } F_2(a, \eta) = 0. \tag{1.30}$$

If the following equation exists

$$\left| \frac{\partial (F_1, F_2)}{\partial (a, \eta)} \right| \neq 0, \tag{1.31}$$

then

$$a(\varepsilon) = \sum_{k=0}^{\infty} \varepsilon^k a_k, \text{ and } \eta(\varepsilon) = \sum_{k=0}^{\infty} \varepsilon^k \eta_k \tag{1.32}$$

and the solution of Equation (1.25) is

$$x(\theta) = a(0) \cos \theta + \sum_{k=0}^{\infty} \varepsilon^k \gamma_k(\theta) \tag{1.33}$$

In the foregoing procedure, the nonlinear solution is based on the variation of linear solution, which may not be adequate. This method is the foundation of multiple-scale method. Introduce

$$\omega = 1 + \sum_{k=0}^{\infty} \varepsilon^k \omega_k. \tag{1.34}$$

The following quantities are assumed as

$$\theta_N = \left(1 + \sum_{k=1}^{N} \varepsilon^k \omega_k \right) t \text{ and } x_M = a_0 \cos \theta + \sum_{k=1}^{M} \varepsilon^k \gamma_k(\theta). \tag{1.35}$$

Such a procedure makes the problem more complicated.

1.1.3 *Method of Averaging*

Based on the Lagrange standard form, one developed the method of averaging. van der Pol (1920) used the averaging method to determine the periodic motions of self-excited systems in circuits, and the presence of natural entrainment frequencies in such a system was observed in van der Pol and van der Mark (1927). Cartwright and Littlewood (1945) discussed the periodic motions of the van der Pol equation and proved the existence of periodic motions. Levinson (1948) used a piecewise linear model to describe the van der Pol equation and determined the existence of periodic motions. Levinson (1949) further developed the structures of periodic solutions in such a second order differential equation through the piecewise linear model, and discovered that infinite periodic solutions exist in such a piecewise linear model.

Since the nonlinear phenomena was observed in engineering, Duffing (1918) used the hardening spring model to investigate the vibration of electro-magnetized vibrating beam, and after that, the Duffing oscillator has been extensively used in structural dynamics. In addition to determining the existence of periodic motions in nonlinear different equations of the second order in mathematics, one has applied the Poincare perturbation methods for periodic motions in nonlinear dynamical systems. Fatou (1928) provided the first proof of asymptotic validity of the method of averaging through the existence of solutions of differential equations. Krylov and Bogolyubov (1935) developed systematically the method of averaging and the detailed discussion can be found in Bogoliubov and Mitropolsky (1961). The method of averaging is presented as follows:

Theorem 1.3 *Consider a dynamical system*

$$\dot{\mathbf{x}} = \varepsilon \mathbf{f}(\mathbf{x}, t) + \varepsilon^2 \mathbf{g}(\mathbf{x}, t, \varepsilon), \mathbf{x}(t_0) = \mathbf{x}_0 \tag{1.36}$$

If the following conditions are satisfied, that is,

i. *the vector functions* $\mathbf{f}, \mathbf{g}$ *and* $\partial \mathbf{f}/\partial \mathbf{x}$ *are defined, continuous and bounded;*
ii. $\mathbf{f}(\mathbf{x}, t)$ *is T-periodic with*

$$\mathbf{f}^0(\mathbf{x}, t) = \frac{1}{T}\int_0^T \mathbf{f}(\mathbf{x}, t)dt \tag{1.37}$$

where T *is constant independent of* ε*, and*

$$\dot{\mathbf{y}} = \varepsilon \mathbf{f}^0(\mathbf{y}), \mathbf{y}(t_0) = \mathbf{y}_0 \tag{1.38}$$

with $\mathbf{x}, \mathbf{y} \in D \subset \mathscr{R}^n$, $|t - t_0| < h$, *and* $0 \le \varepsilon \le \varepsilon_0 < 1$;
iii. $\mathbf{g}$ *is Lipschitz-continuous in* $\mathbf{x}$ *for* $\mathbf{x} \in D$, *and* $\mathbf{y}(t)$ *is in the subset of* D; *then*

$$\mathbf{x}(t) - \mathbf{y}(t) = O(\varepsilon) \tag{1.39}$$

on the time scale $1/\varepsilon$.

Proof. The proof can be referred to Verhulst (1991). ■

The classic perturbation methods for nonlinear oscillators were presented (e.g., Stoker, 1950; Minorsky, 1962; Hayashi, 1964). Hayashi (1964) used the method of averaging and the harmonic balance method (HBM) to discuss the approximate periodic solutions of nonlinear systems and the corresponding stability. Nayfeh (1973) employed the multiple-scale perturbation method to develop approximate solutions of periodic motions in the Duffing oscillators. Holmes and Rand (1976) discussed the stability and bifurcation of periodic motions in the Duffing oscillator. Nayfeh and Mook (1979) applied the perturbation analysis to nonlinear structural vibrations via the Duffing oscillators, and Holmes (1979) demonstrated chaotic motions in nonlinear oscillators through the Duffing oscillator with a twin-well potential. Ueda (1980) numerically simulated chaos via period-doubling of periodic motions of Duffing oscillators. Thus, one continues using the perturbation analysis to determine the approximate analytical solution of periodic motions. Coppola and Rand (1990) determined limit cycles of nonlinear oscillators through elliptic functions in the averaging method. Wang *et al.* (1992) used the harmonic balance method and the Floquet theory to investigate the nonlinear behaviors of the Duffing oscillator with a bounded potential well (also see, Kao, Wang, and Yang, 1992). Luo and Han (1997) determined the stability and bifurcation conditions of periodic motions of the Duffing oscillator. However, only symmetric periodic motions of the Duffing oscillators were investigated. Luo and Han (1999) investigated the analytical prediction of chaos in nonlinear rods through the Duffing oscillator. Peng *et al.* (2008) presented the approximate symmetric solution of period-1 motions in the Duffing oscillator by the harmonic balance method with three harmonic terms. In addition, Buonomo (1998a,b) showed the procedure for periodic solutions of van der Pol oscillator in power series. Kovacic and Mickens (2012) applied the generalized Krylov-Bogoliubov method to the van der Pol oscillator with small nonlinearity for limit cycles.

For parametric oscillators, the Mathieu equation should be mentioned herein. Mathieu (1868) investigated the linear Mathieu equation (also see, Mathieu, 1873; McLachlan, 1947). Whittaker (1913) presented a method to find the unstable solutions for very weak excitation (also see, Whittaker and Watson, 1935). In engineering, Sevin (1961) used the Mathieu equation to investigate the vibration-absorber with parametric excitation. Hsu (1963) discussed the first approximation analysis and stability criteria for a multiple-degree of freedom dynamical system (also see, Hsu, 1965). Tso and Caughey (1965) discussed the stability of parametric, nonlinear systems. Mond *et al.* (1993) presented the stability analysis of nonlinear Mathieu equation. Zounes and Rand (2000) discussed the transient response for the quasi-periodic Mathieu equation. Luo (2004) discussed chaotic motions in the resonant separatrix bands of the Mathieu-Duffing oscillator with a twin-well potential. Shen *et al.* (2008) used the incremental harmonic balance method to investigate the bifurcation and route to chaos in the Mathieu-Duffing oscillator.

The rotor dynamics is about the vibration of rotating shaft with disks. The shaft is supported by bearings with seals. In industrial application, flexible rotors are extensively used, which is relatively long. In 1883, Gustav Delaval manufactured a gas turbine which can operate over the first critical rotation speed. The high performance machines always operate over the first critical speed. Jeffcott (1919) first developed equations of motion for linear rotor dynamics. For such a linear rotor system, it can be easily analyzed. However, the results may not be adequate for flexible rotors with high operation speed. Thus one considered the bearing clearance, squeezing film dampers, seals, and fluid dynamics effects in the flexible rotor systems. Begg (1974) investigated the stability of a friction-induced rotor whirl motion. Childs (1982) applied a perturbation method to investigate subharmonic responses of a rotor with a small nonlinearity. Choi and Noah (1987) used the harmonic balance method and fast Fourier transformation (FFT) to study the subharmonic and superharmonic responses in a rotor with a bearing clearance. Day (1987) used multiple-scale method to show the aperiodic motion. Ehrich (1988) numerically investigated higher order subharmonic responses in such a rotor system under a high operation speed. Kim and Noah (1990) used the harmonic balance method to discuss the bifurcation of periodic motions in a modified Jeffcott rotor with bearing clearings. Choi and Noah (1994) still used the harmonic balance method to investigate mode-locking motion and chaos in such a Jeffcott rotor. The quasi-periodic motions and stability for such a modified Jeffcott rotor was also presented through the harmonic balance method in Kim and Noah (1996). Chu and Zhang (1998) used the harmonic balance method to determine periodic motions and numerically show the bifurcation scenarios. In fact, the modified Jeffcott rotor is discontinuous. Thus, the harmonic balance method may not be an adequate method for periodic motions in such a modified rotor with discontinuity, which can be as a rough prediction. Jiang and Ulbrich (2001) investigated stability of sliding whirl in a nonlinear Jeffcott rotor.

1.1.4 Generalized Harmonic Balance

As mentioned in previous sections, those analytical methods are based on solutions of linear systems to solve the nonlinear dynamical systems. Variation of constants in the solution of a so-called related linear system enforces the original nonlinear system to satisfy under the perturbation expansion with a small parameter. One always thinks the periodic motion as like a circle with a harmonic term. Such a complicated procedure cannot give satisfactory solutions. In such a mathematical treatment, the original vector fields are changed through the perturbation expansion. Thus, the approximate solutions cannot represent the original dynamical systems for a long time period.

To determine periodic solutions in nonlinear systems, we should find a basis of periodic functions to represent the periodic solution in nonlinear dynamical systems instead of perturbation expansion. Luo (2012a) developed a generalized harmonic balance method to get the approximate analytical solutions of periodic motions and chaos in nonlinear dynamical systems. This method used the finite term Fourier series to express periodic motions and the coefficients are time-varying. With the principle of virtual work, a dynamical system of coefficients are obtained from which the steady-state solution are achieved and the corresponding stability and bifurcation are completed. Two theorems will be presented herein, which will be used in other chapters. The detailed description of such a theory can be referred to Luo (2012a, 2013, 2014). Without excitation, the corresponding theorem of a nonlinear vibration system is stated as follows.

Theorem 1.4 *Consider a nonlinear vibration system as*

$$\ddot{\mathbf{x}} = \mathbf{f}(\mathbf{x}, \dot{\mathbf{x}}, \mathbf{p}) \in \mathscr{R}^n \tag{1.40}$$

where $\mathbf{f}(\mathbf{x}, \dot{\mathbf{x}}, \mathbf{p})$ *is a* C^r *– continuous nonlinear function vector* ($r \geq 1$). *If such a dynamical system has a period-*m *motion* $\mathbf{x}^{(m)}(t)$ *with finite norm* $\|\mathbf{x}^{(m)}\|$ *and period* $T = 2\pi/\Omega$, *there is a generalized coordinate transformation with* $\theta = \Omega t$ *for the periodic motion of Equation (1.40) in the form of*

$$\mathbf{x}^{(m)}(t) = \mathbf{a}_0^{(m)}(t) + \sum_{k=1}^{\infty} \mathbf{b}_{k/m}(t) \cos\left(\frac{k}{m}\theta\right) + \mathbf{c}_{k/m}(t) \sin\left(\frac{k}{m}\theta\right) \tag{1.41}$$

with

$$\begin{aligned} \mathbf{a}_0^{(m)} &= (a_{01}^{(m)}, a_{02}^{(m)}, \ldots, a_{0n}^{(m)})^{\mathrm{T}}, \\ \mathbf{b}_{k/m} &= (b_{k/m1}, b_{k/m2}, \ldots, b_{k/mn})^{\mathrm{T}}, \\ \mathbf{c}_{k/m} &= (c_{k/m1}, c_{k/m2}, \ldots, c_{k/mn})^{\mathrm{T}} \end{aligned} \tag{1.42}$$

and

$$\begin{aligned} &\|\mathbf{x}^{(m)}\| = \|\mathbf{a}_0^{(m)}\| + \sum_{k=1}^{\infty} \|\mathbf{A}_{k/m}\|, \text{ and } \lim_{k\to\infty} \|\mathbf{A}_{k/m}\| = 0 \text{ but not uniform} \\ &\text{with } \mathbf{A}_{k/m} = (A_{k/m1}, A_{k/m2} \ldots, A_{k/mn})^{\mathrm{T}} \\ &\text{and } A_{k/mj} = \sqrt{b_{k/mj}^2 + c_{k/mj}^2}\ (j = 1, 2, \ldots, n). \end{aligned} \tag{1.43}$$

For $\|\mathbf{x}^{(m)}(t) - \mathbf{x}^{(m)*}(t)\| < \varepsilon$ *with a prescribed small* $\varepsilon > 0$, *the infinite term transformation* $\mathbf{x}^{(m)}(t)$ *of period-*m *motion of Equation (1.40), given by Equation (1.41), can be approximated by a finite term transformation* $\mathbf{x}^{(m)*}(t)$ *as*

$$\mathbf{x}^{(m)*}(t) = \mathbf{a}_0^{(m)}(t) + \sum_{k=1}^{N} \mathbf{b}_{k/m}(t)\cos\left(\frac{k}{m}\theta\right) + \mathbf{c}_{k/m}(t)\sin\left(\frac{k}{m}\theta\right) \tag{1.44}$$

and the generalized coordinates are determined by

$$\begin{aligned}
\ddot{\mathbf{a}}_0^{(m)} &= \mathbf{F}_0^{(m)}(\mathbf{a}_0^{(m)}, \mathbf{b}^{(m)}, \mathbf{c}^{(m)}, \dot{\mathbf{a}}_0^{(m)}, \dot{\mathbf{b}}^{(m)}, \dot{\mathbf{c}}^{(m)}),\\
\ddot{\mathbf{b}}^{(m)} &= -2\frac{\Omega}{m}\mathbf{k}_1\dot{\mathbf{c}}^{(m)} + \frac{\Omega^2}{m^2}\mathbf{k}_2\mathbf{b}^{(m)} + \mathbf{F}_1^{(m)}(\mathbf{a}_0^{(m)}, \mathbf{b}^{(m)}, \mathbf{c}^{(m)}, \dot{\mathbf{a}}_0^{(m)}, \dot{\mathbf{b}}^{(m)}, \dot{\mathbf{c}}^{(m)}),\\
\ddot{\mathbf{c}}^{(m)} &= 2\frac{\Omega}{m}\mathbf{k}_1\dot{\mathbf{b}}^{(m)} + \frac{\Omega^2}{m^2}\mathbf{k}_2\mathbf{c}^{(m)} + \mathbf{F}_2^{(m)}(\mathbf{a}_0^{(m)}, \mathbf{b}^{(m)}, \mathbf{c}^{(m)}, \dot{\mathbf{a}}_0^{(m)}, \dot{\mathbf{b}}^{(m)}, \dot{\mathbf{c}}^{(m)})
\end{aligned} \tag{1.45}$$

where

$$\begin{aligned}
&\mathbf{k}_1 = diag(\mathbf{I}_{n\times n}, 2\mathbf{I}_{n\times n}, \ldots, N\mathbf{I}_{n\times n}),\\
&\mathbf{k}_2 = diag(\mathbf{I}_{n\times n}, 2^2\mathbf{I}_{n\times n}, \ldots, N^2\mathbf{I}_{n\times n}),\\
&\mathbf{b}^{(m)} = (\mathbf{b}_{1/m}, \mathbf{b}_{2/m}, \ldots, \mathbf{b}_{N/m})^{\mathrm{T}},\\
&\mathbf{c}^{(m)} = (\mathbf{c}_{1/m}, \mathbf{c}_{2/m}, \ldots, \mathbf{c}_{N/m})^{\mathrm{T}},\\
&\mathbf{F}_1^{(m)} = (\mathbf{F}_{11}^{(m)}, \mathbf{F}_{12}^{(m)}, \ldots, \mathbf{F}_{1N}^{(m)})^{\mathrm{T}},\\
&\mathbf{F}_2^{(m)} = (\mathbf{F}_{21}^{(m)}, \mathbf{F}_{22}^{(m)}, \ldots, \mathbf{F}_{2N}^{(m)})^{\mathrm{T}}\\
&\text{for } N = 1, 2, \ldots, \infty
\end{aligned} \tag{1.46}$$

and for $k = 1, 2, \ldots, N$

$$\begin{aligned}
&\mathbf{F}_0^{(m)}(\mathbf{a}_0^{(m)}, \mathbf{b}^{(m)}, \mathbf{c}^{(m)}, \dot{\mathbf{a}}_0^{(m)}, \dot{\mathbf{b}}^{(m)}, \dot{\mathbf{c}}^{(m)})\\
&\quad = \frac{1}{2m\pi}\int_0^{2m\pi} \mathbf{f}(\mathbf{x}^{(m)*}, \dot{\mathbf{x}}^{(m)*}, \mathbf{p})d\theta;\\
&\mathbf{F}_{1k}^{(m)}(\mathbf{a}_0^{(m)}, \mathbf{b}^{(m)}, \mathbf{c}^{(m)}, \dot{\mathbf{a}}_0^{(m)}, \dot{\mathbf{b}}^{(m)}, \dot{\mathbf{c}}^{(m)})\\
&\quad = \frac{1}{m\pi}\int_0^{2m\pi} \mathbf{f}(\mathbf{x}^{(m)*}, \dot{\mathbf{x}}^{(m)*}, \mathbf{p})\cos\left(\frac{k}{m}\theta\right)d\theta,\\
&\mathbf{F}_{2k}^{(m)}(\mathbf{a}_0^{(m)}, \mathbf{b}^{(m)}, \mathbf{c}^{(m)}, \dot{\mathbf{a}}_0^{(m)}, \dot{\mathbf{b}}^{(m)}, \dot{\mathbf{c}}^{(m)})\\
&\quad = \frac{1}{m\pi}\int_0^{2m\pi} \mathbf{f}(\mathbf{x}^{(m)*}, \dot{\mathbf{x}}^{(m)*}, \mathbf{p})\sin\left(\frac{k}{m}\theta\right)d\theta.
\end{aligned} \tag{1.47}$$

The state-space form of Equation (1.45) is

$$\dot{\mathbf{z}}^{(m)} = \mathbf{z}_1^{(m)} \text{ and } \dot{\mathbf{z}}_1^{(m)} = \mathbf{g}^{(m)}(\mathbf{z}^{(m)}, \mathbf{z}_1^{(m)}) \tag{1.48}$$

where

$$\mathbf{z}^{(m)} = (\mathbf{a}_0^{(m)}, \mathbf{b}^{(m)}, \mathbf{c}^{(m)})^{\mathrm{T}}, \dot{\mathbf{z}}^{(m)} = \mathbf{z}_1^{(m)}$$

$$\mathbf{g}^{(m)} = \left(\mathbf{F}_0^{(m)}, -2\frac{\Omega}{m}\mathbf{k}_1\dot{\mathbf{c}}^{(m)} + \frac{\Omega^2}{m^2}\mathbf{k}_2\mathbf{b}^{(m)} + \mathbf{F}_1^{(m)},\right.$$

$$\left. 2\frac{\Omega}{m}\mathbf{k}_1\dot{\mathbf{b}}^{(m)} + \frac{\Omega^2}{m^2}\mathbf{k}_2\mathbf{c}^{(m)} + \mathbf{F}_2^{(m)}\right)^{\mathrm{T}}. \tag{1.49}$$

An equivalent system of Equation (1.48) is

$$\dot{\mathbf{y}}^{(m)} = \mathbf{f}^{(m)}(\mathbf{y}^{(m)}) \tag{1.50}$$

where

$$\mathbf{y}^{(m)} = (\mathbf{z}^{(m)}, \mathbf{z}_1^{(m)})^{\mathrm{T}} \text{ and } \mathbf{f}^{(m)} = (\mathbf{z}_1^{(m)}, \mathbf{g}^{(m)})^{\mathrm{T}}. \tag{1.51}$$

If equilibrium $\mathbf{y}^{(m)*}$ *of Equation (1.50) (i.e.,* $\mathbf{f}^{(m)}(\mathbf{y}^{(m)*}) = \mathbf{0}$*) exists, then the approximate solution of period-*m *motion exists as in Equation (1.44). In vicinity of equilibrium* $\mathbf{y}^{(m)*}$*, with* $\mathbf{y}^{(m)} = \mathbf{y}^{(m)*} + \Delta\mathbf{y}^{(m)}$*, the linearized equation of Equation (1.50) is*

$$\Delta\dot{\mathbf{y}}^{(m)} = D\mathbf{f}^{(m)}(\mathbf{y}^{(m)*})\Delta\mathbf{y}^{(m)} \tag{1.52}$$

and the eigenvalue analysis of the equilibrium $\mathbf{y}^*$ *is given by*

$$|D\mathbf{f}^{(m)}(\mathbf{y}^{(m)*}) - \lambda\mathbf{I}_{2n(2N+1)\times 2n(2N+1)}| = 0 \tag{1.53}$$

where $D\mathbf{f}^{(m)}(\mathbf{y}^{(m)*}) = \partial\mathbf{f}^{(m)}(\mathbf{y}^{(m)})/\partial\mathbf{y}^{(m)}|_{\mathbf{y}^{(m)*}}$*. Thus, the stability and bifurcation period-*m *motion can be classified by the eigenvalues of* $D\mathbf{f}^{(m)}(\mathbf{y}^{(m)*})$ *with*

$$(n_1, n_2, n_3 | n_4, n_5, n_6). \tag{1.54}$$

i. *If all eigenvalues of the equilibrium possess negative real parts, the approximate periodic solution is stable.*
ii. *If at least one of the eigenvalues of the equilibrium possesses a positive real part, the approximate periodic solution is unstable.*
iii. *The boundaries between stable and unstable equilibriums with higher order singularity give bifurcation and stability conditions with higher order singularity.*

Proof. The proof can be referred to Luo (2012a, 2013, 2014). ■

With periodic excitation, the dynamical systems can be stated as follows:

Theorem 1.5 *Consider a periodically forced, nonlinear vibration system as*

$$\ddot{\mathbf{x}} = \mathbf{F}(\mathbf{x}, \dot{\mathbf{x}}, t, \mathbf{p}) \in \mathscr{R}^n \tag{1.55}$$

where $\mathbf{F}(\mathbf{x}, \dot{\mathbf{x}}, t, \mathbf{p})$ *is a* C^r *– continuous nonlinear function vector* ($r \geq 1$) *with forcing period* $T = 2\pi/\Omega$*. If such a vibration system has a period-*m *motion* $\mathbf{x}^{(m)}(t)$ *with finite norm* $\|\mathbf{x}^{(m)}\|$

and period $T = 2\pi/\Omega$, there is a generalized coordinate transformation with $\theta = \Omega t$ for the periodic motion of Equation (1.55) in the form of

$$\mathbf{x}^{(m)}(t) = \mathbf{a}_0^{(m)}(t) + \sum_{k=1}^{\infty} \mathbf{b}_{k/m}(t)\cos\left(\frac{k}{m}\theta\right) + \mathbf{c}_{k/m}(t)\sin\left(\frac{k}{m}\theta\right) \tag{1.56}$$

with

$$\begin{aligned}
\mathbf{a}_0^{(m)} &= (a_{01}^{(m)}, a_{02}^{(m)}, \ldots, a_{0n}^{(m)})^{\mathrm{T}},\\
\mathbf{b}_{k/m} &= (b_{k/m1}, b_{k/m2}, \ldots, b_{k/mn})^{\mathrm{T}},\\
\mathbf{c}_{k/m} &= (c_{k/m1}, c_{k/m2}, \ldots, c_{k/mn})^{\mathrm{T}}
\end{aligned} \tag{1.57}$$

and

$$\begin{aligned}
&\|\mathbf{x}^{(m)}\| = \|\mathbf{a}_0^{(m)}\| + \sum_{k=1}^{\infty} \|\mathbf{A}_{k/m}\|, \text{ and } \lim_{k\to\infty} \|\mathbf{A}_{k/m}\| = 0 \text{ but not uniform}\\
&\text{with } \mathbf{A}_{k/m} = (A_{k/m1}, A_{k/m2} \ldots, A_{k/mn})^{\mathrm{T}}\\
&\text{and } A_{k/mj} = \sqrt{b_{k/mj}^2 + c_{k/mj}^2}\ (j = 1, 2, \ldots, n).
\end{aligned} \tag{1.58}$$

*For $\|\mathbf{x}^{(m)}(t) - \mathbf{x}^{(m)*}(t)\| < \varepsilon$ with a prescribed small $\varepsilon > 0$, the infinite term transformation $\mathbf{x}^{(m)}(t)$ of period-*m *motion of Equation (1.55), given by Equation (1.56), can be approximated by a finite term transformation $\mathbf{x}^{(m)*}(t)$ as*

$$\mathbf{x}^{(m)*}(t) = \mathbf{a}_0^{(m)}(t) + \sum_{k=1}^{N} \mathbf{b}_{k/m}(t)\cos\left(\frac{k}{m}\theta\right) + \mathbf{c}_{k/m}(t)\sin\left(\frac{k}{m}\theta\right) \tag{1.59}$$

and the generalized coordinates are determined by

$$\begin{aligned}
\ddot{\mathbf{a}}_0^{(m)} &= \mathbf{F}_0^{(m)}(\mathbf{a}_0^{(m)}, \mathbf{b}^{(m)}, \mathbf{c}^{(m)}, \dot{\mathbf{a}}_0^{(m)}, \dot{\mathbf{b}}^{(m)}, \dot{\mathbf{c}}^{(m)}),\\
\ddot{\mathbf{b}}^{(m)} &= -2\frac{\Omega}{m}\mathbf{k}_1\dot{\mathbf{c}}^{(m)} + \frac{\Omega^2}{m^2}\mathbf{k}_2\mathbf{b}^{(m)} + \mathbf{F}_1^{(m)}(\mathbf{a}_0^{(m)}, \mathbf{b}^{(m)}, \mathbf{c}^{(m)}, \dot{\mathbf{a}}_0^{(m)}, \dot{\mathbf{b}}^{(m)}, \dot{\mathbf{c}}^{(m)}),\\
\ddot{\mathbf{c}}^{(m)} &= 2\frac{\Omega}{m}\mathbf{k}_1\dot{\mathbf{b}}^{(m)} + \frac{\Omega^2}{m^2}\mathbf{k}_2\mathbf{c}^{(m)} + \mathbf{F}_2^{(m)}(\mathbf{a}_0^{(m)}, \mathbf{b}^{(m)}, \mathbf{c}^{(m)}, \dot{\mathbf{a}}_0^{(m)}, \dot{\mathbf{b}}^{(m)}, \dot{\mathbf{c}}^{(m)})
\end{aligned} \tag{1.60}$$

where for $N = 1, 2, \ldots, \infty$

$$\begin{aligned}
\mathbf{k}_1 &= diag(\mathbf{I}_{n\times n}, 2\mathbf{I}_{n\times n}, \ldots, N\mathbf{I}_{n\times n}),\\
\mathbf{k}_2 &= diag(\mathbf{I}_{n\times n}, 2^2\mathbf{I}_{n\times n}, \ldots, N^2\mathbf{I}_{n\times n}),\\
\mathbf{b}^{(m)} &= (\mathbf{b}_{1/m}, \mathbf{b}_{2/m}, \ldots, \mathbf{b}_{N/m})^{\mathrm{T}},\\
\mathbf{c}^{(m)} &= (\mathbf{c}_{1/m}, \mathbf{c}_{2/m}, \ldots, \mathbf{c}_{N/m})^{\mathrm{T}},\\
\mathbf{F}_1^{(m)} &= (\mathbf{F}_{11}^{(m)}, \mathbf{F}_{12}^{(m)}, \ldots, \mathbf{F}_{1N}^{(m)})^{\mathrm{T}}\\
\mathbf{F}_2^{(m)} &= (\mathbf{F}_{21}^{(m)}, \mathbf{F}_{22}^{(m)}, \ldots, \mathbf{F}_{2N}^{(m)})^{\mathrm{T}};
\end{aligned} \tag{1.61}$$

and for $k = 1, 2, \ldots, N$

$$\begin{aligned}
&\mathbf{F}_0^{(m)}(\mathbf{a}_0^{(m)}, \mathbf{b}^{(m)}, \mathbf{c}^{(m)}, \dot{\mathbf{a}}_0^{(m)}, \dot{\mathbf{b}}^{(m)}, \dot{\mathbf{c}}^{(m)}) \\
&\quad = \frac{1}{2m\pi} \int_0^{2m\pi} \mathbf{F}(\mathbf{x}^{(m)*}, \dot{\mathbf{x}}^{(m)*}, t, \mathbf{p}) d\theta; \\
&\mathbf{F}_{1k}^{(m)}(\mathbf{a}_0^{(m)}, \mathbf{b}^{(m)}, \mathbf{c}^{(m)}, \dot{\mathbf{a}}_0^{(m)}, \dot{\mathbf{b}}^{(m)}, \dot{\mathbf{c}}^{(m)}) \\
&\quad = \frac{1}{m\pi} \int_0^{2m\pi} \mathbf{F}(\mathbf{x}^{(m)*}, \dot{\mathbf{x}}^{(m)*}, t, \mathbf{p}) \cos\left(\frac{k}{m}\theta\right) d\theta, \\
&\mathbf{F}_{2k}^{(m)}(\mathbf{a}_0^{(m)}, \mathbf{b}^{(m)}, \mathbf{c}^{(m)}, \dot{\mathbf{a}}_0^{(m)}, \dot{\mathbf{b}}^{(m)}, \dot{\mathbf{c}}^{(m)}) \\
&\quad = \frac{1}{m\pi} \int_0^{2m\pi} \mathbf{F}(\mathbf{x}^{(m)*}, \dot{\mathbf{x}}^{(m)*}, t, \mathbf{p}) \sin\left(\frac{k}{m}\theta\right) d\theta.
\end{aligned} \tag{1.62}$$

The state-space form of Equation (1.61) is

$$\dot{\mathbf{z}}^{(m)} = \mathbf{z}_1^{(m)} \text{ and } \dot{\mathbf{z}}_1^{(m)} = \mathbf{g}^{(m)}(\mathbf{z}^{(m)}, \mathbf{z}_1^{(m)}) \tag{1.63}$$

where

$$\begin{aligned}
\mathbf{z}^{(m)} &= (\mathbf{a}_0^{(m)}, \mathbf{b}^{(m)}, \mathbf{c}^{(m)})^{\mathrm{T}}, \dot{\mathbf{z}}^{(m)} = \mathbf{z}_1^{(m)}, \\
\mathbf{g}^{(m)} &= \left(\mathbf{F}_0^{(m)}, -2\frac{\Omega}{m}\mathbf{k}_1\dot{\mathbf{c}}^{(m)} + \frac{\Omega^2}{m^2}\mathbf{k}_2\mathbf{b}^{(m)} + \mathbf{F}_1^{(m)},\right. \\
&\qquad \left. 2\frac{\Omega}{m}\mathbf{k}_1\dot{\mathbf{b}}^{(m)} + \frac{\Omega^2}{m^2}\mathbf{k}_2\mathbf{c}^{(m)} + \mathbf{F}_2^{(m)}\right)^{\mathrm{T}}.
\end{aligned} \tag{1.64}$$

An equivalent system of Equation (1.63) is

$$\dot{\mathbf{y}}^{(m)} = \mathbf{f}^{(m)}(\mathbf{y}^{(m)}) \tag{1.65}$$

where

$$\mathbf{y}^{(m)} = (\mathbf{z}^{(m)}, \mathbf{z}_1^{(m)})^{\mathrm{T}} \text{ and } \mathbf{f}^{(m)} = (\mathbf{z}_1^{(m)}, \mathbf{g}^{(m)})^{\mathrm{T}}. \tag{1.66}$$

If equilibrium $\mathbf{y}^{(m)*}$ *of Equation (1.65) exists (i.e.,* $\mathbf{f}^{(m)}(\mathbf{y}^{(m)*}) = \mathbf{0}$*), then the approximate solution of period-*m *motion exists as in Equation (1.59). In vicinity of equilibrium* $\mathbf{y}^{(m)*}$*, with* $\mathbf{y}^{(m)} = \mathbf{y}^{(m)*} + \Delta\mathbf{y}^{(m)}$*, the linearized equation of Equation (1.65) is*

$$\Delta\dot{\mathbf{y}}^{(m)} = D\mathbf{f}^{(m)}(\mathbf{y}^{(m)*})\Delta\mathbf{y}^{(m)} \tag{1.67}$$

and the eigenvalue analysis of equilibrium $\mathbf{y}^*$ *is given by*

$$|D\mathbf{f}^{(m)}(\mathbf{y}^{(m)*}) - \lambda\mathbf{I}_{2n(2N+1)\times 2n(2N+1)}| = 0 \tag{1.68}$$

where $D\mathbf{f}^{(m)}(\mathbf{y}^{(m)*}) = \partial\mathbf{f}^{(m)}(\mathbf{y}^{(m)})/\partial\mathbf{y}^{(m)}|_{\mathbf{y}^{(m)*}}$*. The stability and bifurcation of period-*m *motion can be classified by eigenvalues of* $D\mathbf{f}^{(m)}(\mathbf{y}^{(m)*})$ *are classified by*

$$(n_1, n_2, n_3 | n_4, n_5, n_6). \tag{1.69}$$

i. *If all eigenvalues of the equilibrium possess negative real parts, the approximate periodic solution is stable.*

ii. If at least one of the eigenvalues of the equilibrium possesses positive real part, the approximate periodic solution is unstable.
iii. The boundaries between stable and unstable equilibriums with higher order singularity give bifurcation and stability conditions with higher order singularity.

Proof. The proof can be referred to Luo (2012a, 2013, 2014). ■

As in the aforementioned two theorems for period-m motions, the analytical solution structures of quasi-periodic motions in nonlinear vibration systems will be presented from Luo (2014) as follows.

Theorem 1.6 *Consider a nonlinear vibration system as*

$$\ddot{\mathbf{x}} = \mathbf{f}(\mathbf{x}, \dot{\mathbf{x}}, \mathbf{p}) \in \mathscr{R}^n \tag{1.70}$$

where $\mathbf{f}(\mathbf{x}, \dot{\mathbf{x}}, \mathbf{p})$ *is a* C^r – *continuous nonlinear function vector* ($r \geq 1$).

A. *If such a dynamical system has a period-m motion* $\mathbf{x}^{(m)}(t)$ *with finite norm* $\|\mathbf{x}^{(m)}\|$ *and period* $T = 2\pi/\Omega$, *there is a generalized coordinate transformation with* $\theta = \Omega t$ *for the periodic motion of Equation (1.70) in a form of*

$$\mathbf{x}^{(m)}(t) = \mathbf{a}_0^{(m)}(t) + \sum_{k=1}^{\infty} \mathbf{b}_{k/m}(t) \cos\left(\frac{k}{m}\theta\right) + \mathbf{c}_{k/m}(t) \sin\left(\frac{k}{m}\theta\right) \tag{1.71}$$

with

$$\begin{aligned}
\mathbf{a}_1^{(0)} &\equiv \mathbf{a}_0^{(m)} = (a_{01}^{(m)}, a_{02}^{(m)}, \dots, a_{0n}^{(m)})^{\mathrm{T}}, \\
\mathbf{a}_2^{(k)} &\equiv \mathbf{b}_{k/m} = (b_{k/m1}, b_{k/m2}, \dots, b_{k/mn})^{\mathrm{T}}, \\
\mathbf{a}_3^{(k)} &\equiv \mathbf{c}_{k/m} = (c_{k/m1}, c_{k/m2}, \dots, c_{k/mn})^{\mathrm{T}}
\end{aligned} \tag{1.72}$$

which, under $\|\mathbf{x}^{(m)}(t) - \mathbf{x}^{(m)*}(t)\| < \varepsilon$ *with a prescribed small* $\varepsilon > 0$, *can be approximated by a finite term transformation* $\mathbf{x}^{(m)*}(t)$

$$\mathbf{x}^{(m)*}(t) = \mathbf{a}_0^{(m)}(t) + \sum_{k=1}^{N_0} \mathbf{b}_{k/m}(t) \cos\left(\frac{k}{m}\theta\right) + \mathbf{c}_{k/m}(t) \sin\left(\frac{k}{m}\theta\right) \tag{1.73}$$

and the generalized coordinates are determined by

$$\ddot{\mathbf{a}}_{s_0} = \mathbf{g}_{s_0}(\mathbf{a}_{s_0}, \dot{\mathbf{a}}_{s_0}, \mathbf{p}) \tag{1.74}$$

where

$$\begin{aligned}
\mathbf{k}_0^{(1)} &= diag(\mathbf{I}_{n\times n}, 2\mathbf{I}_{n\times n}, \dots, N_0\mathbf{I}_{n\times n}), \\
\mathbf{k}_0^{(2)} &= diag(\mathbf{I}_{n\times n}, 2^2\mathbf{I}_{n\times n}, \dots, N_0^2\mathbf{I}_{n\times n}), \\
\mathbf{a}_1^{(0)} &\equiv \mathbf{a}_0^{(m)}, \mathbf{a}_2^{(k)} \equiv \mathbf{b}_{k/m}, \mathbf{a}_3^{(k)} \equiv \mathbf{c}_{k/m}; \\
\mathbf{a}_1 &= \mathbf{a}_1^{(0)}, \\
\mathbf{a}_2 &= (\mathbf{a}_2^{(1)}, \mathbf{a}_2^{(2)}, \dots, \mathbf{a}_2^{(N)})^{\mathrm{T}} \equiv \mathbf{b}^{(m)}, \\
\mathbf{a}_3 &= (\mathbf{a}_3^{(1)}, \mathbf{a}_3^{(2)}, \dots, \mathbf{a}_3^{(N)})^{\mathrm{T}} \equiv \mathbf{c}^{(m)},
\end{aligned}$$

$$\begin{aligned}
&\mathbf{F}_1 = \mathbf{F}_0^{(m)},\\
&\mathbf{F}_2 = (\mathbf{F}_{11}^{(m)}, \mathbf{F}_{12}^{(m)}, \dots, \mathbf{F}_{1N_0}^{(m)})^{\mathrm{T}},\\
&\mathbf{F}_3 = (\mathbf{F}_{21}^{(m)}, \mathbf{F}_{22}^{(m)}, \dots, \mathbf{F}_{2N_0}^{(m)})^{\mathrm{T}};\\
&\mathbf{a}_{s_0} = (\mathbf{a}_1, \mathbf{a}_2, \mathbf{a}_3)^{\mathrm{T}},\\
&\mathbf{g}_{s_0} = \left(\mathbf{F}_1^{(m)}, -2\frac{\Omega}{m}\mathbf{k}_0^{(1)}\dot{\mathbf{a}}_3 + \frac{\Omega^2}{m^2}\mathbf{k}_0^{(2)}\mathbf{a}_2 + \mathbf{F}_2,\right.\\
&\qquad\left. 2\frac{\Omega}{m}\mathbf{k}_0^{(1)}\dot{\mathbf{a}}_2 + \frac{\Omega^2}{m^2}\mathbf{k}_0^{(2)}\mathbf{a}_3 + \mathbf{F}_3\right)^{\mathrm{T}}\\
&\text{for } N_0 = 1, 2, \dots, \infty;
\end{aligned} \tag{1.75}$$

and

$$\begin{aligned}
&\mathbf{F}_0^{(m)}(\mathbf{a}_0^{(m)}, \mathbf{b}^{(m)}, \mathbf{c}^{(m)}) = \frac{1}{2m\pi}\int_0^{2m\pi} \mathbf{f}(\mathbf{x}^{(m)*}, \dot{\mathbf{x}}^{(m)*}, \mathbf{p})d\theta;\\
&\mathbf{F}_{1k}^{(m)}(\mathbf{a}_0^{(m)}, \mathbf{b}^{(m)}, \mathbf{c}^{(m)}) = \frac{1}{m\pi}\int_0^{2m\pi} \mathbf{f}(\mathbf{x}^{(m)*}, \dot{\mathbf{x}}^{(m)*}, \mathbf{p})\cos\left(\frac{k}{m}\theta\right)d\theta,\\
&\mathbf{F}_{2k}^{(m)}(\mathbf{a}_0^{(m)}, \mathbf{b}^{(m)}, \mathbf{c}^{(m)}) = \frac{1}{m\pi}\int_0^{2m\pi} \mathbf{f}(\mathbf{x}^{(m)*}, \dot{\mathbf{x}}^{(m)*}, \mathbf{p})\sin\left(\frac{k}{m}\theta\right)d\theta\\
&\text{for } k = 1, 2, \dots, N_0.
\end{aligned} \tag{1.76}$$

B. *If after the* k*th Hopf bifurcation with* $p_k\omega_k = \omega_{k-1}$ $(k = 1, 2, \dots)$ *and* $\omega_0 = \Omega/m$, *there is a dynamical system of coefficients as*

$$\ddot{\mathbf{a}}_{s_0 s_1 \dots s_k} = \mathbf{g}_{s_0 s_1 \dots s_k}(\mathbf{a}_{s_0 s_1 \dots s_k}, \dot{\mathbf{a}}_{s_0 s_1 \dots s_k}, \mathbf{p}) \tag{1.77}$$

where

$$\begin{aligned}
&\mathbf{a}_{s_0 s_1 \dots s_k} = (\mathbf{a}_{s_0 s_1 \dots s_{k-1} 1}, \mathbf{a}_{s_0 s_1 \dots s_{k-1} 2}, \mathbf{a}_{s_0 s_1 \dots s_{k-1} 3})^{\mathrm{T}},\\
&\mathbf{g}_{s_0 s_1 \dots s_k} = (\mathbf{F}_{s_0 s_1 \dots s_{k-1} 1}, -2\omega_k \mathbf{k}_k^{(1)}\dot{\mathbf{a}}_{s_0 s_1 \dots s_{k-1} 3} + \omega_k^2 \mathbf{k}_k^{(2)}\mathbf{a}_{s_0 s_1 \dots s_{k-1} 2} + \mathbf{F}_{s_0 s_1 \dots s_{k-1} 2},\\
&\qquad 2\omega_k \mathbf{k}_k^{(1)}\dot{\mathbf{a}}_{s_0 s_1 \dots s_{k-1} 2} + \omega_k^2 \mathbf{k}_k^{(2)}\mathbf{a}_{s_1 s_2 \dots s_{k-1} 3} + \mathbf{F}_{s_1 s_2 \dots s_{k-1} 3})^{\mathrm{T}},\\
&\mathbf{k}_k^{(1)} = diag(\mathbf{I}_{n_{k-1}\times n_{k-1}}, 2\mathbf{I}_{n_{k-1}\times n_{k-1}}, \dots, N_k \mathbf{I}_{n_{k-1}\times n_{k-1}}),\\
&\mathbf{k}_k^{(2)} = diag(\mathbf{I}_{n_{k-1}\times n_{k-1}}, 2^2\mathbf{I}_{n_{k-1}\times n_{k-1}}, \dots, N_k^2 \mathbf{I}_{n_{k-1}\times n_{k-1}})\\
&n_{k-1} = n(2N_0 + 1)(2N_1 + 1)\dots(2N_{k-1} + 1)
\end{aligned} \tag{1.78}$$

with a periodic solution as

$$\begin{aligned}
\mathbf{a}_{s_0 s_1 \dots s_k} = \mathbf{a}_{s_0 s_1 \dots s_k 1}^{(0)}(t) + \sum_{l_{k+1}=1}^{\infty} \mathbf{a}_{s_0 s_1 \dots s_k 2}^{(l_{k+1})}(t)\cos(l_{k+1}\theta_{k+1})\\
+ \mathbf{a}_{s_0 s_1 \dots s_k 3}^{(l_{k+1})}(t)\sin(l_{k+1}\theta_{k+1})
\end{aligned} \tag{1.79}$$

with

$$
\begin{aligned}
&s_i = 1,2,3\ (i = 0,1,2,\ldots,k),\\
&\mathbf{a}_{s_0s_1\ldots s_k1} = \mathbf{a}^{(0)}_{s_0s_1\ldots s_k1},\\
&\mathbf{a}_{s_0s_1\ldots s_k2} = (\mathbf{a}^{(1)}_{s_0s_1\ldots s_k2}, \mathbf{a}^{(2)}_{s_0s_1\ldots s_k2}, \ldots, \mathbf{a}^{(N_{k+1})}_{s_0s_1\ldots s_k2})^{\mathrm{T}},\\
&\mathbf{a}_{s_0s_1\ldots s_k3} = (\mathbf{a}^{(1)}_{s_0s_1\ldots s_k3}, \mathbf{a}^{(2)}_{s_0s_1\ldots s_k3}, \ldots, \mathbf{a}^{(N_{k+1})}_{s_0s_1\ldots s_k3})^{\mathrm{T}};\\
&\mathbf{a}_{s_0s_1\ldots s_{k-1}1} = \mathbf{a}^{(0)}_{s_0s_1\ldots s_{k-1}1},\\
&\mathbf{a}_{s_0s_1\ldots s_{k-1}2} = (\mathbf{a}^{(1)}_{s_0s_1\ldots s_{k-1}2}, \mathbf{a}^{(2)}_{s_0s_1\ldots s_{k-1}2}, \ldots, \mathbf{a}^{(N_k)}_{s_0s_1\ldots s_{k-1}2})^{\mathrm{T}},\\
&\mathbf{a}_{s_0s_1\ldots s_{k-1}3} = (\mathbf{a}^{(1)}_{s_0s_1\ldots s_{k-1}3}, \mathbf{a}^{(2)}_{s_0s_1\ldots s_{k-1}3}, \ldots, \mathbf{a}^{(N_k)}_{s_0s_1\ldots s_{k-1}3})^{\mathrm{T}};\\
&\vdots\\
&\mathbf{a}_1 = \mathbf{a}^{(0)}_1,\\
&\mathbf{a}_2 = (\mathbf{a}^{(1)}_2, \mathbf{a}^{(2)}_2, \ldots, \mathbf{a}^{(N_0)}_2)^{\mathrm{T}},\\
&\mathbf{a}_3 = (\mathbf{a}^{(1)}_3, \mathbf{a}^{(2)}_3, \ldots, \mathbf{a}^{(N_0)}_3)^{\mathrm{T}};
\end{aligned} \tag{1.80}
$$

which, under $\|\mathbf{a}_{s_0s_1\ldots s_k}(t) - \mathbf{a}^*_{s_0s_1\ldots s_k}(t)\| < \varepsilon$ *with a prescribed small* $\varepsilon > 0$, *can be approximated by a finite term transformation* $\mathbf{a}^*_{s_0s_1\ldots s_k}(t)$

$$
\begin{aligned}
\mathbf{a}^*_{s_0s_1\ldots s_k} &= \mathbf{a}^{(0)}_{s_0s_1\ldots s_k1}(t) + \sum_{l_{k+1}=1}^{N_{k+1}} \mathbf{a}^{(l_{k+1})}_{s_0s_1\ldots s_k2}(t)\cos(l_{k+1}\theta_{k+1})\\
&\quad + \mathbf{a}^{(l_{k+1})}_{s_0s_1\ldots s_k3}(t)\sin(l_{k+1}\theta_{k+1})
\end{aligned} \tag{1.81}
$$

and the generalized coordinates are determined by

$$
\ddot{\mathbf{a}}_{s_0s_1\ldots s_{k+1}} = \mathbf{g}_{s_0s_1\ldots s_{k+1}}(\mathbf{a}_{s_0s_1\ldots s_{k+1}}, \dot{\mathbf{a}}_{s_0s_1\ldots s_{k+1}}, \mathbf{p}) \tag{1.82}
$$

where

$$
\begin{aligned}
\mathbf{a}_{s_0s_1\ldots s_{k+1}} &= (\mathbf{a}_{s_0s_1\ldots s_k1}, \mathbf{a}_{s_0s_1\ldots s_k2}, \mathbf{a}_{s_0s_1\ldots s_k3})^{\mathrm{T}},\\
\mathbf{g}_{s_0s_1\ldots s_{k+1}} &= (\mathbf{F}_{s_1s_2\ldots s_k1}, -2\omega_{k+1}\mathbf{k}^{(1)}_{k+1}\dot{\mathbf{a}}_{s_1s_2\ldots s_k3} + \omega^2_{k+1}\mathbf{k}^{(2)}_{k+1}\mathbf{a}_{s_1s_2\ldots s_k2} + \mathbf{F}_{s_1s_2\ldots s_k2},\\
&\quad 2\omega_{k+1}\mathbf{k}^{(1)}_{k+1}\dot{\mathbf{a}}_{s_1s_2\ldots s_k2} + \omega^2_{k+1}\mathbf{k}^{(2)}_{k+1}\mathbf{a}_{s_1s_2\ldots s_k3} + \mathbf{F}_{s_1s_2\ldots s_k3})^{\mathrm{T}};
\end{aligned} \tag{1.83}
$$

and

$$
\begin{aligned}
\mathbf{k}^{(1)}_{k+1} &= diag(\mathbf{I}_{n_k\times n_k}, 2\mathbf{I}_{n_k\times n_k}, \ldots, N_{k+1}\mathbf{I}_{n_k\times n_k}),\\
\mathbf{k}^{(2)}_{k+1} &= diag(\mathbf{I}_{n_k\times n_k}, 2^2\mathbf{I}_{n_k\times n_k}, \ldots, N^2_{k+1}\mathbf{I}_{n_k\times n_k})\\
n_k &= n(2N_0+1)(2N_1+1)\ldots(2N_k+1);
\end{aligned}
$$

$$
\begin{aligned}
\mathbf{a}_{s_0 s_1 \ldots s_k 1} &= \mathbf{a}^{(0)}_{s_0 s_1 \ldots s_k 1},\\
\mathbf{a}_{s_0 s_1 \ldots s_k 2} &= (\mathbf{a}^{(1)}_{s_0 s_1 \ldots s_k 2}, \mathbf{a}^{(2)}_{s_0 s_1 \ldots s_k 2}, \ldots, \mathbf{a}^{(N_{k+1})}_{s_0 s_1 \ldots s_k 2})^{\mathrm{T}},\\
\mathbf{a}_{s_0 s_1 \ldots s_k 3} &= (\mathbf{a}^{(1)}_{s_0 s_1 \ldots s_k 3}, \mathbf{a}^{(2)}_{s_0 s_1 \ldots s_k 3}, \ldots, \mathbf{a}^{(N_{k+1})}_{s_0 s_1 \ldots s_k 3})^{\mathrm{T}};\\
\mathbf{a}_{s_0 s_1 \ldots s_{k+1}} &= (\mathbf{a}_{s_0 s_1 \ldots s_k 1}, \mathbf{a}_{s_0 s_1 \ldots s_k 2}, \mathbf{a}_{s_0 s_1 \ldots s_k 3})^{\mathrm{T}};\\
\mathbf{F}_{s_0 s_1 \ldots s_k 1} &= \mathbf{F}^{(0)}_{s_0 s_1 \ldots s_k 1},\\
\mathbf{F}_{s_0 s_1 \ldots s_k 2} &= (\mathbf{F}^{(1)}_{s_0 s_1 \ldots s_k 2}, \mathbf{F}^{(2)}_{s_0 s_1 \ldots s_k 2}, \ldots, \mathbf{F}^{(N_{k+1})}_{s_0 s_1 \ldots s_k 2})^{\mathrm{T}},\\
\mathbf{F}_{s_0 s_1 \ldots s_k 3} &= (\mathbf{F}^{(1)}_{s_0 s_1 \ldots s_k 3}, \mathbf{F}^{(2)}_{s_0 s_1 \ldots s_k 3}, \ldots, \mathbf{F}^{(N_{k+1})}_{s_0 s_1 \ldots s_k 3})^{\mathrm{T}}\\
&\text{for } N_{k+1} = 1, 2, \ldots, \infty;
\end{aligned}
\tag{1.84}
$$

and

$$
\begin{aligned}
&\mathbf{F}_{s_0 s_1 \ldots s_k 1}(\mathbf{a}_{s_0 s_1 \ldots s_{k+1}}, \dot{\mathbf{a}}_{s_0 s_1 \ldots s_{k+1}}, \mathbf{p})\\
&\quad = \frac{1}{2\pi}\int_0^{2\pi} \mathbf{g}_{s_0 s_1 \ldots s_k}(\mathbf{a}^{*}_{s_0 s_1 \ldots s_k}, \dot{\mathbf{a}}^{*}_{s_0 s_1 \ldots s_k}, \mathbf{p})\, d\theta_{k+1};\\
&\mathbf{F}^{(l_{k+1})}_{s_0 s_1 \ldots s_k 2}(\mathbf{a}_{s_0 s_1 \ldots s_{k+1}}, \dot{\mathbf{a}}_{s_0 s_1 \ldots s_{k+1}}, \mathbf{p})\\
&\quad = \frac{1}{\pi}\int_0^{2\pi} \mathbf{g}_{s_0 s_1 \ldots s_k}(\mathbf{a}^{*}_{s_0 s_1 \ldots s_k}, \dot{\mathbf{a}}^{*}_{s_0 s_1 \ldots s_k}, \mathbf{p}) \cos(l_{k+1}\theta_{k+1})\, d\theta_{k+1},\\
&\mathbf{F}^{(l_{k+1})}_{s_0 s_1 \ldots s_k 3}(\mathbf{a}_{s_0 s_1 \ldots s_{k+1}}, \dot{\mathbf{a}}_{s_0 s_1 \ldots s_{k+1}}, \mathbf{p})\\
&\quad = \frac{1}{\pi}\int_0^{2\pi} \mathbf{g}_{s_0 s_1 \ldots s_k}(\mathbf{a}^{*}_{s_0 s_1 \ldots s_k}, \dot{\mathbf{a}}^{*}_{s_0 s_1 \ldots s_k}, \mathbf{p}) \sin(l_{k+1}\theta_{k+1})\, d\theta_{k+1}\\
&\text{for } l_{k+1} = 1, 2, \ldots, N_{k+1}.
\end{aligned}
\tag{1.85}
$$

C. *Equation (1.82) becomes*

$$
\dot{\mathbf{z}}_{s_0 s_1 \ldots s_{k+1}} = \mathbf{f}_{s_0 s_1 \ldots s_{k+1}}(\mathbf{z}_{s_0 s_1 \ldots s_{k+1}})
\tag{1.86}
$$

where

$$
\begin{aligned}
\mathbf{z}_{s_0 s_1 \ldots s_{k+1}} &= (\mathbf{a}_{s_0 s_1 \ldots s_{k+1}}, \dot{\mathbf{a}}_{s_0 s_1 \ldots s_{k+1}})^{\mathrm{T}},\\
\mathbf{f}_{s_0 s_1 \ldots s_{k+1}} &= (\dot{\mathbf{a}}_{s_0 s_1 \ldots s_{k+1}}, \mathbf{g}_{s_0 s_1 \ldots s_{k+1}})^{\mathrm{T}}.
\end{aligned}
\tag{1.87}
$$

If equilibrium $\mathbf{z}^{*}_{s_0 s_1 \ldots s_{k+1}}$ *of Equation (1.86) (i.e.,* $\mathbf{f}_{s_0 s_1 \ldots s_{k+1}}(\mathbf{z}^{*}_{s_0 s_1 \ldots s_{k+1}}) = \mathbf{0}$*) exists, then the approximate solution of the periodic motion of the* kth *generalized coordinates for the period-*m *motion exists as in Equation (1.81). In the vicinity of equilibrium* $\mathbf{z}^{*}_{s_0 s_1 \ldots s_{k+1}}$*, with*

$$
\mathbf{z}_{s_0 s_1 \ldots s_{k+1}} = \mathbf{z}^{*}_{s_0 s_1 \ldots s_{k+1}} + \Delta\mathbf{z}_{s_0 s_1 \ldots s_{k+1}},
\tag{1.88}
$$

the linearized equation of Equation (1.86) is

$$\Delta \dot{\mathbf{z}}_{s_0 s_1 \ldots s_{k+1}} = D\mathbf{f}_{s_0 s_1 \ldots s_{k+1}}(\mathbf{z}^*_{s_0 s_1 \ldots s_{k+1}})\Delta \mathbf{z}_{s_0 s_1 \ldots s_{k+1}} \tag{1.89}$$

and the eigenvalue analysis of equilibrium $\mathbf{z}^*$ *is given by*

$$|D\mathbf{f}_{s_0 s_1 \ldots s_{k+1}}(\mathbf{z}^*_{s_0 s_1 \ldots s_{k+1}}) - \lambda \mathbf{I}_{2n_k(2N_{k+1}+1)\times 2n_k(2N_{k+1}+1)}| = 0 \tag{1.90}$$

where

$$D\mathbf{f}_{s_0 s_1 \ldots s_{k+1}}(\mathbf{z}^*_{s_0 s_1 \ldots s_{k+1}}) = \left.\frac{\partial \mathbf{f}_{s_0 s_1 \ldots s_{k+1}}\left(\mathbf{z}_{s_0 s_1 \ldots s_{k+1}}\right)}{\partial \mathbf{z}_{s_0 s_1 \ldots s_{k+1}}}\right|_{\mathbf{z}^*_{s_1 s_2 \ldots s_{k+1}}}. \tag{1.91}$$

The stability and bifurcation of such a periodic motion of the kth *generalized coordinates can be classified by the eigenvalues of* $D\mathbf{f}_{s_0 s_1 \ldots s_{k+1}}(\mathbf{z}^*_{s_0 s_1 \ldots s_{k+1}})$ *with*

$$(n_1, n_2, n_3 | n_4, n_5, n_6). \tag{1.92}$$

i. *If all eigenvalues of the equilibrium possess negative real parts, the approximate quasi-periodic solution is stable.*
ii. *If at least one of the eigenvalues of the equilibrium possesses positive real part, the approximate quasi-periodic solution is unstable.*
iii. *The boundaries between stable and unstable equilibriums with higher order singularity give bifurcation and stability conditions with higher order singularity.*

D. *For the* kth *order Hopf bifurcation of period-*m *motion, a relation exists as*

$$p_k \omega_k = \omega_{k-1}. \tag{1.93}$$

i. *If* p_k *is an irrational number, the* kth*-order Hopf bifurcation of the period-*m *motion is called the quasi-period-*p_k *Hopf bifurcation, and the corresponding solution of the* kth *generalized coordinates is* p_k*-quasi-periodic to the system of the* $(k-1)$th *generalized coordinates.*
ii. *If* $p_k = 2$*, the* kth*-order Hopf bifurcation of the period-*m *motion is called a period-doubling Hopf bifurcation (or a period-2 Hopf bifurcation), and the corresponding solution of the* kth *generalized coordinates is period-doubling to the system of the* $(k-1)$th *generalized coordinates.*
iii. *If* $p_k = q$ *with an integer* q*, the* kth*-order Hopf bifurcation of the period-*m *motion is called a period-*q *Hopf bifurcation, and the corresponding solution of the* kth *generalized coordinates is of* q*-times period to the system of the* $(k-1)$th *generalized coordinates.*
iv. *If* $p_k = p/q$ $(p, q$ *are irreducible integer), the* kth*-order Hopf bifurcation of the period-*m *motion is called a period-*p/q *Hopf bifurcation, and the corresponding solution of the* kth *generalized coordinates is* p/q*-periodic to the system of the* $(k-1)$th *generalized coordinates.*

Proof. The proof of this theorem can be referred to Luo (2014). ■

Similarly, for periodically forced vibration systems, the analytical solution of quasi-periodic motions can be presented as follows.

Theorem 1.7 *Consider a periodically forced, nonlinear vibration system as*

$$\ddot{\mathbf{x}} = \mathbf{F}(\mathbf{x}, \dot{\mathbf{x}}, t, \mathbf{p}) \in \mathscr{R}^n \tag{1.94}$$

where $\mathbf{F}(\mathbf{x}, \dot{\mathbf{x}}, t, \mathbf{p})$ *is a* C^r *– continuous nonlinear function vector* ($r \geq 1$) *with forcing period* $T = 2\pi/\Omega$.

A. *If such a vibration system has a period-*m *motion* $\mathbf{x}^{(m)}(t)$ *with finite norm* $\|\mathbf{x}^{(m)}\|$, *there is a generalized coordinate transformation with* $\theta = \Omega t$ *for the period-*m *motion of Equation (1.94) in a form of*

$$\mathbf{x}^{(m)}(t) = \mathbf{a}_0^{(m)}(t) + \sum_{k=1}^{\infty} \mathbf{b}_{k/m}(t) \cos\left(\frac{k}{m}\theta\right) + \mathbf{c}_{k/m}(t) \sin\left(\frac{k}{m}\theta\right) \tag{1.95}$$

with

$$\begin{aligned}
\mathbf{a}_1^{(0)} &\equiv \mathbf{a}_0^{(m)} = (a_{01}^{(m)}, a_{02}^{(m)}, \ldots, a_{0n}^{(m)})^{\mathrm{T}}, \\
\mathbf{a}_2^{(k)} &\equiv \mathbf{b}_{k/m} = (b_{k/m1}, b_{k/m2}, \ldots, b_{k/mn})^{\mathrm{T}}, \\
\mathbf{a}_3^{(k)} &\equiv \mathbf{c}_{k/m} = (c_{k/m1}, c_{k/m2}, \ldots, c_{k/mn})^{\mathrm{T}}
\end{aligned} \tag{1.96}$$

which, under $\|\mathbf{x}^{(m)}(t) - \mathbf{x}^{(m)*}(t)\| < \varepsilon$ *with a prescribed small* $\varepsilon > 0$, *can be approximated by a finite term transformation* $\mathbf{x}^{(m)*}(t)$

$$\mathbf{x}^{(m)*}(t) = \mathbf{a}_0^{(m)}(t) + \sum_{k=1}^{N} \mathbf{b}_{k/m}(t) \cos\left(\frac{k}{m}\theta\right) + \mathbf{c}_{k/m}(t) \sin\left(\frac{k}{m}\theta\right) \tag{1.97}$$

and the generalized coordinates are determined by

$$\ddot{\mathbf{a}}_{s_0} = \mathbf{g}_{s_0}(\mathbf{a}_{s_0}, \dot{\mathbf{a}}_{s_0}, \mathbf{p}) \tag{1.98}$$

where

$$\begin{aligned}
\mathbf{k}_0^{(1)} &= diag(\mathbf{I}_{n\times n}, 2\mathbf{I}_{n\times n}, \ldots, N_0\mathbf{I}_{n\times n}), \\
\mathbf{k}_0^{(2)} &= diag(\mathbf{I}_{n\times n}, 2^2\mathbf{I}_{n\times n}, \ldots, N_0^2\mathbf{I}_{n\times n}), \\
\mathbf{a}_1^{(0)} &\equiv \mathbf{a}_0^{(m)}, \mathbf{a}_2^{(k)} \equiv \mathbf{b}_{k/m}, \mathbf{a}_3^{(k)} \equiv \mathbf{c}_{k/m}; \\
\mathbf{a}_1 &= \mathbf{a}_1^{(0)}, \\
\mathbf{a}_2 &= (\mathbf{a}_2^{(1)}, \mathbf{a}_2^{(2)}, \ldots, \mathbf{a}_2^{(N_0)})^{\mathrm{T}} \equiv \mathbf{b}^{(m)}, \\
\mathbf{a}_3 &= (\mathbf{a}_3^{(1)}, \mathbf{a}_3^{(2)}, \ldots, \mathbf{a}_3^{(N_0)})^{\mathrm{T}} \equiv \mathbf{c}^{(m)}, \\
\mathbf{F}_2 &= (\mathbf{F}_{11}^{(m)}, \mathbf{F}_{12}^{(m)}, \ldots, \mathbf{F}_{1N_0}^{(m)})^{\mathrm{T}}, \\
\mathbf{F}_3 &= (\mathbf{F}_{21}^{(m)}, \mathbf{F}_{22}^{(m)}, \ldots, \mathbf{F}_{2N_0}^{(m)})^{\mathrm{T}}; \\
\mathbf{a}_{s_0} &= (\mathbf{a}_1, \mathbf{a}_2, \mathbf{a}_3)^{\mathrm{T}},
\end{aligned}$$

$$\mathbf{g}_{s_0} = \left(\mathbf{F}_1^{(m)}, -2\frac{\Omega}{m}\mathbf{k}_0^{(1)}\dot{\mathbf{a}}_3 + \frac{\Omega^2}{m^2}\mathbf{k}_0^{(2)}\mathbf{a}_2 + \mathbf{F}_2,\right.$$

$$\left. 2\frac{\Omega}{m}\mathbf{k}_0^{(1)}\dot{\mathbf{a}}_2 + \frac{\Omega^2}{m^2}\mathbf{k}_0^{(2)}\mathbf{a}_3 + \mathbf{F}_3\right)^{\mathrm{T}}$$

$$\text{for } N_0 = 1, 2, \ldots, \infty; \tag{1.99}$$

and

$$\mathbf{F}_0^{(m)}(\mathbf{a}_0^{(m)}, \mathbf{b}^{(m)}, \mathbf{c}^{(m)}, \dot{\mathbf{a}}_0^{(m)}, \dot{\mathbf{b}}^{(m)}, \dot{\mathbf{c}}^{(m)})$$
$$= \frac{1}{2m\pi}\int_0^{2m\pi} \mathbf{F}(\mathbf{x}^{(m)*}, \dot{\mathbf{x}}^{(m)*}, t, \mathbf{p})d\theta;$$
$$\mathbf{F}_{1k}^{(m)}(\mathbf{a}_0^{(m)}, \mathbf{b}^{(m)}, \mathbf{c}^{(m)}, \dot{\mathbf{a}}_0^{(m)}, \dot{\mathbf{b}}^{(m)}, \dot{\mathbf{c}}^{(m)})$$
$$= \frac{1}{m\pi}\int_0^{2m\pi} \mathbf{F}(\mathbf{x}^{(m)*}, \dot{\mathbf{x}}^{(m)*}, t, \mathbf{p})\cos\left(\frac{k}{m}\theta\right)d\theta,$$
$$\mathbf{F}_{2k}^{(m)}(\mathbf{a}_0^{(m)}, \mathbf{b}^{(m)}, \mathbf{c}^{(m)}, \dot{\mathbf{a}}_0^{(m)}, \dot{\mathbf{b}}^{(m)}, \dot{\mathbf{c}}^{(m)})$$
$$= \frac{1}{m\pi}\int_0^{2m\pi} \mathbf{F}(\mathbf{x}^{(m)*}, \dot{\mathbf{x}}^{(m)*}, t, \mathbf{p})\sin\left(\frac{k}{m}\theta\right)d\theta. \tag{1.100}$$

B. *For the* kth *Hopf bifurcation with $p_k\omega_k = \omega_{k-1}$ $(k = 1, 2, \ldots)$ and $\omega_0 = \Omega/m$, there is a dynamical system of coefficients as*

$$\ddot{\mathbf{a}}_{s_0s_1\ldots s_k} = \mathbf{g}_{s_0s_1\ldots s_k}(\mathbf{a}_{s_0s_1\ldots s_k}, \dot{\mathbf{a}}_{s_0s_1\ldots s_k}, \mathbf{p}) \tag{1.101}$$

where

$$\mathbf{a}_{s_0s_1\ldots s_k} = (\mathbf{a}_{s_0s_1\ldots s_{k-1}1}, \mathbf{a}_{s_0s_1\ldots s_{k-1}2}, \mathbf{a}_{s_0s_1\ldots s_{k-1}3})^{\mathrm{T}},$$
$$\mathbf{g}_{s_0s_1\ldots s_k} = (\mathbf{F}_{s_1s_2\ldots s_{k-1}1}, -2\omega_k\mathbf{k}_k^{(1)}\dot{\mathbf{a}}_{s_1s_2\ldots s_k3} + \omega_k^2\mathbf{k}_k^{(2)}\mathbf{a}_{s_1s_2\ldots s_{k-1}2} + \mathbf{F}_{s_1s_2\ldots s_{k-1}2},$$
$$2\omega_k\mathbf{k}_k^{(1)}\dot{\mathbf{a}}_{s_1s_2\ldots s_k2} + \omega_k^2\mathbf{k}_k^{(2)}\mathbf{a}_{s_1s_2\ldots s_{k-1}3} + \mathbf{F}_{s_1s_2\ldots s_{k-1}3})^{\mathrm{T}},$$
$$\mathbf{k}_k^{(1)} = diag(\mathbf{I}_{n_{k-1}\times n_{k-1}}, 2\mathbf{I}_{n_{k-1}\times n_{k-1}}, \ldots, N_k\mathbf{I}_{n_{k-1}\times n_{k-1}}),$$
$$\mathbf{k}_k^{(2)} = diag(\mathbf{I}_{n_{k-1}\times n_{k-1}}, 2^2\mathbf{I}_{n_{k-1}\times n_{k-1}}, \ldots, N_k^2\mathbf{I}_{n_{k-1}\times n_{k-1}}),$$
$$n_{k-1} = n(2N_0 + 1)(2N_1 + 1)\ldots(2N_{k-1} + 1) \tag{1.102}$$

with a periodic solution as

$$\mathbf{a}_{s_0s_1\ldots s_k} = \mathbf{a}_{s_0s_1\ldots s_k1}^{(0)}(t) + \sum_{l_{k+1}=1}^{\infty} \mathbf{a}_{s_0s_1\ldots s_k2}^{(l_{k+1})}(t)\cos(l_{k+1}\theta_{k+1})$$
$$+ \mathbf{a}_{s_0s_1\ldots s_k3}^{(l_{k+1})}(t)\sin(l_{k+1}\theta_{k+1}) \tag{1.103}$$

with

$$
\begin{aligned}
& s_i = 1,2,3 \; (i = 0,1,2,\ldots,k), \\
& \mathbf{a}_{s_0 s_1 \ldots s_k 1} = \mathbf{a}^{(0)}_{s_0 s_1 \ldots s_k 1}, \\
& \mathbf{a}_{s_0 s_1 \ldots s_k 2} = (\mathbf{a}^{(1)}_{s_0 s_1 \ldots s_k 2}, \mathbf{a}^{(2)}_{s_0 s_1 \ldots s_k 2}, \ldots, \mathbf{a}^{(N_{k+1})}_{s_0 s_1 \ldots s_k 2})^{\mathrm{T}}, \\
& \mathbf{a}_{s_0 s_1 \ldots s_k 3} = (\mathbf{a}^{(1)}_{s_0 s_1 \ldots s_k 3}, \mathbf{a}^{(2)}_{s_0 s_1 \ldots s_k 3}, \ldots, \mathbf{a}^{(N_{k+1})}_{s_0 s_1 \ldots s_k 3})^{\mathrm{T}}; \\
& \mathbf{a}_{s_0 s_1 \ldots s_{k-1} 1} = \mathbf{a}^{(0)}_{s_0 s_1 \ldots s_{k-1} 1}, \\
& \mathbf{a}_{s_0 s_1 \ldots s_{k-1} 2} = (\mathbf{a}^{(1)}_{s_0 s_1 \ldots s_{k-1} 2}, \mathbf{a}^{(2)}_{s_0 s_1 \ldots s_{k-1} 2}, \ldots, \mathbf{a}^{(N_k)}_{s_0 s_1 \ldots s_{k-1} 2})^{\mathrm{T}}, \\
& \mathbf{a}_{s_0 s_1 \ldots s_{k-1} 3} = (\mathbf{a}^{(1)}_{s_0 s_1 \ldots s_{k-1} 3}, \mathbf{a}^{(2)}_{s_0 s_1 \ldots s_{k-1} 3}, \ldots, \mathbf{a}^{(N_k)}_{s_0 s_1 \ldots s_{k-1} 3})^{\mathrm{T}}; \\
& \vdots \\
& \mathbf{a}_1 = \mathbf{a}^{(0)}_1, \\
& \mathbf{a}_2 = (\mathbf{a}^{(1)}_2, \mathbf{a}^{(2)}_2, \ldots, \mathbf{a}^{(N_0)}_2)^{\mathrm{T}}, \\
& \mathbf{a}_3 = (\mathbf{a}^{(1)}_3, \mathbf{a}^{(2)}_3, \ldots, \mathbf{a}^{(N_0)}_3)^{\mathrm{T}};
\end{aligned} \tag{1.104}
$$

which, under $\|\mathbf{a}_{s_0 s_1 \ldots s_k}(t) - \mathbf{a}^*_{s_0 s_1 \ldots s_k}(t)\| < \varepsilon$ *with a prescribed small* $\varepsilon > 0$*, can be approximated by a finite term transformation* $\mathbf{a}^*_{s_0 s_1 \ldots s_k}(t)$

$$
\begin{aligned}
\mathbf{a}^*_{s_0 s_1 \ldots s_k} = {} & \mathbf{a}^{(0)}_{s_0 s_1 \ldots s_k 1}(t) + \sum_{l_{k+1}=1}^{N_{k+1}} \mathbf{a}^{(l_{k+1})}_{s_0 s_1 \ldots s_k 2}(t) \cos(l_{k+1}\theta_{k+1}) \\
& + \mathbf{a}^{(l_{k+1})}_{s_0 s_1 \ldots s_k 3}(t) \sin(l_{k+1}\theta_{k+1})
\end{aligned} \tag{1.105}
$$

and the generalized coordinates are determined by

$$
\ddot{\mathbf{a}}_{s_0 s_1 \ldots s_{k+1}} = \mathbf{g}_{s_0 s_1 \ldots s_{k+1}}(\mathbf{a}_{s_0 s_1 \ldots s_{k+1}}, \dot{\mathbf{a}}_{s_0 s_1 \ldots s_{k+1}}, \mathbf{p}) \tag{1.106}
$$

where

$$
\begin{aligned}
\mathbf{a}_{s_0 s_1 \ldots s_{k+1}} &= (\mathbf{a}_{s_0 s_1 \ldots s_k 1}, \mathbf{a}_{s_0 s_1 \ldots s_k 2}, \mathbf{a}_{s_0 s_1 \ldots s_k 3})^{\mathrm{T}}, \\
\mathbf{g}_{s_0 s_1 \ldots s_{k+1}} &= (\mathbf{F}_{s_0 s_1 \ldots s_k 1}, -2\omega_{k+1}\mathbf{k}^{(1)}_{k+1}\dot{\mathbf{a}}_{s_0 s_1 \ldots s_k 3} + \omega^2_{k+1}\mathbf{k}^{(2)}_{k+1}\mathbf{a}_{s_0 s_1 \ldots s_k 2} + \mathbf{F}_{s_0 s_1 \ldots s_k 2}, \\
& \quad 2\omega_{k+1}\mathbf{k}^{(1)}_{k+1}\dot{\mathbf{a}}_{s_0 s_1 \ldots s_k 2} + \omega^2_{k+1}\mathbf{k}^{(2)}_{k+1}\mathbf{a}_{s_0 s_1 \ldots s_k 3} + \mathbf{F}_{s_0 s_1 \ldots s_k 3})^{\mathrm{T}};
\end{aligned} \tag{1.107}
$$

and

$$
\begin{aligned}
& \mathbf{k}^{(1)}_{k+1} = diag(\mathbf{I}_{n_k \times n_k}, 2\mathbf{I}_{n_k \times n_k}, \ldots, N_{k+1}\mathbf{I}_{n_k \times n_k}), \\
& \mathbf{k}^{(2)}_{k+1} = diag(\mathbf{I}_{n_k \times n_k}, 2^2\mathbf{I}_{n_k \times n_k}, \ldots, N^2_{k+1}\mathbf{I}_{n_k \times n_k}) \\
& n_k = n(2N_0 + 1)(2N_1 + 1) \ldots (2N_k + 1); \\
& \mathbf{a}_{s_0 s_1 \ldots s_k 1} = \mathbf{a}^{(0)}_{s_0 s_1 \ldots s_k 1},
\end{aligned}
$$

$$\begin{aligned}
\mathbf{a}_{s_0 s_1 \ldots s_k 2} &= (\mathbf{a}^{(1)}_{s_0 s_1 \ldots s_k 2}, \mathbf{a}^{(2)}_{s_0 s_1 \ldots s_k 2}, \ldots, \mathbf{a}^{(N_{k+1})}_{s_0 s_1 \ldots s_k 2})^{\mathrm{T}},\\
\mathbf{a}_{s_0 s_1 \ldots s_k 3} &= (\mathbf{a}^{(1)}_{s_0 s_1 \ldots s_k 3}, \mathbf{a}^{(2)}_{s_0 s_1 \ldots s_k 3}, \ldots, \mathbf{a}^{(N_{k+1})}_{s_0 s_1 \ldots s_k 3})^{\mathrm{T}};\\
\mathbf{a}_{s_0 s_1 \ldots s_{k+1}} &= (\mathbf{a}_{s_0 s_1 \ldots s_k 1}, \mathbf{a}_{s_0 s_1 \ldots s_k 2}, \mathbf{a}_{s_0 s_1 \ldots s_k 3})^{\mathrm{T}};\\
\mathbf{F}_{s_0 s_1 \ldots s_k 1} &= \mathbf{F}^{(0)}_{s_0 s_1 \ldots s_k 1},\\
\mathbf{F}_{s_0 s_1 \ldots s_k 2} &= (\mathbf{F}^{(1)}_{s_0 s_1 \ldots s_k 2}, \mathbf{F}^{(2)}_{s_0 s_1 \ldots s_k 2}, \ldots, \mathbf{F}^{(N_{k+1})}_{s_0 s_1 \ldots s_k 2})^{\mathrm{T}},\\
\mathbf{F}_{s_0 s_1 \ldots s_k 3} &= (\mathbf{F}^{(1)}_{s_0 s_1 \ldots s_k 3}, \mathbf{F}^{(2)}_{s_0 s_1 \ldots s_k 3}, \ldots, \mathbf{F}^{(N_{k+1})}_{s_0 s_1 \ldots s_k 3})^{\mathrm{T}}\\
&\text{for } N_{k+1} = 1, 2, \ldots, \infty;
\end{aligned} \tag{1.108}$$

and

$$\begin{aligned}
&\mathbf{F}_{s_0 s_1 \ldots s_k 1}(\mathbf{a}_{s_0 s_1 \ldots s_{k+1}}, \dot{\mathbf{a}}_{s_0 s_1 \ldots s_{k+1}}, \mathbf{p})\\
&\quad = \frac{1}{2\pi}\int_0^{2\pi} \mathbf{g}_{s_0 s_1 \ldots s_k}(\mathbf{a}^*_{s_0 s_1 \ldots s_k}, \dot{\mathbf{a}}^*_{s_1 s_2 \ldots s_k}, \mathbf{p}) d\theta_{k+1};\\
&\mathbf{F}^{(l_{k+1})}_{s_0 s_1 \ldots s_k 2}(\mathbf{a}_{s_0 s_1 \ldots s_{k+1}}, \dot{\mathbf{a}}_{s_0 s_1 \ldots s_{k+1}}, \mathbf{p})\\
&\quad = \frac{1}{\pi}\int_0^{2\pi} \mathbf{g}_{s_0 s_1 \ldots s_k}(\mathbf{a}^*_{s_0 s_1 \ldots s_k}, \dot{\mathbf{a}}^*_{s_1 s_2 \ldots s_k}, \mathbf{p}) \cos(l_{k+1}\theta_{k+1}) d\theta_{k+1},\\
&\mathbf{F}^{(l_{k+1})}_{s_0 s_1 \ldots s_k 3}(\mathbf{a}_{s_0 s_1 \ldots s_{k+1}}, \dot{\mathbf{a}}_{s_0 s_1 \ldots s_{k+1}}, \mathbf{p})\\
&\quad = \frac{1}{\pi}\int_0^{2\pi} \mathbf{g}_{s_0 s_1 \ldots s_k}(\mathbf{a}^*_{s_0 s_1 \ldots s_k}, \dot{\mathbf{a}}^*_{s_1 s_2 \ldots s_k}, \mathbf{p}) \sin(l_{k+1}\theta_{k+1}) d\theta_{k+1}\\
&\text{for } l_{k+1} = 1, 2, \ldots, N_{k+1}.
\end{aligned} \tag{1.109}$$

C. *Equation (1.106) becomes*

$$\dot{\mathbf{z}}_{s_0 s_1 \ldots s_{k+1}} = \mathbf{f}_{s_0 s_1 \ldots s_{k+1}}(\mathbf{z}_{s_0 s_1 \ldots s_{k+1}}) \tag{1.110}$$

where

$$\begin{aligned}
\mathbf{z}_{s_0 s_1 \ldots s_{k+1}} &= (\mathbf{a}_{s_0 s_1 \ldots s_{k+1}}, \dot{\mathbf{a}}_{s_0 s_1 \ldots s_{k+1}})^{\mathrm{T}},\\
\mathbf{f}_{s_0 s_1 \ldots s_{k+1}} &= (\dot{\mathbf{a}}_{s_0 s_1 \ldots s_{k+1}}, \mathbf{g}_{s_0 s_1 \ldots s_{k+1}})^{\mathrm{T}}.
\end{aligned} \tag{1.111}$$

If equilibrium $\mathbf{z}^*_{s_1 s_2 \ldots s_{k+1}}$ *of Equation (1.110) (i.e.,* $\mathbf{f}_{s_1 s_2 \ldots s_{k+1}}(\mathbf{z}^*_{s_1 s_2 \ldots s_{k+1}}) = \mathbf{0}$*) exists, then the approximate solution of the periodic motion of the* kth *generalized coordinates for the period-*m *motion exists as in Equation (1.105). In the vicinity of equilibrium* $\mathbf{z}^*_{s_1 s_2 \ldots s_{k+1}}$, *with*

$$\mathbf{z}_{s_0 s_1 \ldots s_{k+1}} = \mathbf{z}^*_{s_0 s_1 \ldots s_{k+1}} + \Delta\mathbf{z}_{s_0 s_1 \ldots s_{k+1}}, \tag{1.112}$$

the linearized equation of Equation (1.110) is

$$\Delta\dot{\mathbf{z}}_{s_0 s_1 \ldots s_{k+1}} = D\mathbf{f}_{s_0 s_1 \ldots s_{k+1}}(\mathbf{z}^*_{s_0 s_1 \ldots s_{k+1}})\Delta\mathbf{z}_{s_0 s_1 \ldots s_{k+1}} \tag{1.113}$$

and the eigenvalue analysis of equilibrium $\mathbf{z}^*$ *is given by*

$$|D\mathbf{f}_{s_0 s_1 \ldots s_{k+1}}(\mathbf{z}^*_{s_0 s_1 \ldots s_{k+1}}) - \lambda\mathbf{I}_{2n_k(2N_{k+1}+1)\times 2n_k(2N_{k+1}+1)}| = 0 \tag{1.114}$$

where

$$D\mathbf{f}_{s_0 s_1 \ldots s_{k+1}}(\mathbf{z}^*_{s_0 s_1 \ldots s_{k+1}}) = \left.\frac{\partial\mathbf{f}_{s_0 s_1 \ldots s_{k+1}}\left(\mathbf{z}_{s_0 s_1 \ldots s_{k+1}}\right)}{\partial\mathbf{z}_{s_0 s_1 \ldots s_{k+1}}}\right|_{\mathbf{z}^*_{s_0 s_1 \ldots s_{k+1}}}. \tag{1.115}$$

The stability and bifurcation of such a periodic motion of the kth *generalized coordinates can be classified by the eigenvalues of* $D\mathbf{f}_{s_0 s_1 \ldots s_{k+1}}(\mathbf{z}^*_{s_0 s_1 \ldots s_{k+1}})$ *with*

$$(n_1, n_2, n_3 | n_4, n_5, n_6). \tag{1.116}$$

i. *If all eigenvalues of the equilibrium possess negative real parts, the approximate quasi-periodic solution is stable.*
ii. *If at least one of the eigenvalues of the equilibrium possesses positive real part, the approximate quasi-periodic solution is unstable.*
iii. *The boundaries between stable and unstable equilibriums with higher order singularity give bifurcation and stability conditions with higher order singularity.*

D. *For the* kth *order Hopf bifurcation of period-*m *motion, a relation exists as*

$$p_k\omega_k = \omega_{k-1}. \tag{1.117}$$

i. *If* p_k *is an irrational number, the* kth*-order Hopf bifurcation of the period-*m *motion is called the quasi-period-*p_k *Hopf bifurcation, and the corresponding solution of the* kth *generalized coordinates is* p_k*-quasi-periodic to the system of the* $(k-1)$th *generalized coordinates.*
ii. *If* $p_k = 2$, *the* kth*-order Hopf bifurcation of the period-*m *motion is called a period-doubling Hopf bifurcation (or a period-2 Hopf bifurcation), and the corresponding solution of the* kth *generalized coordinates is period-doubling to the system of the* $(k-1)$th *generalized coordinates.*
iii. *If* $p_k = q$ *with an integer q, the* kth*-order Hopf bifurcation of the period-*m *motion is called a period-q Hopf bifurcation, and the corresponding solution of the* kth *generalized coordinates is of q-times period to the system of the* $(k-1)$th *generalized coordinates.*
iv. *If* $p_k = p/q$ *(p, q are irreducible integer), the* kth*-order Hopf bifurcation of the period-*m *motion is called a period-*p/q *Hopf bifurcation, and the corresponding solution of the* kth *generalized coordinates is* p/q*-periodic to the system of the* $(k-1)$th *generalized coordinates.*

Proof. The proof of this theorem can be referred to Luo (2014). ■

The general theory for the general nonlinear dynamical systems was found in Luo (2012a, 2014), and the analytical solutions for nonlinear dynamical systems with time-delay was presented as well. The generalized harmonic balance method is different from the traditional harmonic balance method. This generalized harmonic balance method provides a theoretic framework to analytically express all possible periodic motions in nonlinear dynamical systems. The procedure for different periodic solutions in different dynamical systems is *of the same*, as presented in Luo (2012a, 2013, 2014). However, the analytical expressions for different periodic solutions in the same dynamical systems are *distinguishing*, which should be obtained through the different, transformed, nonlinear dynamical systems. For instance, the period-1, period-2, and period-m solutions possess the completely different solution expressions. Even for the same period-m solutions with different parameters and/or locations of initial conditions, the analytical solutions in the same nonlinear system are completely different. One needs to work on them to obtain the *complete* pictures (*dynamics*) of stable and unstable periodic solutions plus chaos.

The detailed mathematical theory of the generalized harmonic balance method with the vigorous proof was presented in Luo (2012a, 2013, 2014). In fact, this method provides a finite-harmonic-term transformation with different time scales to obtain an autonomous nonlinear system of coefficients in the Fourier series form with finite harmonic terms. The dynamical behaviors of such an autonomous nonlinear system will determine the solution behaviors of original dynamical systems. For periodic solutions, the Fourier series forms of the finite harmonic terms are convergent. For transient solutions, such Fourier series forms of the finite harmonic terms may not be convergent. For different periodic solutions in a nonlinear dynamical system, the Fourier series solution forms are different, which are determined by how many finite harmonic terms with time-varying coefficients in the Fourier series form. To determine different periodic solutions in the same dynamical system and the corresponding dynamical behaviors, the different, transformed, nonlinear dynamical systems relative to the prescribed finite harmonic terms should be employed. Of course, periodic solutions in different dynamical systems are different, and the corresponding investigation should be carried out individually because the transformed, nonlinear dynamical systems are totally different. In summary, the generalized harmonic balance method provides a possibility of finding all possible periodic solutions plus chaos analytically. For the current stage, this method is the best way to analytically determine the complete dynamics of periodic solutions in nonlinear dynamical systems. In addition, the generalized harmonic balance method is also *a small-parameter-free method* to determine the periodic solutions in nonlinear dynamical systems.

Luo and Huang (2012a) the generalized harmonic balance method with finite terms to obtain the analytical solution of period-1 motion of the Duffing oscillator with a twin-well potential. Luo and Huang (2012b) presented a generalized harmonic balance method to find analytical solutions of period-m motions in such a Duffing oscillator. The analytical bifurcation trees of periodic motions in the Duffing oscillator to chaos are obtained (also see, Luo and Huang, 2012c,d, 2013a,b,c, 2014a). Such analytical bifurcation trees show the connection from periodic solution to chaos analytically. To better understand nonlinear behaviors in nonlinear dynamical systems, the analytical solutions for the bifurcation trees from period-1 motion to chaos in a periodically forced oscillator with quadratic nonlinearity were presented in Luo and Yu (2013a,b,c), and period-m motions in the periodically forced, van der Pol equation was presented in Luo and Lakeh (2013a). The analytical solutions for the van der Pol oscillator can be used to verify the conclusions in Cartwright and Littlewood (1945) and Levinson (1949).

The results for the quadratic nonlinear oscillator in Luo and Yu (2013a,b,c) analytically show the complicated period-1 motions and the corresponding bifurcation structures. In this book, the generalized harmonic balance method will be used to develop the analytical solutions.

1.2 Book Layout

This book consists of five chapters. Chapter 1 gave the brief literature review on analytical methods, including perturbation methods, the method of averaging, and generalized harmonic balance methods. Other chapters are briefly summarized as follows.

In Chapter 2, analytical bifurcation trees from period-m motions to chaos in periodically forced, Duffing oscillators will be presented. The analytical solutions of period-m motions in Duffing oscillators will be discussed because the Duffing oscillators are extensively applied in structural vibrations and physical problems. The bifurcation trees of period-1 motions to chaos for the Duffing oscillators will be discussed and the bifurcation trees of period-3 motions to chaos will also be presented for the Duffing oscillators. Different types of Duffing oscillators possess completely different bifurcation trees.

In Chapter 3, analytical solutions for period-m motions in periodically forced, self-excited oscillators will be presented in the Fourier series form with finite harmonic terms, and the stability and bifurcation of the corresponding period-m motions will be completed. The period-m motions in the periodically forced, van der Pol oscillator will be discussed, and the limit cycles for the van der Pol oscillator without any excitation will be discussed as well. The period-m motions are in independent periodic solution windows embedded in quasi-periodic and chaotic motions. The period-m motions for the van der Pol-Duffing oscillator will be presented, and bifurcation tree of period-m motion will be discussed. For a better understanding of complex period-m motions in such a van der Pol-Duffing oscillator, trajectories and amplitude spectrums will be illustrated numerically.

In Chapter 4, analytical solutions for period-m motions in parametrically forced, nonlinear oscillators are discussed. The bifurcation trees of periodic motions to chaos in a parametric oscillator with quadratic nonlinearity will be discussed analytically. Nonlinear behaviors of such periodic motions will be characterized through frequency-amplitude curves. This investigation shows that period-1 motions exist in parametric nonlinear systems and the corresponding bifurcation trees to chaos exist as well. In addition, analytical solutions for periodic motions in a Mathieu-Duffing oscillator are presented. The frequency-amplitude characteristics of asymmetric period-1 and symmetric period-2 motions will be discussed. Period-1 asymmetric and period-2 symmetric motions will be illustrated for a better understanding of periodic motions in the Mathieu-Duffing oscillator.

In Chapter 5, analytical solutions for period-m motions in a nonlinear rotor system will be discussed. This rotor system with two degrees of freedom is a simple rotor dynamical system and periodic excitations are from the rotor eccentricity. The analytical expressions of periodic solutions will be developed. The corresponding stability and bifurcation analyses of period-m motions will be carried out. Analytical bifurcation trees of period-1 motions to chaos will be presented. The Hopf bifurcation of periodic motion can cause not only the bifurcation tree but quasi-periodic motions. Displacement orbits of periodic motions in nonlinear rotor systems show motion complexity, and harmonic amplitude spectrums gives harmonic effects on periodic motions.

2

Bifurcation Trees in Duffing Oscillators

In this chapter, analytical bifurcation trees from period-m motions to chaos in periodically forced, Duffing oscillators will be presented. The analytical solutions of period-m motions in Duffing oscillators will be discussed because the Duffing oscillators are applied in structural vibrations and physical problems. The bifurcation trees of period-1 motions to chaos for the Duffing oscillators will be discussed and the bifurcation trees of period-3 motions to chaos will be presented for the Duffing oscillators as well. Different types of Duffing oscillators possess completely different bifurcation trees.

2.1 Analytical Solutions

In this section, analytical solutions for period-m motions in periodically forced, Duffing oscillators will be discussed with finite harmonic terms based on the prescribed accuracy of harmonic amplitudes.

Consider a periodically forced Duffing oscillator as

$$\ddot{x} + \delta\dot{x} + \alpha x + \beta x^3 = Q_0 \cos \Omega t \tag{2.1}$$

where $\dot{x} = dx/dt$ is velocity, Q_0 and Ω are excitation amplitude and frequency, respectively. The damping coefficient δ, linear and nonlinear terms α and β are for the Duffing oscillator. Equation (2.1) can be expressed in a standard form of

$$\ddot{x} = F(x, \dot{x}, t) \tag{2.2}$$

where

$$F(x, \dot{x}, t) = -\delta\dot{x} - \alpha x - \beta x^3 + Q_0 \cos \Omega t. \tag{2.3}$$

The analytical solution of period-1 motion for the hardening Duffing oscillator is

$$x^*(t) = a_0(t) + \sum_{k=1}^{N} b_k(t) \cos(k\Omega t) + c_k(t) \sin(k\Omega t). \tag{2.4}$$

Analytical Routes to Chaos in Nonlinear Engineering, First Edition. Albert C. J. Luo.

In Luo (2012), the analytical solution of period-m motion with $\theta = \Omega t$ can be written as

$$x^{(m)*}(t) = a_0^{(m)}(t) + \sum_{k=1}^{N} b_{k/m}(t)\cos\left(\frac{k}{m}\theta\right) + c_{k/m}(t)\sin\left(\frac{k}{m}\theta\right). \tag{2.5}$$

The first and second order derivatives of Equation (2.5) with respect to time give

$$\begin{aligned}
\dot{x}^{(m)*}(t) = \dot{a}_0^{(m)} + \sum_{k=1}^{N} \Bigg[& \left(\dot{b}_{k/m} + \frac{k\Omega}{m}c_{k/m}\right)\cos\left(\frac{k\theta}{m}\right) \\
& + \left(\dot{c}_{k/m} - \frac{k\Omega}{m}b_{k/m}\right)\sin\left(\frac{k\theta}{m}\right)\Bigg],
\end{aligned} \tag{2.6}$$

$$\begin{aligned}
\ddot{x}^{(m)*}(t) = \ddot{a}_0^{(m)} + \sum_{k=1}^{N} \Bigg[& \left(\ddot{b}_{k/m} + 2\frac{k\Omega}{m}\dot{c}_{k/m} - \left(\frac{k\Omega}{m}\right)^2 b_{k/m}\right)\cos\left(\frac{k}{m}\theta\right) \\
& + \left(\ddot{c}_{k/m} - 2\frac{k\Omega}{m}\dot{b}_{k/m} - \left(\frac{k\Omega}{m}\right)^2 c_{k/m}\right)\sin\left(\frac{k}{m}\theta\right)\Bigg].
\end{aligned} \tag{2.7}$$

Substitution of Equations (2.5)–(2.7) to Equation (2.2) and application of the virtual work principle for a basis of constant, $\cos(k\theta/m)$ and $\sin(k\theta/m)$ ($k = 1, 2, \ldots$) as a set of virtual displacements gives

$$\begin{aligned}
& \ddot{a}_0^{(m)} = F_0^{(m)}(a_0^{(m)}, \mathbf{b}^{(m)}, \mathbf{c}^{(m)}, \dot{a}_0^{(m)}, \dot{\mathbf{b}}^{(m)}, \dot{\mathbf{c}}^{(m)}), \\
& \ddot{b}_{k/m} + 2\frac{k\Omega}{m}\dot{c}_{k/m} - \left(\frac{k\Omega}{m}\right)^2 b_{k/m} \\
& \quad = F_{1k}^{(m)}(a_0^{(m)}, \mathbf{b}^{(m)}, \mathbf{c}^{(m)}, \dot{a}_0^{(m)}, \dot{\mathbf{b}}^{(m)}, \dot{\mathbf{c}}^{(m)}), \\
& \ddot{c}_{k/m} - 2\frac{k\Omega}{m}\dot{b}_{k/m} - \left(\frac{k\Omega}{m}\right)^2 c_{k/m} \\
& \quad = F_{2k}^{(m)}(a_0^{(m)}, \mathbf{b}^{(m)}, \mathbf{c}^{(m)}, \dot{a}_0^{(m)}, \dot{\mathbf{b}}^{(m)}, \dot{\mathbf{c}}^{(m)}) \\
& \text{for } k = 1, 2, \ldots, N
\end{aligned} \tag{2.8}$$

The coefficients of constant, $\cos(k\theta/m)$ and $\sin(k\theta/m)$ for the function of $F(x, \dot{x}, t)$ in the Fourier series are

$$\begin{aligned}
& F_0^{(m)}(a_0^{(m)}, \mathbf{b}^{(m)}, \mathbf{c}^{(m)}, \dot{a}_0^{(m)}, \dot{\mathbf{b}}^{(m)}, \dot{\mathbf{c}}^{(m)}) \\
& \quad = \frac{1}{mT}\int_0^{mT} F(x^{(m)*}, \dot{x}^{(m)*}, t)dt \\
& \quad = -\delta\dot{a}_0^{(m)} - \alpha a_0^{(m)} - \beta f^{(0)}, \\
& F_{1k}^{(m)}(a_0^{(m)}, \mathbf{b}^{(m)}, \mathbf{c}^{(m)}, \dot{a}_0^{(m)}, \dot{\mathbf{b}}^{(m)}, \dot{\mathbf{c}}^{(m)}) \\
& \quad = \frac{2}{mT}\int_0^{mT} F(x^{(m)*}, \dot{x}^{(m)*}, t)\cos\left(\frac{k}{m}\Omega t\right)dt \\
& \quad = -\delta\left(\dot{b}_{k/m} + \frac{k\Omega}{m}c_{k/m}\right) - \alpha b_{k/m} - \beta f^{(c)} + Q_0\delta_k^m,
\end{aligned}$$

$$\begin{aligned}
&F_{2k}^{(m)}(a_0^{(m)},\mathbf{b}^{(m)},\mathbf{c}^{(m)},\dot{a}_0^{(m)},\dot{\mathbf{b}}^{(m)},\dot{\mathbf{c}}^{(m)})\\
&=\frac{2}{mT}\int_0^{mT}F(x^{(m)*},\dot{x}^{(m)*},t)\sin\left(\frac{k}{m}\Omega t\right)dt\\
&=-\delta(\dot{c}_{k/m}-\frac{k\Omega}{m}b_{k/m})-\alpha c_{k/m}-\beta f^{(s)}
\end{aligned} \tag{2.9}$$

where

$$\begin{aligned}
f^{(0)}=&(a_0^{(m)})^3+\sum_{l=1}^{N}\sum_{j=1}^{N}\sum_{i=1}^{N}\frac{3a_0^{(m)}}{2N}(b_{i/m}b_{j/m}\delta_{i-j}^0+c_{i/m}c_{j/m}\delta_{i-j}^0)\\
&+\frac{1}{4}b_{i/m}b_{j/m}b_{l/m}(\delta_{i-j-l}^0+\delta_{i-j+l}^0+\delta_{i+j-l}^0)\\
&+\frac{3}{4}b_{i/m}c_{j/m}c_{l/m}(\delta_{i+j-l}^0+\delta_{i-j+l}^0-\delta_{i-j-l}^0),
\end{aligned} \tag{2.10}$$

$$\begin{aligned}
f^{(c)}=&\sum_{l=1}^{N}\sum_{j=1}^{N}\sum_{i=1}^{N}3\left(\frac{a_0^{(m)}}{N}\right)^2 b_{l/m}\delta_l^k+\frac{3a_0^{(m)}}{2N}b_{l/m}b_{j/m}(\delta_{|l-j|}^k+\delta_{l+j}^k)\\
&+\frac{3a_0^{(m)}}{2N}c_{l/m}c_{j/m}(\delta_{|l-j|}^k-\delta_{l+j}^k)\\
&+\frac{1}{4}b_{l/m}b_{j/m}b_{i/m}(\delta_{|l-j-i|}^k+\delta_{l+j+i}^k+\delta_{|l-j+i|}^k+\delta_{|l+j-i|}^k)\\
&+\frac{3}{4}b_{l/m}c_{j/m}c_{i/m}(\delta_{|l+j-i|}^k-\delta_{l+j+i}^k+\delta_{|l-j+i|}^k-\delta_{|l-j-i|}^k),
\end{aligned} \tag{2.11}$$

$$\begin{aligned}
f^{(s)}=&\sum_{l=1}^{N}\sum_{j=1}^{N}\sum_{i=1}^{N}3\left(\frac{a_0^{(m)}}{N}\right)^2 c_{l/m}\delta_l^k+\frac{3a_0^{(m)}}{N}b_{l/m}c_{j/m}[\delta_{l+j}^k-\mathrm{sgn}(l-j)\delta_{|l-j|}^k]\\
&+\frac{1}{4}c_{l/m}c_{j/m}c_{i/m}\Big[\mathrm{sgn}\,(l-j+i)\,\delta_{|l-j+i|}^k-\delta_{l+j+i}^k\\
&+\mathrm{sgn}\,(l+j-i)\,\delta_{|l+j-i|}^k-\mathrm{sgn}(l-j-i)\delta_{|l-j-i|}^k\Big]\\
&+\frac{3}{4}b_{l/m}b_{j/m}c_{i/m}\Big[\mathrm{sgn}\,(l-j+i)\,\delta_{|l-j+i|}^k+\delta_{l+j+i}^k\\
&-\mathrm{sgn}\,(l+j-i)\,\delta_{|l+j-i|}^k-\mathrm{sgn}(l-j-i)\delta_{|l-j-i|}^k\Big].
\end{aligned} \tag{2.12}$$

Introduce vectors to express the unknown time-varying coefficients as

$$\begin{aligned}
\mathbf{z}^{(m)}&\triangleq(a_0^{(m)},\mathbf{b}^{(m)},\mathbf{c}^{(m)})^{\mathrm{T}}\\
&=(a_0^{(m)},b_{1/m},\ldots,b_{N/m},c_{1/m},\ldots,c_{N/m})^{\mathrm{T}}\\
&\equiv(z_0^{(m)},z_1^{(m)},\ldots,z_{2N}^{(m)})^{\mathrm{T}},
\end{aligned}$$

$$\begin{aligned}\mathbf{z}_1 &= \dot{\mathbf{z}} = (\dot{a}_0^{(m)}, \dot{\mathbf{b}}^{(m)}, \dot{\mathbf{c}}^{(m)})^{\mathrm{T}} \\ &= (\dot{a}_0^{(m)}, \dot{b}_{1/m}, \ldots, \dot{b}_{N/m}, \dot{c}_{1/m}, \ldots, \dot{c}_{N/m})^{\mathrm{T}} \\ &\equiv (\dot{z}_0^{(m)}, \dot{z}_1^{(m)}, \ldots, \dot{z}_{2N}^{(m)})^{\mathrm{T}}\end{aligned} \tag{2.13}$$

where

$$\begin{aligned}\mathbf{b}^{(m)} &= (b_{1/m}, b_{2/m}, \ldots, b_{N/m})^{\mathrm{T}}, \\ \mathbf{c}^{(m)} &= (c_{1/m}, c_{2/m}, \ldots, c_{N/m})^{\mathrm{T}}.\end{aligned} \tag{2.14}$$

Equation (2.8) can be expressed in the form of vector field as

$$\dot{\mathbf{z}}^{(m)} = \mathbf{z}_1^{(m)} \text{ and } \dot{\mathbf{z}}_1^{(m)} = \mathbf{g}^{(m)}(\mathbf{z}^{(m)}, \mathbf{z}_1^{(m)}), \tag{2.15}$$

where

$$\mathbf{g}^{(m)}(\mathbf{z}^{(m)}, \mathbf{z}_1^{(m)}) = \begin{pmatrix} F_0^{(m)}(\mathbf{z}^{(m)}, \mathbf{z}_1^{(m)}) \\ \mathbf{F}_1^{(m)}(\mathbf{z}^{(m)}, \mathbf{z}_1^{(m)}) - 2\mathbf{k}_1 \frac{\Omega}{m}\dot{\mathbf{c}}^{(m)} + \mathbf{k}_2\left(\frac{\Omega}{m}\right)^2 \mathbf{b}^{(m)} \\ \mathbf{F}_2^{(m)}(\mathbf{z}^{(m)}, \mathbf{z}_1^{(m)}) + 2\mathbf{k}_1 \frac{\Omega}{m}\dot{\mathbf{b}}^{(m)} + \mathbf{k}_2\left(\frac{\Omega}{m}\right)^2 \mathbf{c}^{(m)} \end{pmatrix}, \tag{2.16}$$

$$\begin{aligned}\mathbf{k}_1 &= diag(1, 2, \ldots, N), \\ \mathbf{k}_2 &= diag(1, 2^2, \ldots, N^2), \\ \mathbf{F}_1^{(m)} &= (F_{11}^{(m)}, F_{12}^{(m)}, \ldots, F_{1N}^{(m)})^{\mathrm{T}}, \\ \mathbf{F}_2^{(m)} &= (F_{21}^{(m)}, F_{22}^{(m)}, \ldots, F_{2N}^{(m)})^{\mathrm{T}} \\ &\text{for } N = 1, 2, \ldots, \infty;\end{aligned} \tag{2.17}$$

and

$$\mathbf{y}^{(m)} \equiv (\mathbf{z}^{(m)}, \mathbf{z}_1^{(m)}) \text{ and } \mathbf{f}^{(m)} = (\mathbf{z}_1^{(m)}, \mathbf{g}^{(m)})^{\mathrm{T}}. \tag{2.18}$$

Thus, Equation (2.15) becomes

$$\dot{\mathbf{y}}^{(m)} = \mathbf{f}^{(m)}(\mathbf{y}^{(m)}). \tag{2.19}$$

The solutions of steady-state periodic motion can be obtained by setting $\dot{\mathbf{y}}^{(m)} = \mathbf{0}$, that is,

$$\begin{aligned} F_0^{(m)}(a_0^{(m)*}, \mathbf{b}^{(m)*}, \mathbf{c}^{(m)*}, 0, \mathbf{0}, \mathbf{0}) &= 0, \\ \mathbf{F}_1^{(m)}(a_0^{(m)*}, \mathbf{b}^{(m)*}, \mathbf{c}^{(m)*}, 0, \mathbf{0}, \mathbf{0}) + \frac{\Omega^2}{m^2}\mathbf{k}_2\mathbf{b}^{(m)*} &= \mathbf{0}, \\ \mathbf{F}_2^{(m)}(a_0^{(m)*}, \mathbf{b}^{(m)*}, \mathbf{c}^{(m)*}, 0, \mathbf{0}, \mathbf{0}) + \frac{\Omega^2}{m^2}\mathbf{k}_2\mathbf{c}^{(m)*} &= \mathbf{0}. \end{aligned} \tag{2.20}$$

The solutions of the $(2N+1)$ nonlinear equations in Equation (2.20) are computed from the Newton–Raphson method. The linearized equation at the equilibrium point $\mathbf{y}^{(m)*} = (\mathbf{z}^{(m)*}, \mathbf{0})^{\mathrm{T}}$ is

$$\Delta\dot{\mathbf{y}}^{(m)} = D\mathbf{f}^{(m)}(\mathbf{y}^{*(m)})\Delta\mathbf{y}^{(m)} \tag{2.21}$$

where

$$D\mathbf{f}^{(m)}(\mathbf{y}^{*(m)}) = \partial \mathbf{f}^{(m)}(\mathbf{y}^{(m)})/\partial \mathbf{y}^{(m)}|_{\mathbf{y}^{(m)*}}. \tag{2.22}$$

The Jacobian matrix is

$$D\mathbf{f}^{(m)}(\mathbf{y}^{(m)}) = \begin{bmatrix} \mathbf{0}_{(2N+1)\times(2N+1)} & \mathbf{I}_{(2N+1)\times(2N+1)} \\ \mathbf{G}_{(2N+1)\times(2N+1)} & \mathbf{H}_{(2N+1)\times(2N+1)} \end{bmatrix} \tag{2.23}$$

and

$$\mathbf{G} = \frac{\partial \mathbf{g}^{(m)}}{\partial \mathbf{z}^{(m)}} = (\mathbf{G}^{(0)}, \mathbf{G}^{(c)}, \mathbf{G}^{(s)})^{\mathrm{T}} \tag{2.24}$$

with

$$\begin{aligned} \mathbf{G}^{(0)} &= (G_0^{(0)}, G_1^{(0)}, \ldots, G_{2N}^{(0)}), \\ \mathbf{G}^{(c)} &= (\mathbf{G}_1^{(c)}, \mathbf{G}_2^{(c)}, \ldots, \mathbf{G}_N^{(c)})^{\mathrm{T}}, \\ \mathbf{G}^{(s)} &= (\mathbf{G}_1^{(s)}, \mathbf{G}_2^{(s)}, \ldots, \mathbf{G}_N^{(s)})^{\mathrm{T}} \end{aligned} \tag{2.25}$$

for $N = 1, 2, \ldots \infty$ with

$$\begin{aligned} \mathbf{G}_k^{(c)} &= (G_{k0}^{(c)}, G_{k1}^{(c)}, \ldots, G_{k(2N)}^{(c)}), \\ \mathbf{G}_k^{(s)} &= (G_{k0}^{(s)}, G_{k1}^{(s)}, \ldots, G_{k(2N)}^{(s)}) \end{aligned} \tag{2.26}$$

for $k = 1, 2, \ldots N$. The corresponding components are

$$\begin{aligned} G_r^{(0)} &= -\alpha\delta_r^0 - \beta g_{2r}^{(0)}, \\ G_{kr}^{(c)} &= \left(\frac{k\Omega}{m}\right)^2 \delta_k^r - \delta\frac{k\Omega}{m}\delta_{k+N}^r - \alpha\delta_k^r - \beta g_{2kr}^{(c)}, \\ G_{kr}^{(s)} &= \left(\frac{k\Omega}{m}\right)^2 \delta_{k+N}^r + \delta\frac{k\Omega}{m}\delta_k^r - \alpha\delta_{k+N}^r - \beta g_{2kr}^{(s)} \end{aligned} \tag{2.27}$$

where

$$\begin{aligned} g_{2r}^{(0)} = {} & 3(a_0^{(m)})^2\delta_r^0 + \sum_{l=1}^{N}\sum_{j=1}^{N}\sum_{i=1}^{N}\frac{3}{2N}(b_{i/m}b_{j/m}\delta_r^0 + 2a_0^{(m)}b_{i/m}\delta_j^r)\delta_{i-j}^0 \\ & + \frac{3}{2N}(c_{i/m}c_{j/m}\delta_r^0 + 2a_0^{(m)}c_{i/m}\delta_{j+N}^r)\delta_{i-j}^0 \\ & + \frac{3}{4}b_{i/m}b_{j/m}\delta_l^r(\delta_{i-j-l}^0 + \delta_{i-j+l}^0 + \delta_{i+j-l}^0) \\ & + \frac{3}{4}(c_{j/m}c_{l/m}\delta_i^r + 2b_{i/m}c_{j/m}\delta_{l+N}^r)(\delta_{i+j-l}^0 + \delta_{i-j+l}^0 - \delta_{i-j-l}^0) \end{aligned} \tag{2.28}$$

$$\begin{aligned} g_{2kr}^{(c)} = {} & \sum_{l=1}^{N}\sum_{j=1}^{N}\sum_{i=1}^{N} 3\frac{a_0^{(m)}}{N^2}\left[a_0^{(m)}\delta_l^r + 2b_{l/m}\delta_0^r\right]\delta_l^k \\ & + \frac{3}{2N}(b_{l/m}b_{j/m}\delta_0^r + 2a_0^{(m)}b_{j/m}\delta_l^r)(\delta_{|l-j|}^k + \delta_{l+j}^k) \end{aligned}$$

$$
\begin{aligned}
&+\frac{3}{2N}(c_{l/m}c_{j/m}\delta_0^r + a_0^{(m)}c_{j/m}\delta_{l+N}^r)(\delta_{|l-j|}^k - \delta_{l+j}^k)\\
&+\frac{3}{4}b_{j/m}b_{i/m}\delta_l^r(\delta_{|l-j-i|}^k + \delta_{l+j+i}^k + \delta_{|l-j+i|}^k + \delta_{|l+j-i|}^k)\\
&+\frac{3}{4}(c_{j/m}c_{i/m}\delta_l^r + 2b_{l/m}c_{i/m}\delta_{j+N}^r)(\delta_{|l+j-i|}^k - \delta_{l+j+i}^k + \delta_{|l-j+i|}^k - \delta_{|l-j-i|}^k);
\end{aligned}
\tag{2.29}
$$

$$
\begin{aligned}
g_{2kr}^{(s)} = &\sum_{l=1}^{N}\sum_{j=1}^{N}\sum_{i=1}^{N} 3\frac{a_0^{(m)}}{N^2}[a_0^{(m)}\delta_{l+N}^r + 2c_{l/m}\delta_0^r]\delta_l^k\\
&+\frac{3}{N}(a_0^{(m)}c_{j/m}\delta_l^r + a_0^{(m)}b_{l/m}\delta_{j+N}^r + b_{l/m}c_{j/m}\delta_0^r)[\delta_{l+j}^k - \mathrm{sgn}(l-j)\delta_{|l-j|}^k]\\
&+\frac{3}{4}c_{j/m}c_{i/m}\delta_{l+N}^r[\mathrm{sgn}(l-j+i)\delta_{|l-j+i|}^k - \delta_{l+j+i}^k\\
&+\mathrm{sgn}(l+j-i)\delta_{|l+j-i|}^k - \mathrm{sgn}(l-j-i)\delta_{|l-j-i|}^k]\\
&+\frac{3}{4}(b_{l/m}b_{j/m}\delta_{i+N}^r + 2b_{j/m}c_{i/m}\delta_l^r)[\mathrm{sgn}(l-j+i)\delta_{|l-j+i|}^k + \delta_{l+j+i}^k\\
&-\mathrm{sgn}(l+j-i)\delta_{|l+j-i|}^k - \mathrm{sgn}(l-j-i)\delta_{|l-j-i|}^k)]
\end{aligned}
\tag{2.30}
$$

for $r = 0, 1, \ldots 2N$.

$$
\mathbf{H} = \frac{\partial \mathbf{g}^{(m)}}{\partial \mathbf{z}_1{}^{(m)}} = (\mathbf{H}^{(0)}, \mathbf{H}^{(c)}, \mathbf{H}^{(s)})^{\mathrm{T}} \tag{2.31}
$$

where

$$
\begin{aligned}
\mathbf{H}^{(0)} &= (H_0^{(0)}, H_1^{(0)}, \ldots, H_{2N}^{(0)}),\\
\mathbf{H}^{(c)} &= (\mathbf{H}_1^{(c)}, \mathbf{H}_2^{(c)}, \ldots, \mathbf{H}_N^{(c)})^{\mathrm{T}},\\
\mathbf{H}^{(s)} &= (\mathbf{H}_1^{(s)}, \mathbf{H}_2^{(s)}, \ldots, \mathbf{H}_N^{(s)})^{\mathrm{T}}
\end{aligned}
\tag{2.32}
$$

for $N = 1, 2, \ldots \infty$, with

$$
\begin{aligned}
\mathbf{H}_k^{(c)} &= (H_{k0}^{(c)}, H_{k1}^{(c)}, \ldots, H_{k(2N)}^{(c)}),\\
\mathbf{H}_k^{(s)} &= (H_{k0}^{(s)}, H_{k1}^{(s)}, \ldots, H_{k(2N)}^{(s)})
\end{aligned}
\tag{2.33}
$$

for $k = 1, 2, \ldots N$.

The corresponding components are

$$
\begin{aligned}
H_r^{(0)} &= -\delta\delta_0^r,\\
H_{kr}^{(c)} &= -2\frac{k\Omega}{m}\delta_{k+N}^r - \delta\delta_k^r,\\
H_{kr}^{(s)} &= 2\frac{k\Omega}{m}\delta_k^r - \delta\delta_{k+N}^r
\end{aligned}
\tag{2.34}
$$

for $r = 0, 1, \ldots, 2N$.

The corresponding eigenvalues are given by

$$|D\mathbf{f}^{(m)}(\mathbf{y}^{*(m)}) - \lambda \mathbf{I}_{2(2N+1)\times 2(2N+1)}| = 0. \tag{2.35}$$

If $\mathrm{Re}(\lambda_k) < 0$ $(k = 1, 2, \ldots, 2(2N + 1))$, the approximate, steady-state, periodic solution $\mathbf{y}^{(m)*}$ with truncation of $\cos(N\Omega t/m)$ and $\sin(N\Omega t/m)$ is stable. If $\mathrm{Re}(\lambda_k) < 0$ $(k \in \{1, 2, \ldots, 2(2N + 1)\})$, the truncated approximate steady-state solution is unstable. The boundary between the stable and unstable solutions is given by the bifurcation conditions.

For the symmetric motion, $a_0^{(m)} = 0$ is obtained. For one harmonic term balance, setting $m = k = 1$, Equation (2.9) becomes

$$\begin{aligned} &F_0^{(1)}(a_0, b_1, c_1, \dot{a}_0, \dot{b}_1, \dot{c}_1) = 0, \\ &F_{11}^{(1)}(a_0, b_1, c_1, \dot{a}_0, \dot{b}_1, \dot{c}_1) = -\delta(\dot{b}_1 + \Omega c_1) - \alpha b_1 - \beta f_1^{(c)} + Q_0; \\ &F_{21}^{(1)}(a_0, b_1, c_1, \dot{a}_0, \dot{b}_1, \dot{c}_1) = -\delta(\dot{c}_1 - \Omega b_1) - \alpha c_1 - \beta f_1^{(s)}. \end{aligned} \tag{2.36}$$

where for $i = j = l = 1$ Equations (2.10)–(2.12) gives

$$f_0^{(1)} = 0,\ f_1^{(c)} = \frac{3}{4} b_1 (b_1^2 + c_1^2),\ f_1^{(s)} = \frac{3}{4} c_1 (b_1^2 + c_1^2). \tag{2.37}$$

Thus for $m = k = 1$, Equation (2.8) becomes

$$\begin{aligned} &\ddot{a}_0 = 0, \\ &\ddot{b}_1 + 2\Omega \dot{c}_1 - \Omega^2 b_1 = -\delta(\dot{b}_1 + \Omega c_1) - \alpha b_1 - \frac{3}{4}\beta b_1 (b_1^2 + c_1^2) + Q_0, \\ &\ddot{c}_1 - 2\Omega \dot{b}_1 - \Omega^2 c_1 = -\delta(\dot{c}_1 - \Omega b_1) - \alpha c_1 - \frac{3}{4}\beta c_1 (c_1^2 + b_1^2). \end{aligned} \tag{2.38}$$

The algebraic equations for the traditional harmonic balance with one term is given by the equilibrium point of Equation (2.38), that is,

$$\begin{aligned} &-\Omega^2 b_1^* = -\delta\Omega c_1^* - \alpha b_1^* - \frac{3}{4}\beta b_1^* (b_1^{*2} + c_1^{*2}) + Q_0, \\ &-\Omega^2 c_1^* = \delta\Omega b_1^* - \alpha c_1^* - \frac{3}{4}\beta c_1^* (b_1^{*2} + c_1^{*2}). \end{aligned} \tag{2.39}$$

Setting $A_1^2 = c_1^{*2} + b_1^{*2}$, and deformation of Equation (2.39) produces

$$(\delta\Omega)^2 A_1^2 + A_1^2 \left[(\alpha - \Omega^2) + \frac{3}{4}\beta A_1^2 \right]^2 = Q_0^2. \tag{2.40}$$

From Equation (2.40), the first harmonic amplitude can be determined. Further, from Equation (2.39), the coefficient b_1^* and c_1^* are determined. The corresponding stability and bifurcations can be determined from the eigenvalue analysis of the linearized equation of Equation (2.38). At the equilibrium point (b_1^*, c_1^*), the linearized equation is

$$\ddot{\mathbf{u}} + \mathbf{C}\dot{\mathbf{u}} + \mathbf{K}\mathbf{u} = \mathbf{0}, \tag{2.41}$$

where

$$\mathbf{u} = (\Delta b_1, \Delta c_1)^{\mathrm{T}}, \dot{\mathbf{u}} = (\Delta \dot{b}_1, \Delta \dot{c}_1)^{\mathrm{T}}, \ddot{\mathbf{u}} = (\Delta \ddot{b}_1, \Delta \ddot{c}_1)^{\mathrm{T}}$$

$$\mathbf{C} = \begin{bmatrix} \delta & 2\Omega \\ -2\Omega & \delta \end{bmatrix}, \mathbf{K} = \begin{bmatrix} K_{11} & K_{12} \\ K_{21} & K_{22} \end{bmatrix};$$

$$K_{11} = \alpha - \Omega^2 + \frac{3}{4}\beta\left(3b_1^{*2} + c_1^{*2}\right), K_{12} = \delta\Omega + \frac{3}{2}\beta b_1^* c_1^*,$$

$$K_{21} = -\delta\Omega + \frac{3}{2}\beta b_1^* c_1^*, K_{22} = \alpha - \Omega^2 + \frac{3}{4}\beta\left(b_1^{*2} + 3c_1^{*2}\right). \tag{2.42}$$

The eigenvalues of the linearized equation is determined by

$$|\lambda^2 \mathbf{I} + \lambda \mathbf{C} + \mathbf{K}| = \mathbf{0}. \tag{2.43}$$

In other words,

$$\begin{vmatrix} \lambda^2 + \delta\lambda + K_{11} & 2\Omega\lambda + K_{12} \\ -2\Omega\lambda + K_{21} & \lambda^2 + \delta\lambda + K_{22} \end{vmatrix} = 0. \tag{2.44}$$

From the eigenvalues, the stability and bifurcation of approximate symmetric period-1 motion are determined. For one harmonic term, the symmetric period-1 motion cannot be approximated well.

The truncated harmonic series solutions will be used as an approximate solution for period-m motions. The harmonic amplitudes varying with excitation frequency Ω are presented with the kth order harmonic amplitude and phase as

$$A_{k/m} \equiv \sqrt{b_{k/m}^2 + c_{k/m}^2}, \varphi_{k/m} = \arctan\frac{c_{k/m}}{b_{k/m}} \tag{2.45}$$

and the corresponding solution in Equation (2.5) is

$$x^*(t) = a_0^{(m)} + \sum_{k=1}^{N} A_{k/m}\cos\left(\frac{k}{m}\Omega t - \varphi_{k/m}\right). \tag{2.46}$$

As in Luo and Huang (2013a,2013b), analytical bifurcation trees of period-m motions to chaos in a periodically forced, hardening Duffing oscillator will be presented. The bifurcation trees from period-1 motions to chaos will be discussed first. The bifurcation tree from period-3 motion to chaos will be presented as an example for period-m motions to chaos. In numerical illustrations, without losing generality, the selected system parameters are

$$\delta = 0.2, \alpha = 1.0, \beta = 4.0, Q_0 = 100.0. \tag{2.47}$$

2.2 Period-1 Motions to Chaos

In this section, analytical bifurcation trees of period-1 motions to chaos in a periodically forced, hardening Duffing oscillator will be discussed. The approximate solution with a few harmonic terms for the period-1 motion is presented first, which is appropriate for large excitation frequency as other parameters are specified. To obtain adequate analytical solution, more harmonic terms in the approximate solutions will be considered. In the vicinity of $\Omega = 0$, infinite harmonic terms should be used to determine approximate solutions of period-1 motions.

2.2.1 Period-1 Motions

For symmetric period-1 motion, the first three harmonic terms of the Fourier series expansion (HB3) will be used to obtain the approximate periodic solutions. The first three harmonic amplitudes (A_k) and phases (φ_k) ($k = 1, 3$) versus excitation frequency are plotted in Figure 2.1(a)–(d), respectively. The solid and dashed curves represent the stable and unstable periodic solutions based on the three terms of the harmonic balance (HB3), respectively. The acronyms "SN" and "HB" represent the saddle-node bifurcation and Hopf bifurcation, respectively. The acronyms "S" and "A" represent the symmetric and asymmetric periodic motions, accordingly. In Figure 2.1(a) and (b), the frequency-amplitude curves are presented. For symmetric period-1 motion, $a_0 = A_2 = 0$. From the approximate analysis, the saddle-node bifurcations occur at $\Omega \approx 2.145, 2.245, 2.575, 7.745, 29.52$. $\Omega = 4.25$ is for Hopf bifurcation. The frequency-amplitude curve (Ω, A_1) in Figure 2.1(a) is similar to the one harmonic term. The upper and lower stable branches of solutions exist. The unstable solution is in the middle branch, which is similar to the traditional analysis. The upper branch of symmetric period-1 solution is in $A_1 \in (1.0, 20)$ for $\Omega \in (0, 29.52)$. The lower branch of symmetric period-1 solution is in $A_1 \in (0.0, 2.0)$ for $\Omega \in (7.745, 35)$. From the frequency-amplitude curve (Ω, A_3) in Figure 2.1(b), the higher order harmonics contribution to the upper and lower branches are less than 10% and 1% for $\Omega > 5$, respectively. However, for $\Omega < 5$, the higher order harmonic contribution to the upper branch solution are the same quantity level. So many higher order harmonic terms should be taken into account. The corresponding phase are presented in Figure 2.1(c) and (d). To make illustrations clear, the asymmetric period-1 motion based on the three harmonic terms is presented in Figure 2.2(i)–(vii). For the asymmetric period-1 motion, $a_0 \neq 0$ and $A_2 \neq 0$. The asymmetric period-1 motion exists in about $\Omega \in (2.24, 4.14)$ and there are four parts of stable motion and four parts of unstable motion. In Figure 2.2(i) the constant term coefficient is presented, and the symmetric period-1 motion with $a_0 = 0$ is observed. The eigenvalue analysis gives the saddle-node bifurcation at $\Omega = 2.34, 3.73, 4.05$ and Hopf bifurcation at $\Omega = 2.28, 2.79, 2.96, 3.61, 3.75$. The saddle-node bifurcations of the symmetric and asymmetric period-1 motion are not the intersected points. In Figure 2.2(ii), the frequency-amplitude curve (Ω, A_1) for asymmetric period-1 motion is presented. The frequency-amplitude curve (Ω, A_2) for asymmetric period-1 motion is presented in Figure 2.2(iii) and the symmetric period-1 motion with $A_2 = 0$ is presented as well. The frequency-amplitude curve (Ω, A_3) for asymmetric and symmetric period-1 motion is presented in Figure 2.2(iv). The phase varying with excitation frequency relative to the first, second, and third harmonic terms are presented in Figure 2.2(v)–(vii). For symmetric motion, the phase is $\varphi_2 = 2\pi$.

Using the three harmonic terms, the parameter map (Ω, Q_0) are presented in Figure 2.3 for the period-1 motion, and the corresponding domain are labeled. In the parameter map, the acronyms "U" and "S" are for unstable and stable period-1 motions, respectively. $S^m U^n$ means that m stable period-1 motions and n unstable period-1 motions co-exist. For $m = 0$, U^n means that n unstable period-1 motions co-exist. For $n = 0$, S^m means that m stable period-1 motions co-exist. The Hopf bifurcation boundaries are given by dashed curves. The saddle-node bifurcation boundaries are given by solid curves. The dash-dot curves give the saddle switching. In Figure 2.3(a), the global view of the parameter map is given and the zoomed view of the local details is presented in Figure 2.3(b) and (c). Again the parameter map for the lower excitation frequency region may not be accurate, and a more comprehensive investigation should be completed.

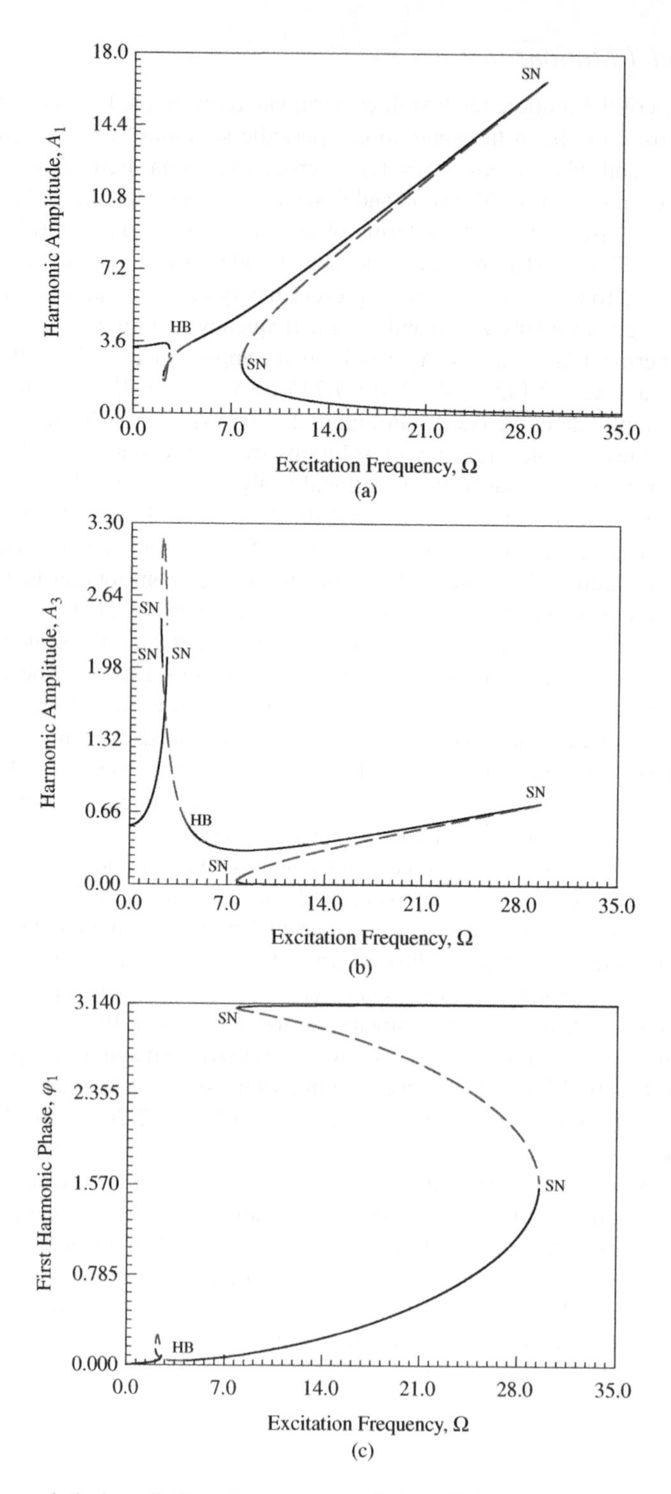

Figure 2.1 The analytical prediction of symmetric period-1 solutions based on three harmonic terms (HB3): (a) and (b) harmonic amplitudes A_k $(k = 1, 3)$; (c) and (d) harmonic phases φ_k $(k = 1, 3)$. ($\delta = 0.2$, $\alpha = 1.0$, $\beta = 4$, $Q_0 = 100.0$)

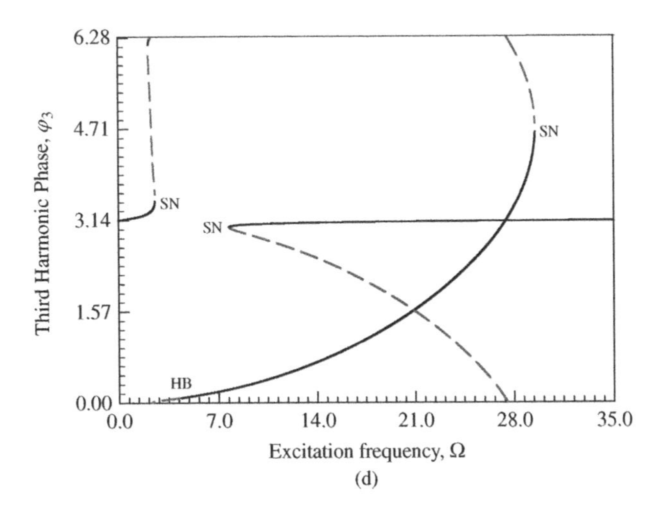

Figure 2.1 *(continued)*

2.2.2 Period-1 to Period-4 Motions

From the discussion in the previous section, it is observed that the approximate solutions are not accurate enough. For $\Omega < 5$, the higher order harmonic contribution to the upper branch solution are the same quantity level. To obtain the appropriate analytical solution for period-1 motion, 10 harmonic terms are used here to give the period-1 motions. From the new solutions of period-1 motions, the corresponding stability and bifurcation will be determined, and period-2 motions with 20 harmonic terms in the approximate solutions will be determined from the Hopf bifurcation of period-1 motions. In a similar fashion, period-4 motions with 40 harmonic terms in the approximate solutions will be determined from the Hopf bifurcation of period-2 motions. Continuously, from this bifurcation trees, chaos relative to period-1 motions can be found, and the analytical bifurcation trees of period-1 motion to chaos can be determined.

The symmetric and asymmetric period-1 motions are presented in Figure 2.4(i)–(xi). In Figure 2.4(i) constant term coefficients varying with excitation frequency are presented. The asymmetric period-1 motion exists in $\Omega \in (2.306, 4.354)$ different from $\Omega \in (2.24, 4.14)$ given by three harmonic terms. The stability ranges and bifurcation points are different because the approximate, period-1 solution with three harmonic terms cannot provide the accurate period-1 solutions for $\Omega < 5$. For the asymmetric period-1 motion, the range of excitation frequency is in $\Omega \in (2.0, 5.0)$. In Figure 2.4(ii), the frequency-amplitude curves (Ω, A_1) for period-1 motions are presented. The solutions of symmetric period-1 motion given by ten harmonic terms are almost the same as by the three harmonic terms for $\Omega > 5$. However, for $\Omega < 5$, the solutions based on the three and ten harmonic terms are different because the higher order harmonic terms have significant contributions on the solutions. In Figure 2.4(iii), the frequency-amplitude curves (Ω, A_2) for asymmetric period-1 motions are presented because of $A_2 = 0$ for this symmetric period-1 motion. Comparing this ten harmonic term solution with the three harmonic term solution, the frequency amplitude curves for asymmetric period-1 motion are modified with $a_0, A_2 \sim 4$ for $\Omega \in (1.5, 5.0)$. In Figure 2.4(iv), the frequency-amplitude curves (Ω, A_3) for symmetric and asymmetric period-1 motions are presented. The amplitude A_3 based on the ten and three harmonic terms are almost the

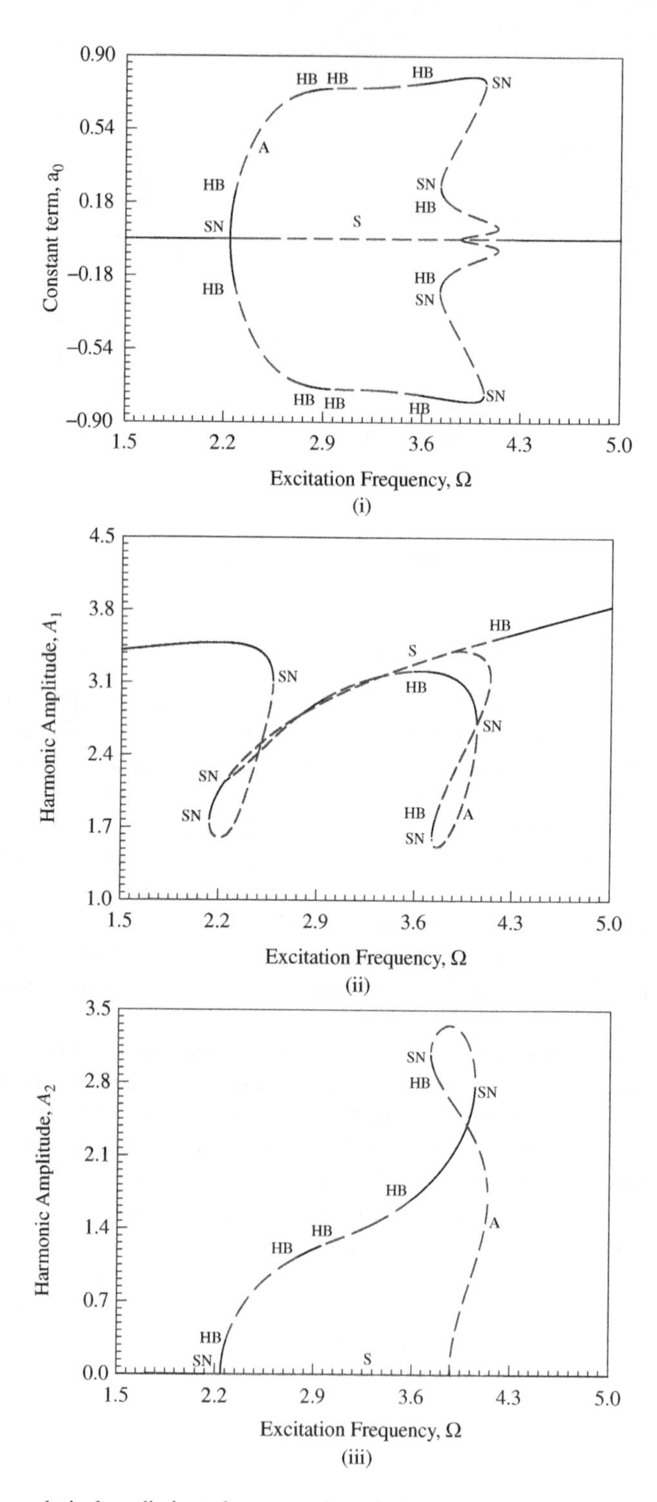

Figure 2.2 The analytical prediction of asymmetric period-1 solutions based on three harmonic terms (HB3): (i) constant term a_0, (ii)–(iv) harmonic amplitudes A_k ($k = 1, 2, 3$); (v)–(vii) harmonic phases φ_k ($k = 1, 3$) ($\delta = 0.2$, $\alpha = 1.0$, $\beta = 4$, $Q_0 = 100.0$)

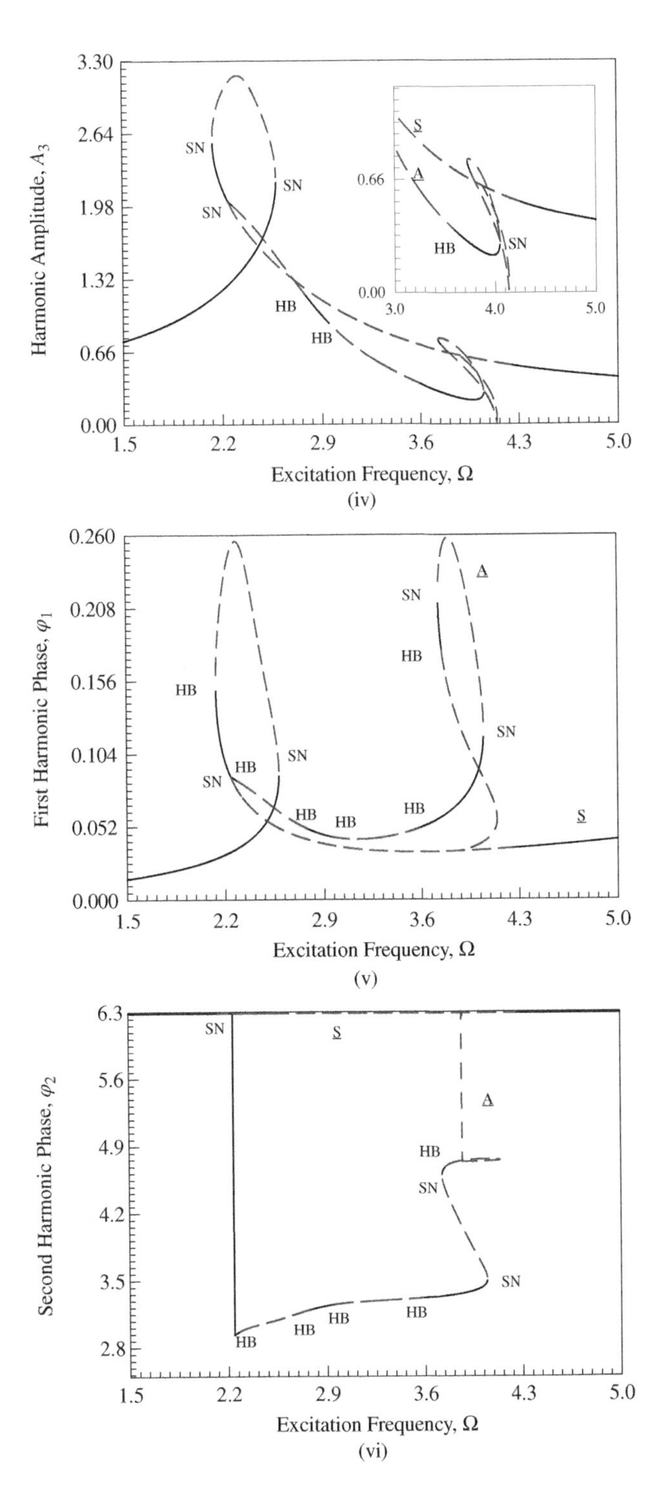

Figure 2.2 *(continued)*

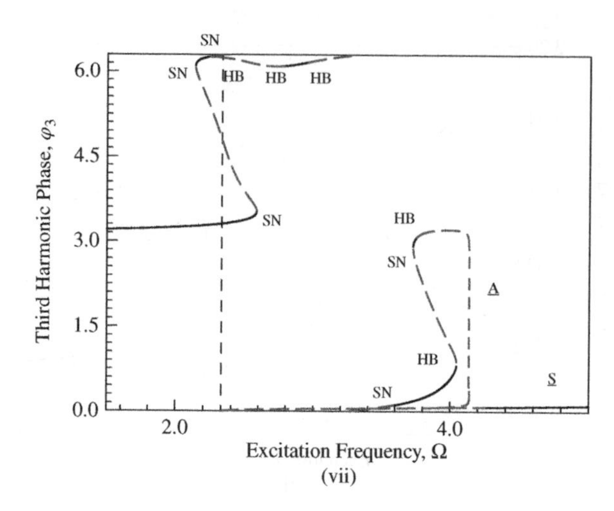

Figure 2.2 (*continued*)

same for $\Omega > 5$. However, for $\Omega < 5$, the symmetric and asymmetric harmonic amplitudes for symmetric and asymmetric period-1 motion are quite different because the higher order harmonic terms possess more effects on the period-1 motion solutions. For asymmetric motion, $A_4 \sim 10^{-1}$, and $A_5 \sim 2$ in Figure 2.4(v) and (vi). In Figure 2.4(vii)–(xi), $A_k \sim 10^{-2}$ $(k = 6, 8, 10)$ and $A_{7,9} \sim 10^{-1}$ for $\Omega < 3$. However, for $\Omega > 3$, $A_5 \leq 0.04$, $A_7 \leq 0.002$, and $A_9 \leq 10^{-4}$. Thus, for symmetric period-1 motion, the Fourier series solution with three harmonic terms can give a good approximation for $\Omega > 5$. From the quantity level of harmonic response amplitudes, effects of the harmonic terms on the solutions can be observed.

The asymmetric period-1 motion with ten harmonic terms has a Hopf bifurcation. Thus a period-2 motion will be formed from such an asymmetric period-1 motion. If this period-2 motion has a Hopf bifurcation, the period-4 motion will appear. The analytical bifurcation route of an asymmetric period-1 motion to period-4 motions $(m = 4)$ is presented in Figure 2.5(i)–(xxiv) through the constant terms $(a_0^{(m)})$ for the left and right sides of the symmetric motion and the harmonic amplitude $A_{k/m}$ $(k = 1, 2, \ldots, 12)$, $A_{k/m}$ $(k = 16, 20, \ldots, 36)$ and $A_{k/m}$ $(k = 37, 38, 39, 40)$. From asymmetric period-1 motion, at $\Omega \approx 3.447$, the approximate solution for asymmetric period-2 motions is obtained. From the asymmetric period-2 motion, at $\Omega \approx 3.321$, the approximate solution for asymmetric period-4 motions is obtained. From the asymmetric period-4 motion, at $\Omega \approx 3.306$, the approximate solutions for asymmetric period-8 motion can be obtained. Continuously, the chaotic motions for such a hardening Duffing oscillator can be achieved. $a_0^{(m)} \sim 1$ for $\Omega \in (3.0, 3.5)$. In Figure 2.5(iii), the harmonic amplitude $A_{1/4} \sim 10^{-1}$ for period-4 motion is presented and $A_{1/4} = 0$ for period-2 and period-1 motions. In Figure 2.5(iv), the harmonic amplitude $A_{1/2} \sim 10^{-1}$ for period-2 and period-4 motions are presented and $A_{1/2} = 0$ for period-1 motion. In Figure 2.5(v), the harmonic amplitude $A_{3/4} \sim 10^{-2}$ for period-4 motion is presented and $A_{3/4} = 0$ for period-2 and period-1 motions. In Figure 2.5(vi), the harmonic amplitude A_1 for period-1, period-2, period-4 motions are presented and $A_1 \in (2.8, 3.13)$ for $\Omega \in (3.0, 3.5)$. In Figure 2.5(vii), the harmonic amplitude $A_{5/4} \sim 10^{-2}$ for period-4 motion is presented and $A_{5/4} = 0$ for period-2 and period-1 motions. In Figure 2.5(viii), the harmonic amplitude $A_{3/2} \sim 10^{-1}$ for period-2 and period-4 motions are presented and $A_{3/2} = 0$ for

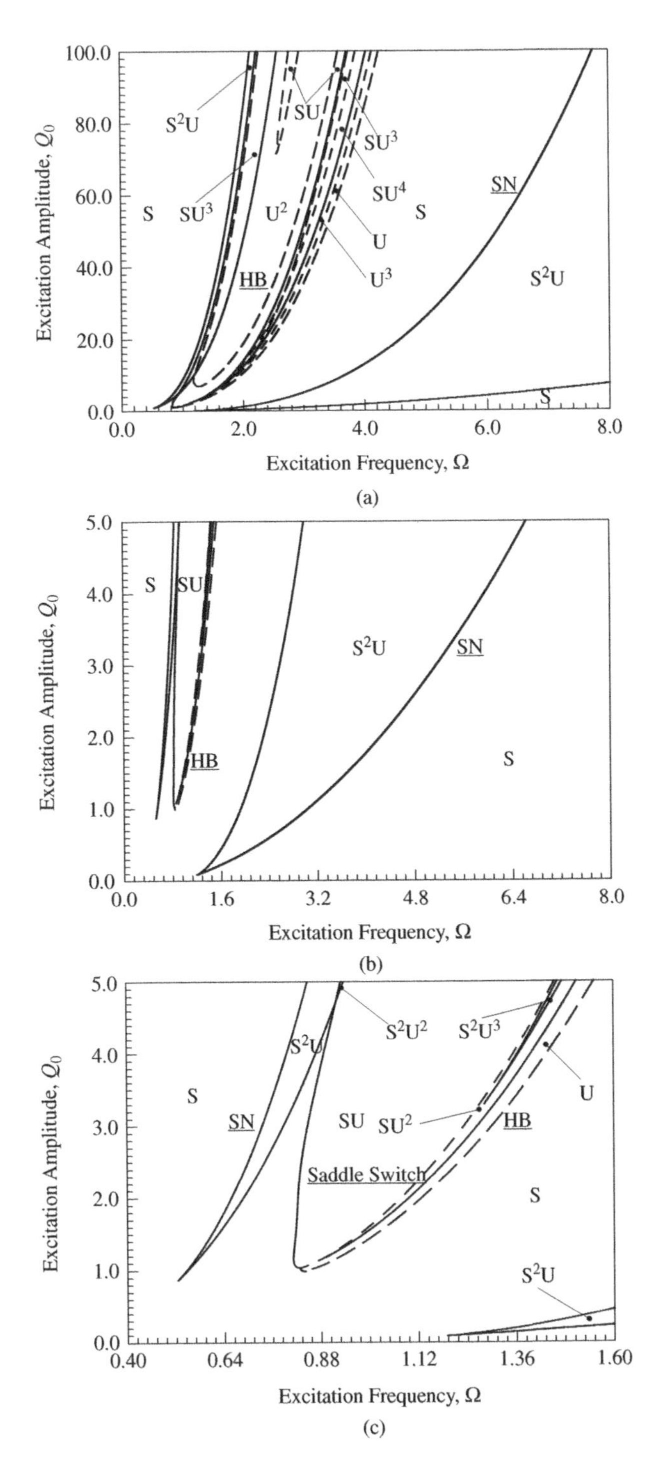

Figure 2.3 A parameter map from the analytical prediction of periodic solutions based on three harmonic terms (HB3): (a) global view, (b) zoomed view, and (c) zoomed view ($\delta = 0.2, \alpha = 1.0, \beta = 4.0$)

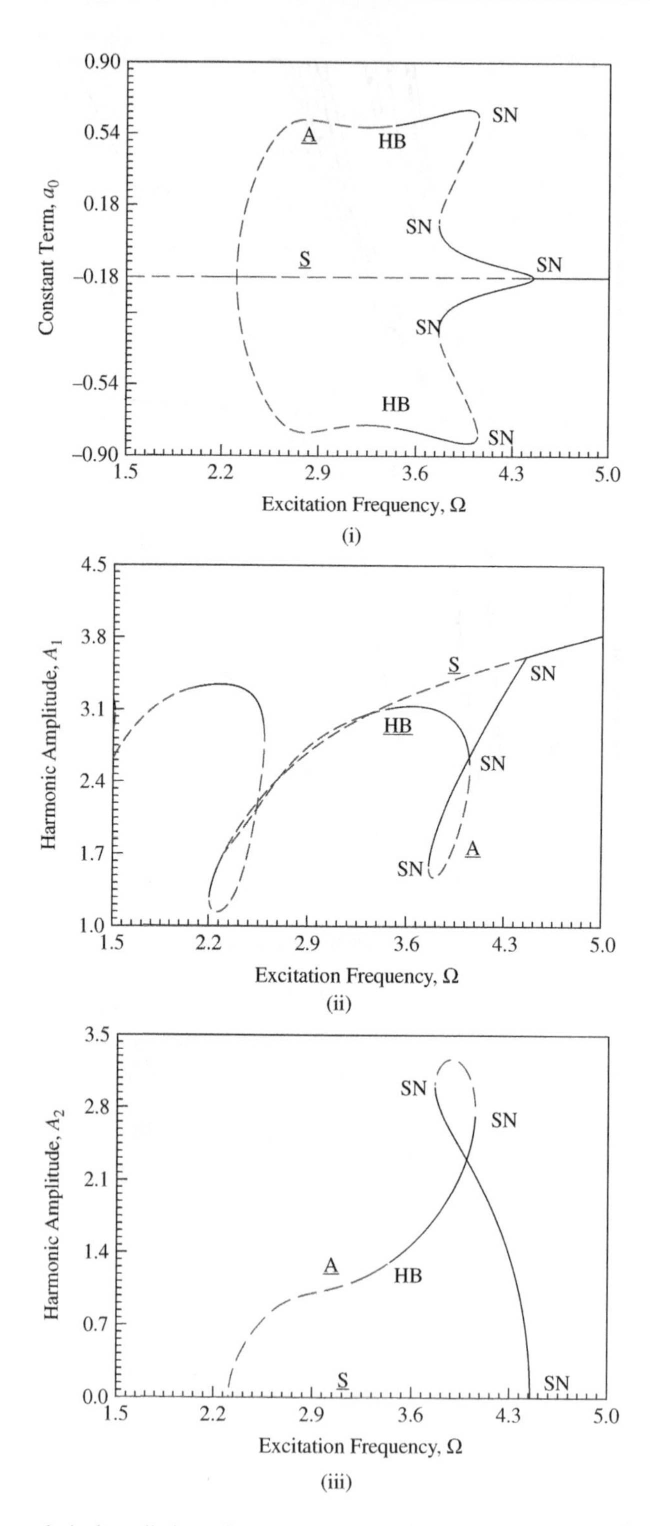

Figure 2.4 The analytical prediction of asymmetric period-1 solutions based on ten harmonic terms (HB10): (i) constant term a_0 and zoomed view, (ii)–(xi) harmonic amplitudes and zoomed views A_k ($k = 1, 2, \ldots, 10$) ($\delta = 0.2$, $\alpha = 1.0$, $\beta = 4$, $Q_0 = 100.0$)

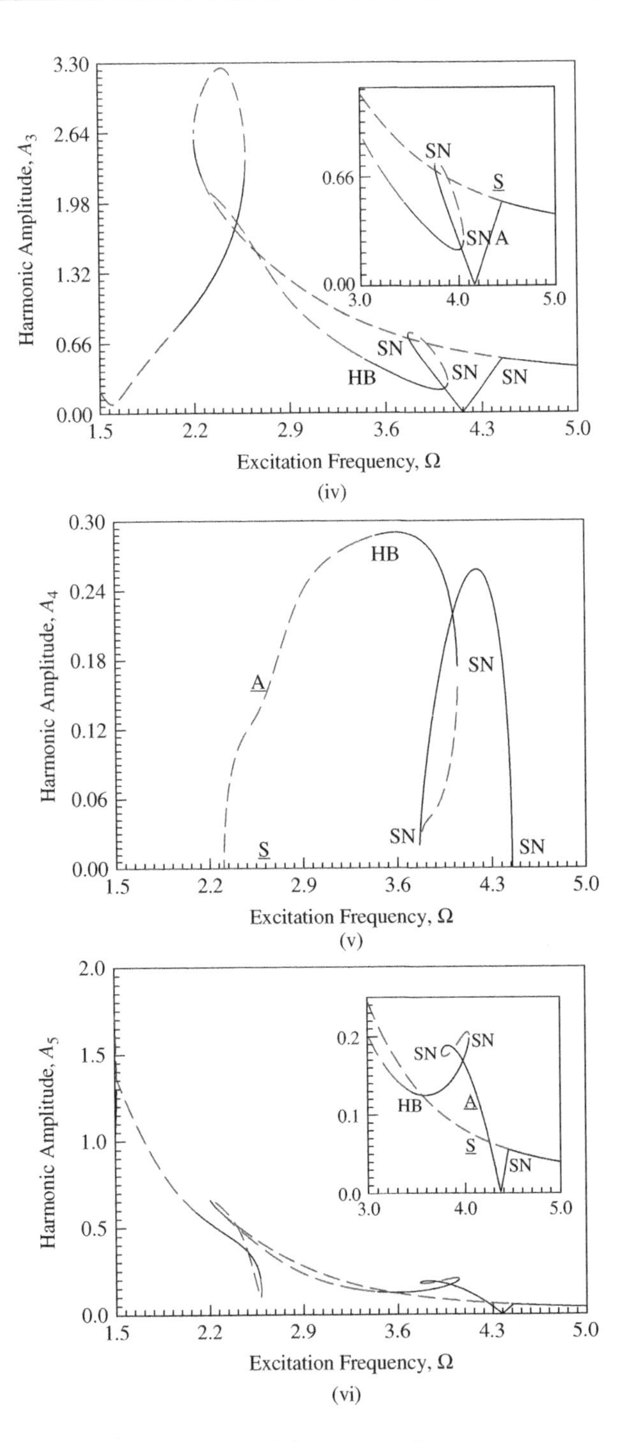

Figure 2.4 (*continued*)

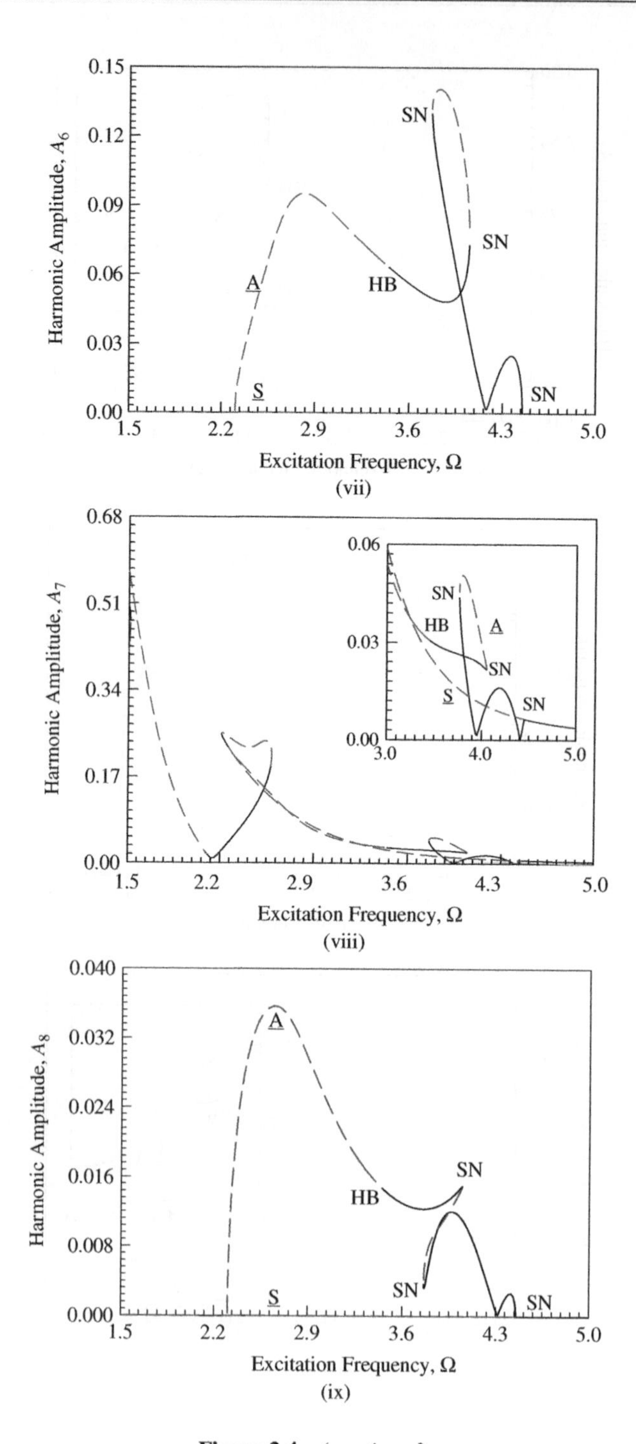

Figure 2.4 (*continued*)

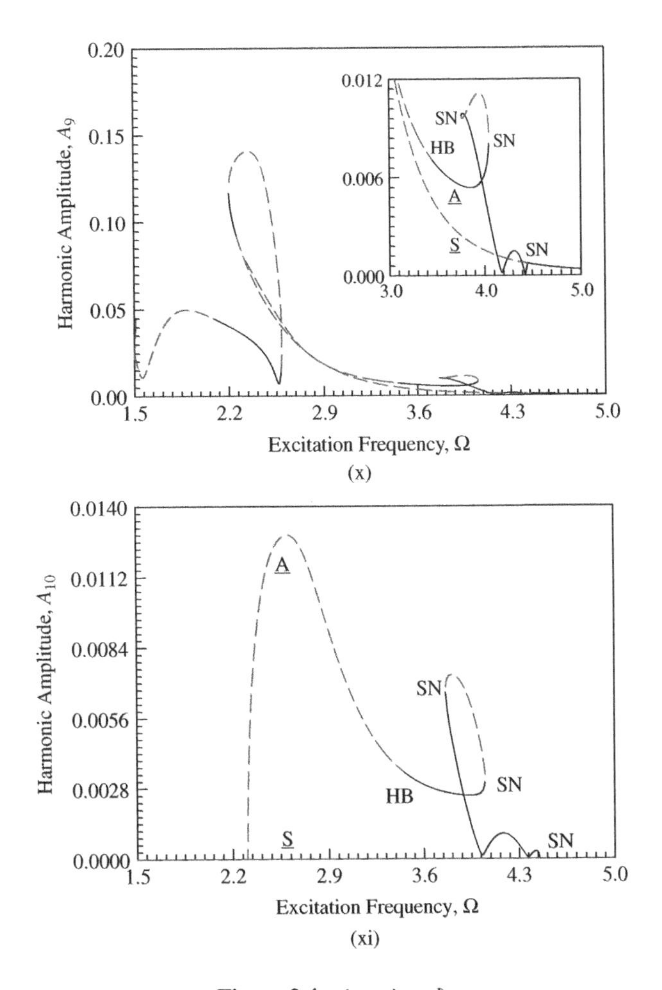

Figure 2.4 *(continued)*

period-1 motion. In Figure 2.5(ix), the harmonic amplitude $A_{7/4} \sim 10^{-1}$ for period-4 motion is presented and $A_{7/4} = 0$ for period-2 and period-1 motions. In Figure 2.5(x), the harmonic amplitude A_2 for period-1, period-2, period-4 motions are presented and $A_2 \in (0.75, 1.40)$ for $\Omega \in (3.0, 3.5)$. In Figure 2.5(xi), the harmonic amplitude ($A_{9/4} \sim 10^{-1}$) for period-4 motion is presented and $A_{9/4} = 0$ for period-2 and period-1 motions. In Figure 2.5(xii), the harmonic amplitude $A_{5/2} \sim 10^0$ for period-2 and period-4 motions are presented and $A_{5/2} = 0$ for period-1 motion. In Figure 2.5(xiii), the harmonic amplitude ($A_{11/4} \sim 10^{-2}$) for period-4 motion is presented and $A_{11/4} = 0$ for period-2 and period-1 motions. In Figure 2.5(xiv), the harmonic amplitude A_3 for period-1, period-2, and period-4 motions are presented and $A_3 \in (0.45, 1.0)$ for $\Omega \in (3.0, 3.5)$. To avoid abundant illustrations, only harmonic terms $A_{k/m}$ ($m = 4$, $k = 16, 20, \ldots, 36$) are presented in Figure 2.5(xvi)–(xx) or period-1, period-2, and period-4 motions. $A_{4,5,6} \sim 10^{-1}$ and $A_{7,8,9} \sim 10^{-2}$. To show convergence, in Figure 2.5(xxi), the harmonic amplitude ($A_{37/4} \sim 10^{-3}$) for period-4 motion is presented and $A_{37/4} = 0$ for period-2 and period-1 motions. In Figure 2.5(xxii), the harmonic amplitude $A_{19/2} \sim 10^{-3}$ for period-2 and period-4 motions are presented and $A_{19/2} = 0$ for period-1

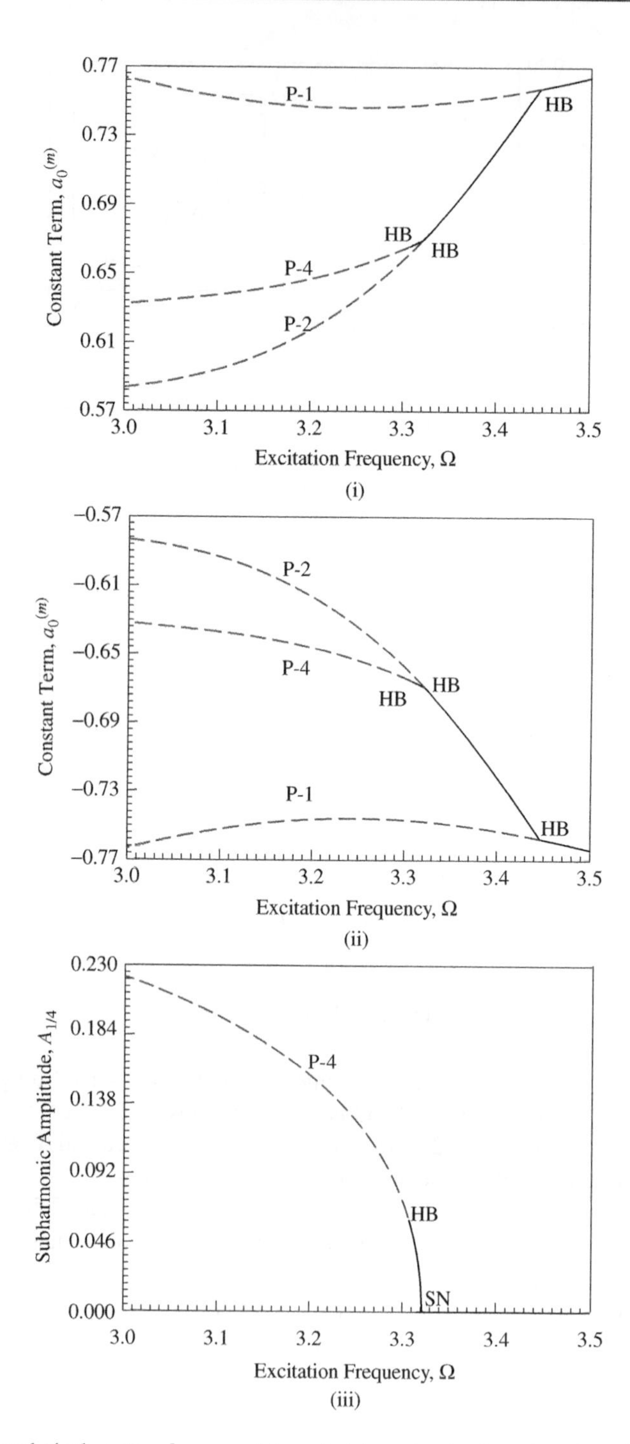

Figure 2.5 The analytical routes of asymmetric period-1 motion to chaos based on 40 harmonic terms (HB40): (i) and (ii) constant term $a_0^{(m)}$, (iii)–(xxiv) harmonic amplitudes $A_{k/m}$ ($k = 1, 2, \ldots, 12, m = 4$), ($k = 16, 20, \ldots, 36, m = 4$), and ($k = 37, 38, 39, 40, m = 4$) ($\delta = 0.2$, $\alpha = 1.0$, $\beta = 4$, $Q_0 = 100.0$)

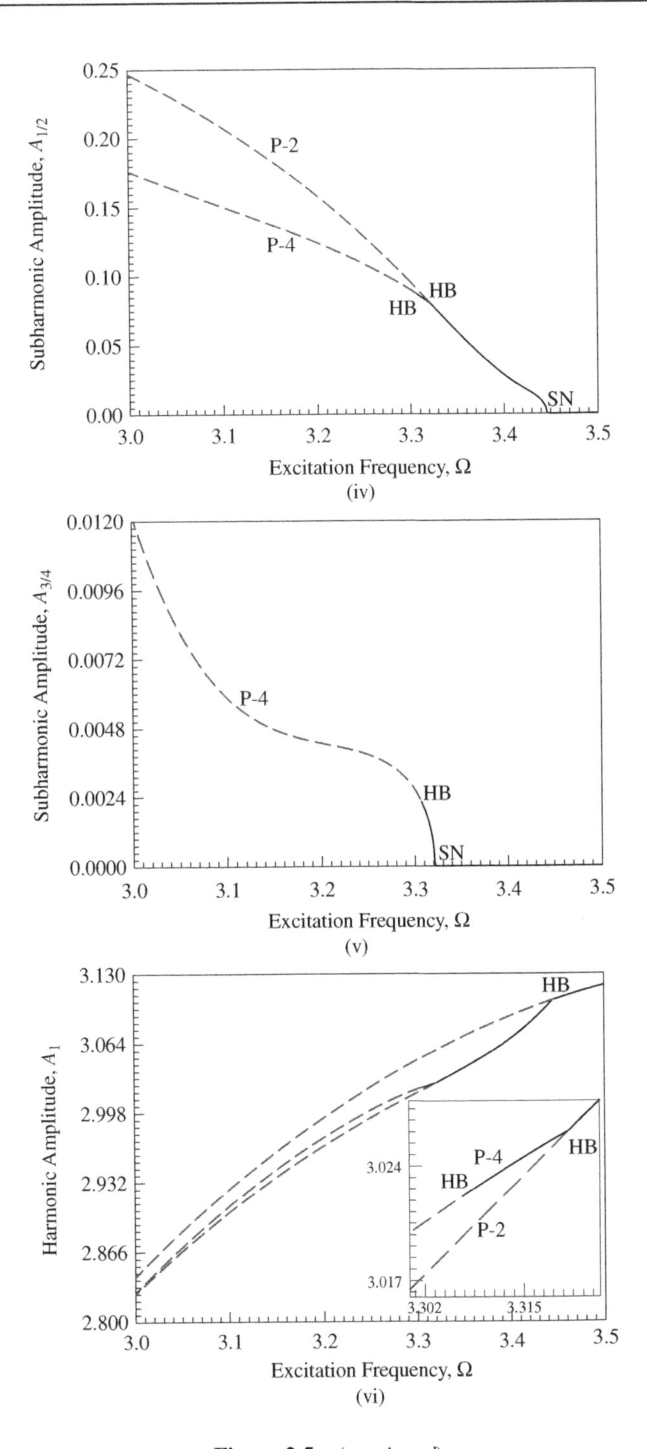

Figure 2.5 *(continued)*

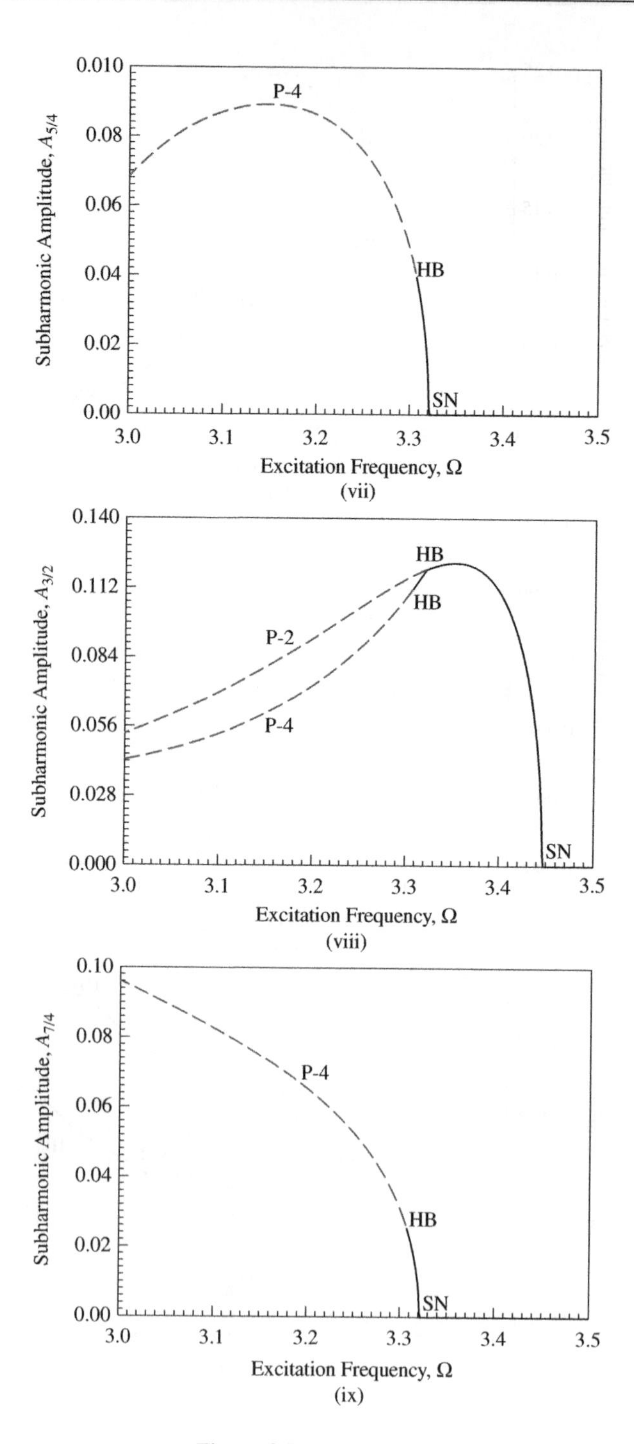

Figure 2.5 (*continued*)

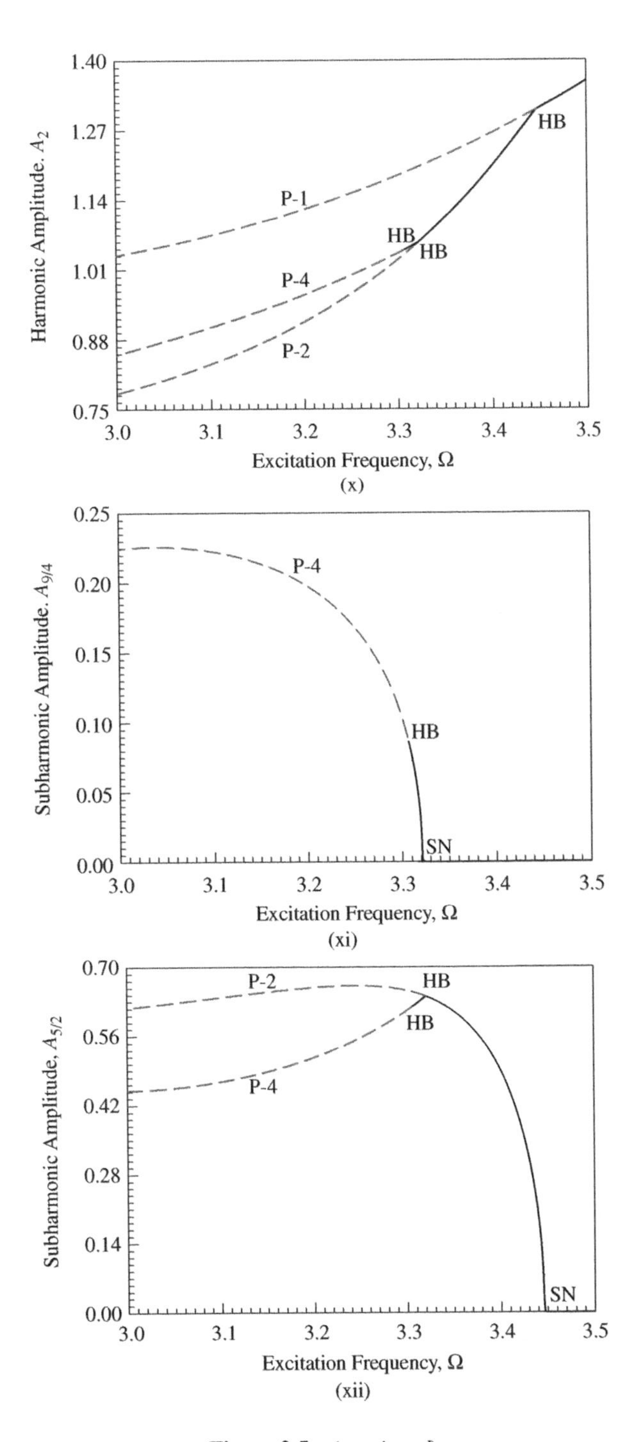

Figure 2.5 (*continued*)

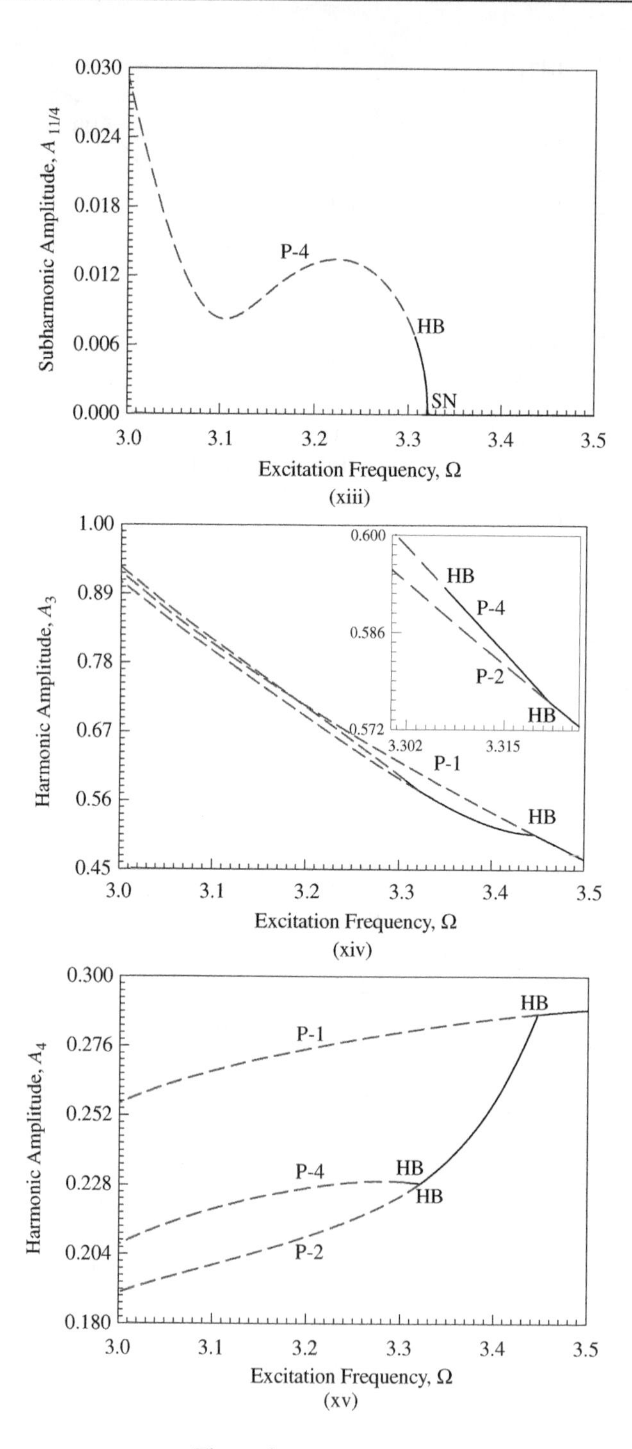

Figure 2.5 *(continued)*

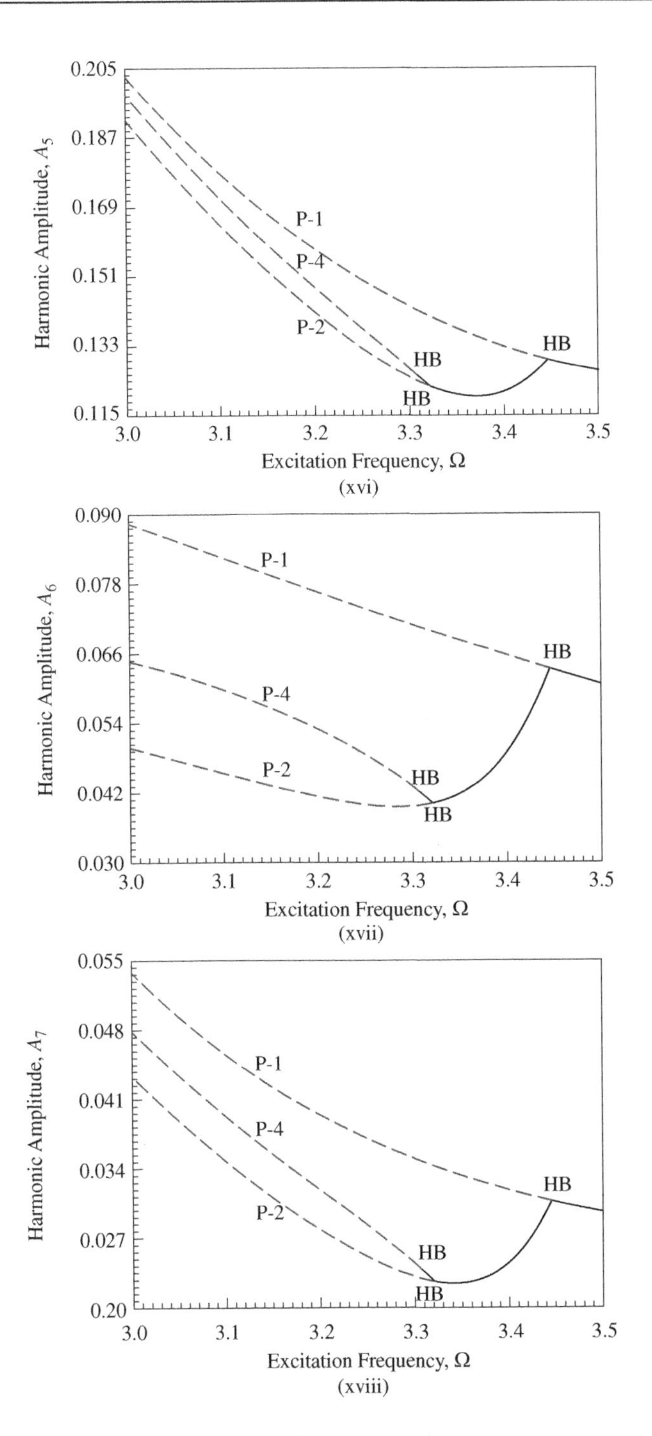

Figure 2.5 *(continued)*

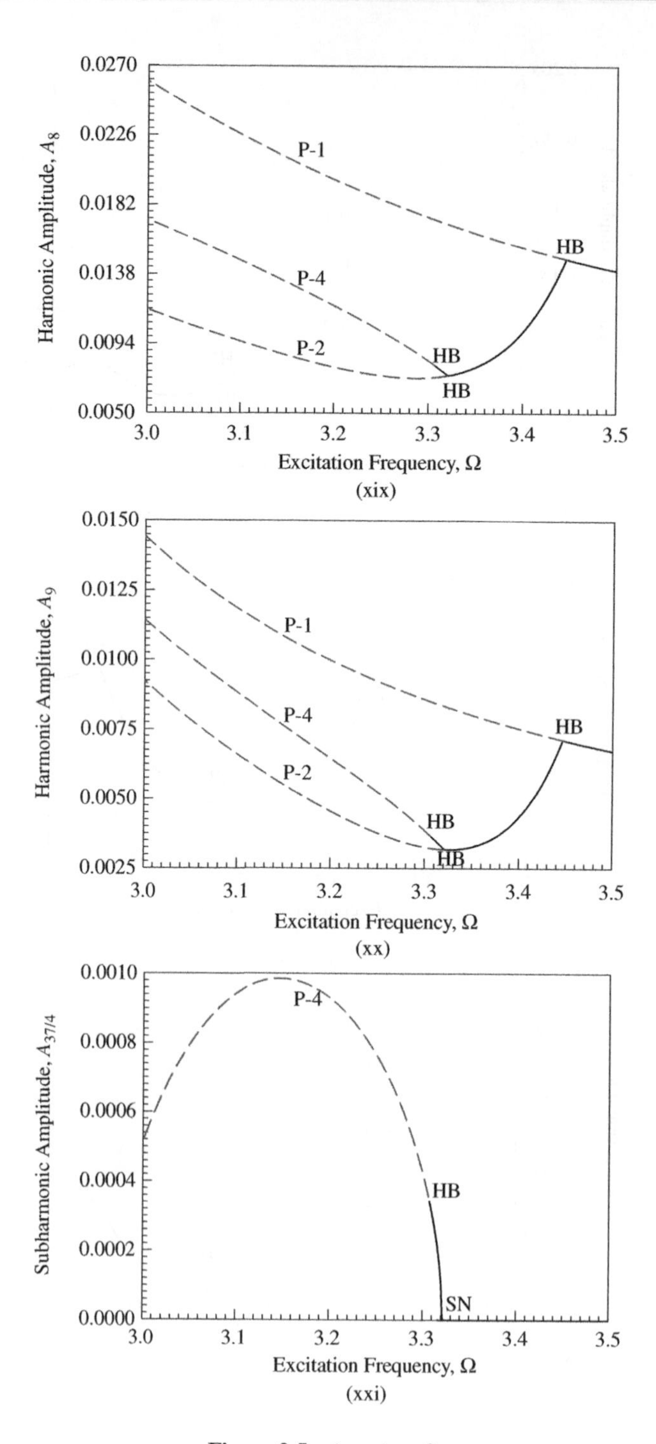

Figure 2.5 (*continued*)

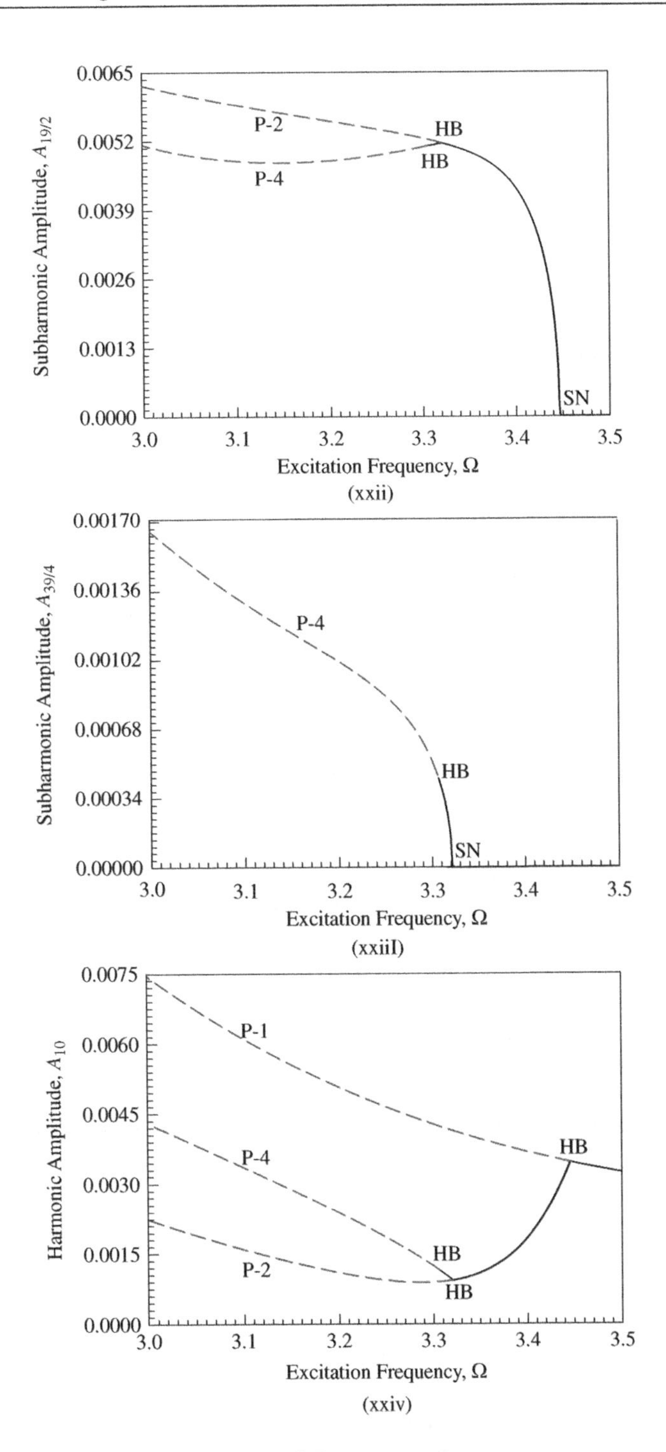

Figure 2.5 *(continued)*

motion. In Figure 2.5(xxiii), the harmonic amplitude ($A_{39/4} \sim 10^{-3}$) for period-4 motion is presented and $A_{39/4} = 0$ for period-2 and period-1 motions. In Figure 2.5(xxiv), the harmonic amplitude A_{10} for period-1, period-2, and period-4 motions are presented and $A_{10} \in (0.0015, 0.0075)$ for $\Omega \in (3.0, 3.5)$. The harmonic phases for right and left asymmetry have a relation like $\varphi^L_{k/m} = \text{mod}(\varphi^R_{k/m} + (k+1)\pi, 2\pi)$, which is not presented herein.

2.2.3 Numerical Simulations

In this section, the initial conditions for numerical simulations are computed from approximate analytical solutions of periodic solutions. In all plots, circular symbols gives approximate solutions, and solid curves give numerical simulation results. The acronym "I.C." with a large circular symbol represents initial condition for all plots. The numerical solutions of periodic motions are generated via the symplectic scheme.

The displacement and trajectory in phase plane for the approximate solutions of stable and unstable symmetric period-1 motion are illustrated in Figure 2.6(i)–(vi). In addition, the numerical simulations are superimposed, and the initial conditions are obtained from the approximate solutions of period-1 motion. The analytical solution of period-1 motion is given by the Fourier series with the 10 harmonic terms (HB10). For the upper branch of stable symmetric period-1 motion, the initial conditions are $t_0 = 0.0$, $x_0 \approx 6.296950$, and $y_0 \approx 11.422100$ for $\Omega = 10$ with other parameters in Equation (2.47). For the lower branch of stable symmetric period-1 motion with the same parameters, the initial conditions are $t_0 = 0.0$, $x_0 \approx -1.045500$, and $y_0 \approx 0.220845$ for $\Omega = 10$. For the middle branch of unstable symmetric period-1 motion with the same parameters, the initial conditions are $t_0 = 0.0$, $x_0 \approx -5.192940$, and $y_0 \approx 6.918470$ for $\Omega = 10$.

In Figure 2.6(i) and (ii), the analytical and numerical solutions overlap each other for the displacement and trajectory of the symmetric period-1 motion. The symmetry of displacement is observed. For 40 periods, the analytical and numerical trajectories of the symmetric period-1 motion in phase plane are plotted and both analytical and numerical results match very well. The motion for one period is labeled. The analytical and numerical solutions for the displacement of the symmetric period-1 motion on the lower branch are presented in Figure 2.6(iii) and (iv), respectively. The analytical and numerical results are in good agreement. For the unstable symmetric period-1 motion, the numerical and analytical displacement and trajectory are presented in Figure 2.6(v) and (vi), respectively. For the first few periods, analytical and numerical unstable period-1 motions match very well. However, after a few periods, the numerical unstable period-1 motion moves away and arrives to a new periodic motion. Such a new stable periodic motion is symmetric period-3 motion, which will be discussed in the next section.

In this chapter, the asymmetric periodic motion is of great interest. The displacement, velocity, and trajectory in the phase plane will be illustrated. Taking account of $\Omega = 4.1$, a stable asymmetric motion and an unstable symmetric motion coexist. For the stable asymmetric period-1 motion, the initial condition is $t_0 = 0.0$, $x_0 \approx 2.735740$, and $y_0 \approx -19.407900$. For the unstable symmetric period-1 motion, the initial condition is $t_0 = 0.0$, $x_0 \approx 4.137450$, and $y_0 \approx 1.295130$. Numerical and analytical solutions match very well. After 40 excitation periods, the numerical and analytical solutions of the stable asymmetric period-1 motion perfectly match as shown in Figure 2.7. Compared to the unstable symmetric period-1 motion in Figure 2.6(v) and (vi), this unstable symmetric period-1 motion possesses a different trajectory shape and its numerical solutions move away to a stable asymmetric period-1 motion.

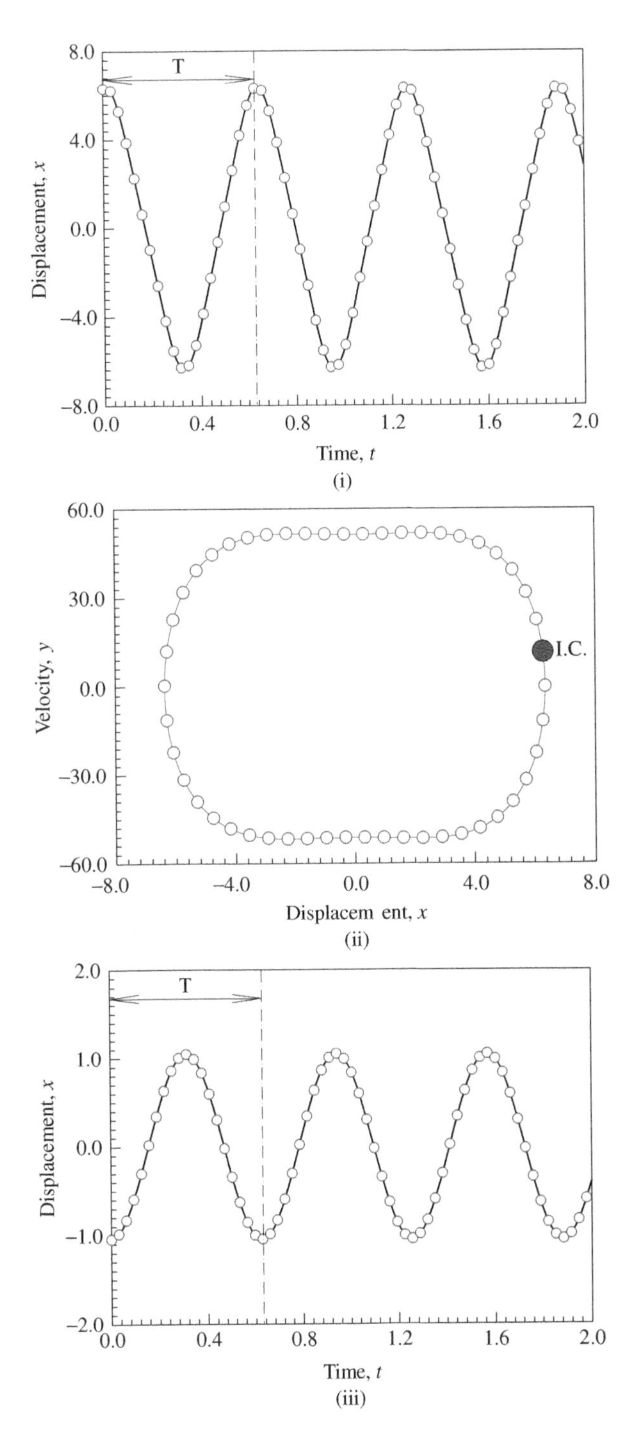

Figure 2.6 A stable, symmetric period-1 motions (HB10) (upper branch): (i) displacement and (ii) phase plane ($x_0 \approx 6.296950$, $y_0 \approx 11.422100$). A stable symmetric period-1 motion (HB10) (lower branch): (iii) displacement and (iv) phase plane ($x_0 \approx -1.045500$, $y_0 \approx 0.220845$). An unstable symmetric period-1 motion (HB10) (middle branch): (v) displacement and (vi) phase plane ($x_0 \approx -5.192940$, $y_0 \approx 6.918470$) ($\Omega = 10$, $\delta = 0.2$, $\alpha = 1.0$, $\beta = 4$, $Q_0 = 100.0$)

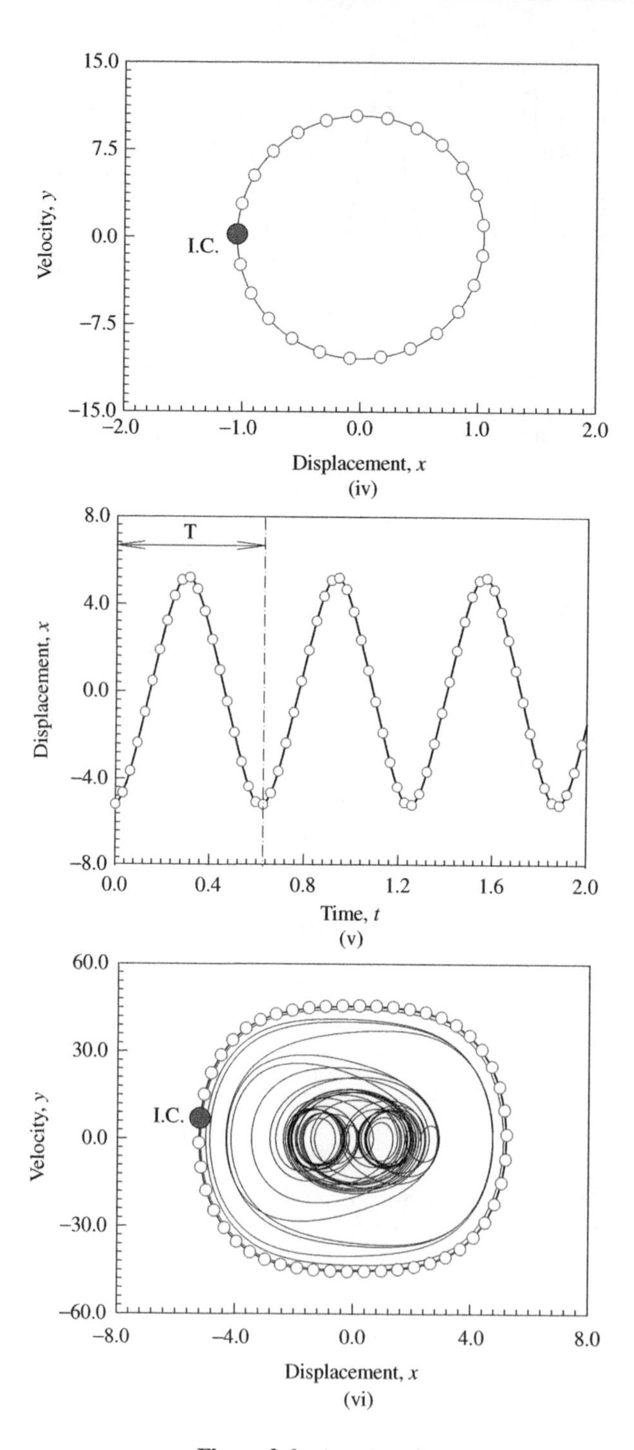

Figure 2.6 (*continued*)

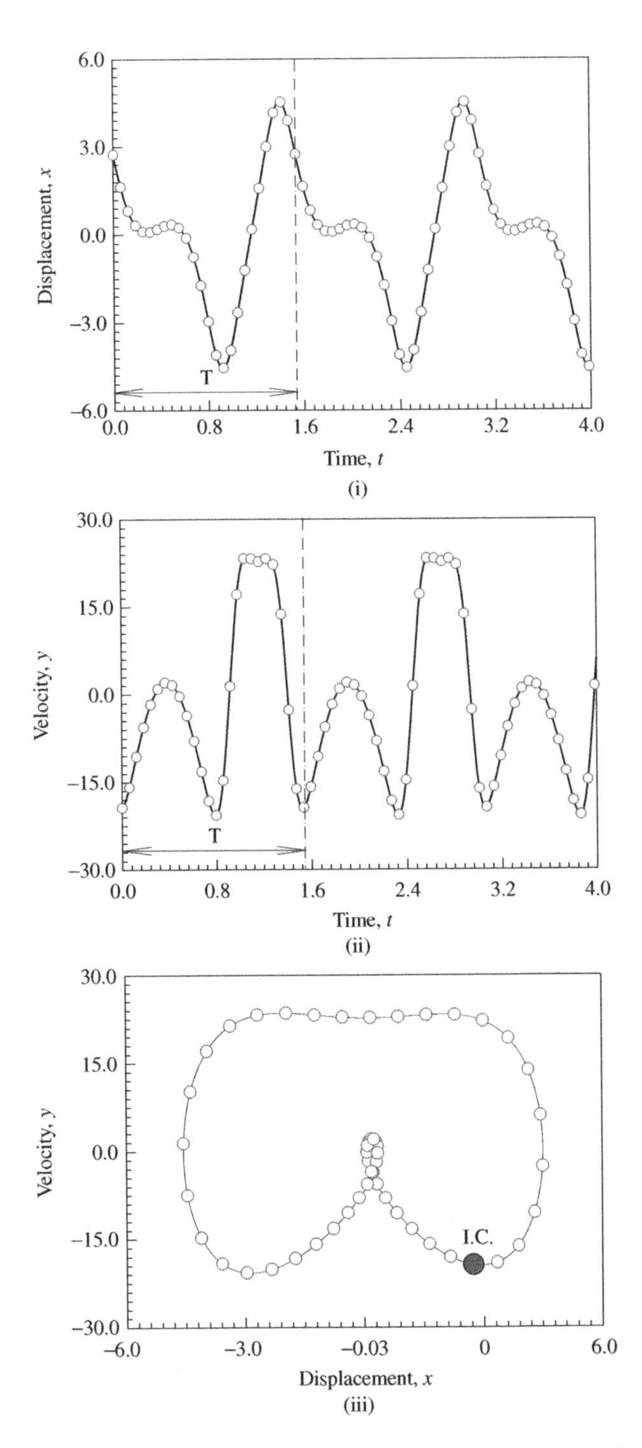

Figure 2.7 A stable, asymmetric period-1 motions (HB10): (i) displacement, (ii) velocity, and (iii) phase plane ($x_0 \approx 2.735740$, $y_0 \approx -19.407900$). An unstable, symmetric period-1 motion (HB10): (iv) displacement, (v) velocity, and (vi) phase plane ($x_0 \approx 4.960370$, $y_0 \approx 3.734470$) ($\Omega = 4.1$, $\delta = 0.2$, $\alpha = 1.0$, $\beta = 4$, $Q_0 = 100.0$)

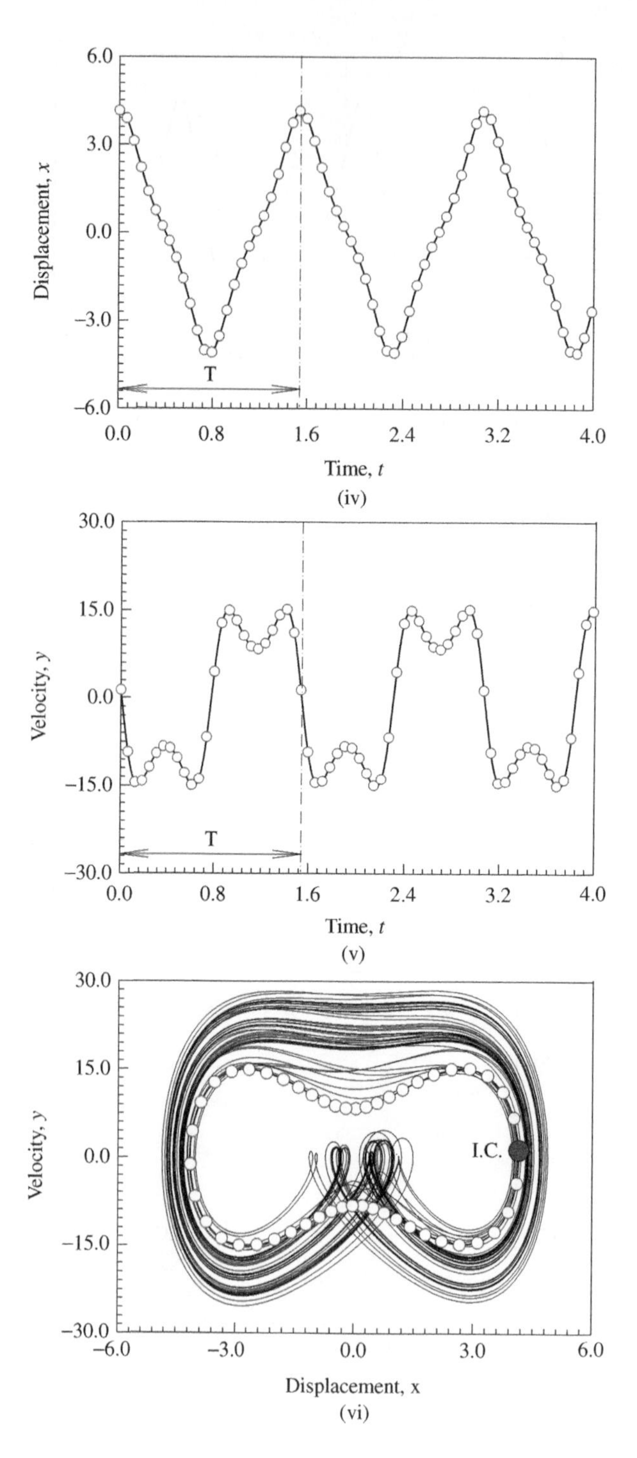

Figure 2.7 (*continued*)

In Figure 2.8, the unstable asymmetric period-1 motions and the stable asymmetric period-2 motions are presented for $\Omega = 3.4$. The initial conditions are $t_0 = 0.0$, $x_0 \approx 2.936730$, and $y_0 \approx -1.173750$ (unstable period-1 motion) and $t_0 = 0.0, x_0 \approx 3.358240$, and $y_0 \approx -1.225180$ (stable period-2 motion). In Figure 2.8(i)–(iii), displacement, velocity, and trajectory for the unstable asymmetric period-1 motions are presented. The unstable asymmetric period-1 motion moves to the stable asymmetric period-2 motion. The analytical solution of the unstable period-1 motion is given by the Fourier series solution with 10 harmonic terms. For the stable asymmetric period-2 motion, the corresponding analytical solution is given by the Fourier series solution with 20 harmonic terms, and the analytical and numerical solutions of displacement, velocity, and trajectories are presented in Figure 2.8(iv)–(vi), respectively. For further demonstration of the analytical tree of periodic motions from asymmetric period-1 motion, consider an excitation frequency of $\Omega = 3.32$ for which the unstable period-1, unstable period-2, and stable period-4 motions coexist on the analytical bifurcation tree from the asymmetric period-1 motion. The displacements and trajectories for three periodic motions are presented in Figure 2.9(i)–(vi). The analytical solutions for period-1, period-2, and period-4 motions are expressed by the Fourier series with 10, 20, and 40 harmonic terms. The numerical solutions for the unstable period-1 and unstable period-2 motions go away from the corresponding analytical solutions to the stable period-4 motion. However, the analytical and numerical solutions for period-4 motions match very well.

2.3 Period-3 Motions to Chaos

In this section, the bifurcation tree from period-3 motion to chaos will be presented as an example for period-m motions to chaos. A symmetric period-3 motion will be presented analytically, which is without any bifurcation tree. Two asymmetric period-3 motions will be predicted analytically and the corresponding analytical bifurcation tree to chaos will be presented.

2.3.1 *Independent, Symmetric Period-3 Motions*

In the previous section, the analytical routes of period-1 motions to chaos were presented through the harmonic amplitude varying with excitation frequency Ω. Herein, the analytical prediction of unstable and stable period-3 motions will be presented through the frequency-amplitude curves. The independent symmetric period-3 motion will be discussed. For the symmetric period-3 motion, $a_0^{(m)} = 0$ and $A_{k/m} = 0$ for $k = 2l$ $(l = 0, 1, 2, \ldots, m = 3)$ are obtained. Therefore, only $A_{k/m}$ for $k = 2l + 1$ $(l = 0, 1, 2, \ldots, m = 3)$ are plotted. From the parameters in Equation (2.47), the independent period-3 motion is analytically predicted with 12 harmonic terms (HB12), as shown in Figure 2.10. The solid and dashed curves represent the stable and unstable period-3 solutions, respectively. The acronym SN represents saddle-node bifurcation. Since $a_0^{(3)} = 0$ of the symmetric period-3 motion exists in the entire frequency range, in Figure 2.10(i), the harmonic amplitude of $A_{1/3} \sim 10^1$ increases with excitation frequency Ω, which is like period-1 motion in the hardening Duffing oscillator. The stable and unstable symmetric period-3 motion formed a closed loop, and only saddle-node bifurcations are observed, which gives the stability switching between the stable and unstable solutions. For the symmetric period-3 motion, $A_{2/3} = 0$ exists in the entire frequency range. The harmonic amplitude of $A_1 \sim 10^0$ is plotted in Figure 2.10(ii). The frequency-amplitude curves for the stable and unstable period-1 motions form an “8” closed loop. Because of $A_{4/3} = 0$,

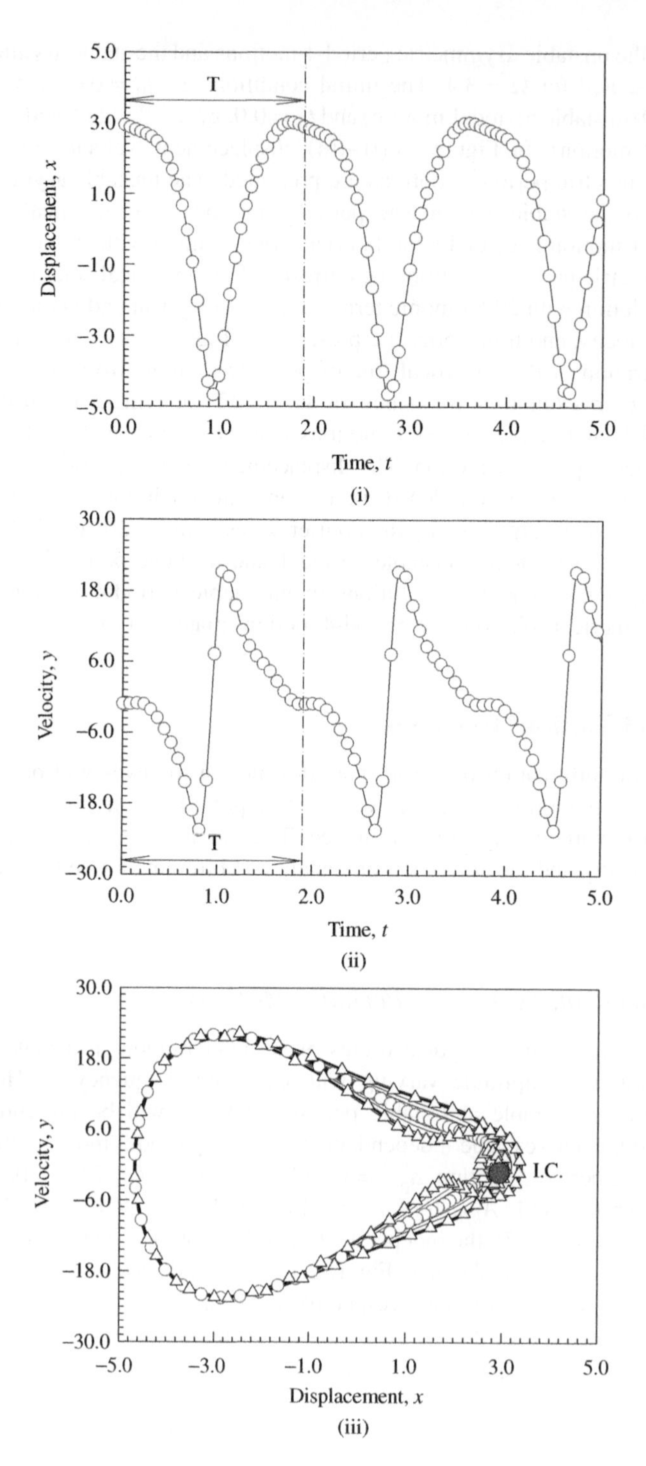

Figure 2.8 An unstable, asymmetric period-1 motions (HB10): (i) displacement, (ii) velocity, and (iii) phase plane ($x_0 \approx 2.936730$, $y_0 \approx -1.173750$). A stable asymmetric period-2 motion (HB20): (iv) displacement, (v) velocity, and (vi) phase plane ($x_0 \approx 3.358240$, $y_0 \approx -1.225180$) ($\Omega = 3.4$, $\delta = 0.2$, $\alpha = 1.0$, $\beta = 4$, $Q_0 = 100.0$)

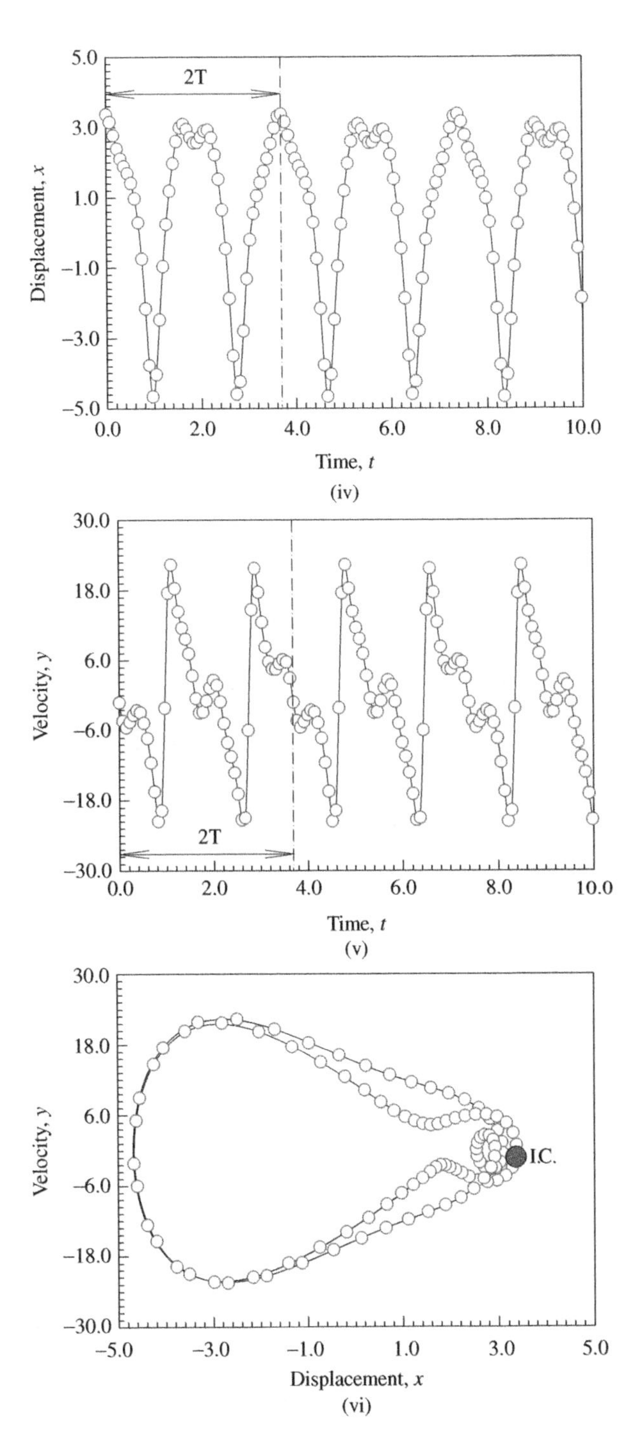

Figure 2.8 *(continued)*

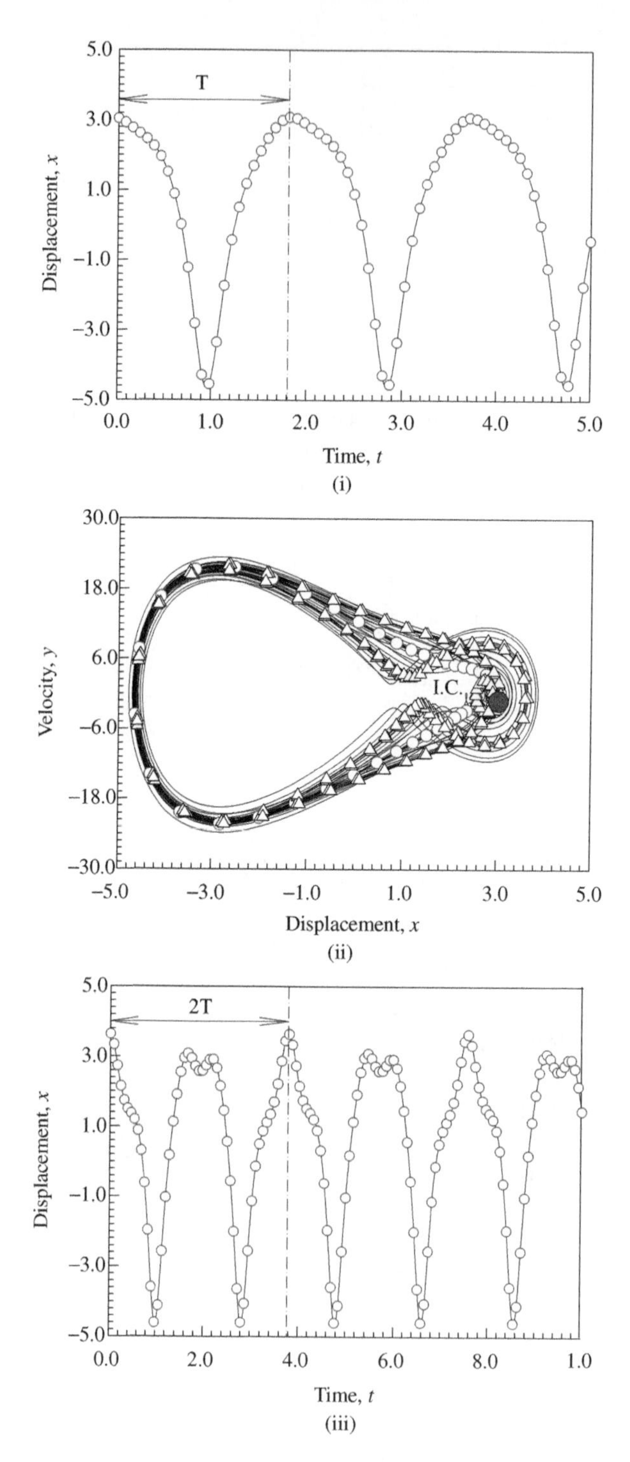

Figure 2.9 An unstable, asymmetric period-1 motions (HB10): (i) displacement and (ii) phase plane ($x_0 \approx 3.034700$, $y_0 \approx -1.134240$). An unstable asymmetric period-2 motion (HB20): (iii) displacement and (iv) phase plane ($x_0 \approx 3.629530$, $y_0 \approx -0.926901$). A stable period-4 motion (HB40): (v) displacement and (vi) phase plane ($x_0 \approx 3.654920$, $y_0 \approx -0.869122$) ($\Omega = 3.32$, $\delta = 0.2$, $\alpha = 1.0$, $\beta = 4$, $Q_0 = 100.0$)

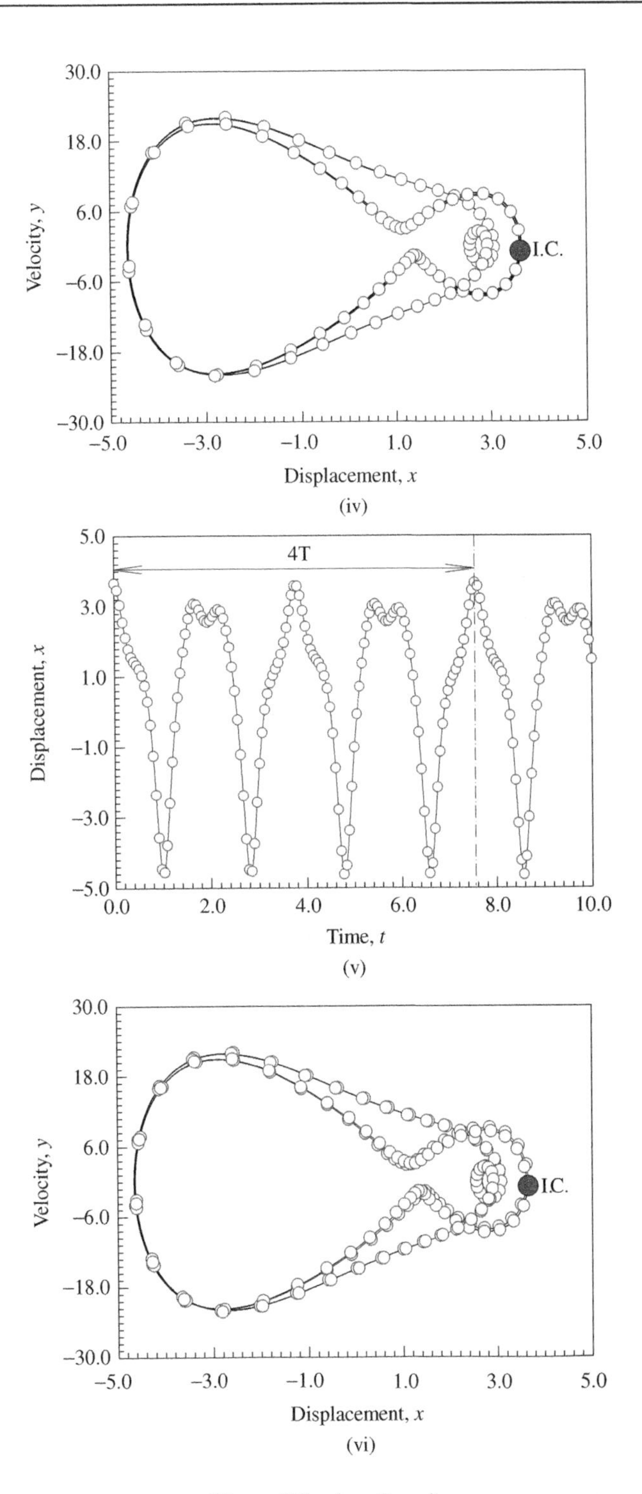

Figure 2.9 (*continued*)

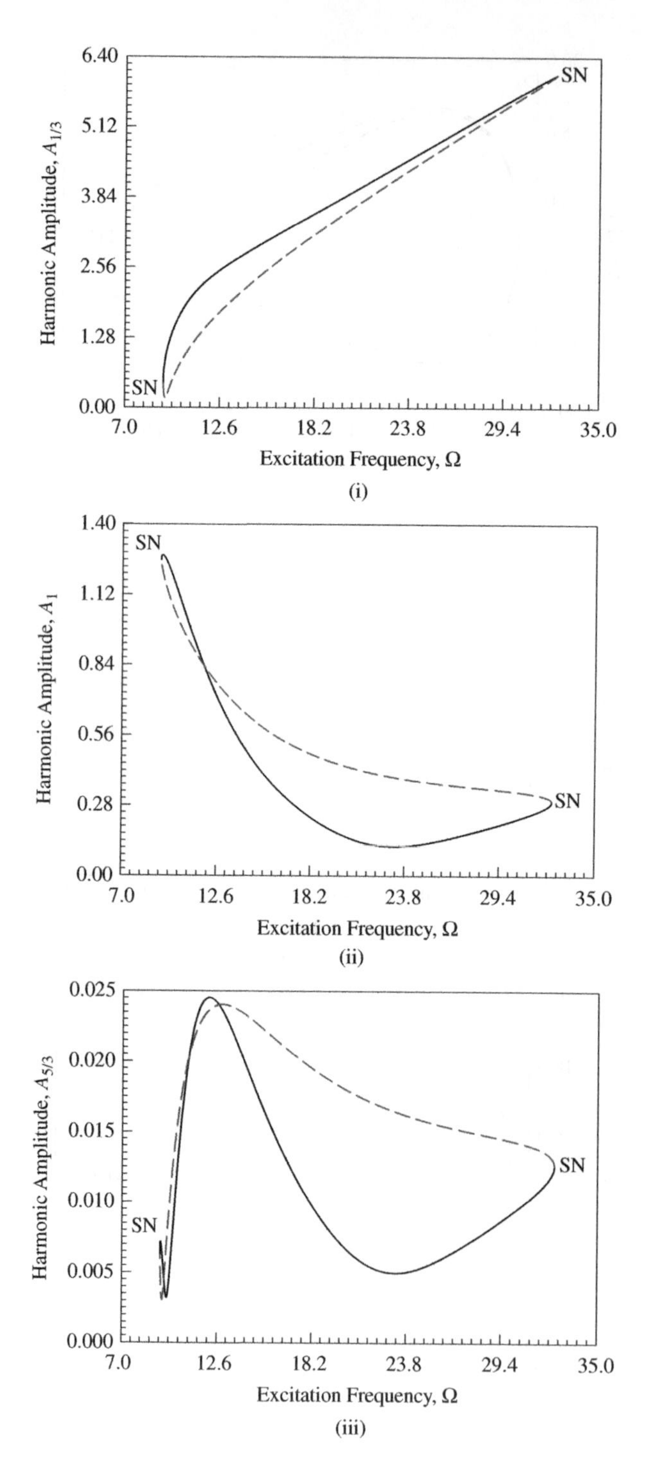

Figure 2.10 Analytical predictions of independent symmetric period-3 motions based on 12 harmonic terms (HB12): (i)–(vi) harmonic amplitudes $A_{k/m}$ ($k = 1, 3, \ldots, 11, m = 3$). Parameters ($\delta = 0.2, \alpha = 1.0, \beta = 4.0, Q_0 = 100.0$)

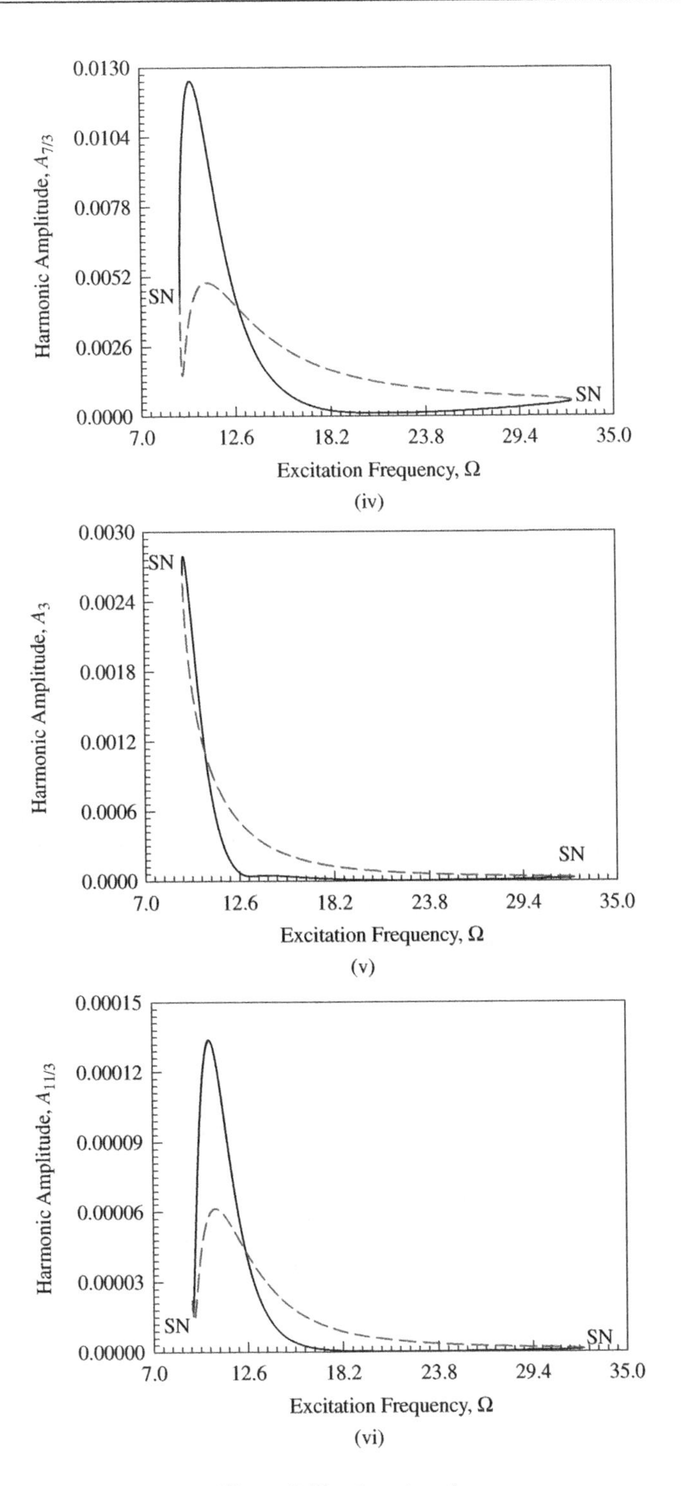

Figure 2.10 (*continued*)

the harmonic amplitude $A_{5/3}$ is depicted in Figure 2.10(iii). $A_{5/3} \sim 2.5 \times 10^{-2}$ is much less than $A_{1/3}$ and A_1 (i.e., $A_{5/3} << A_{1/3}$ and A_1). Due to $A_2 = 0$, the harmonic amplitude of $A_{7/3} \sim 10^{-2}$ is plotted in Figure 2.10(iv), and $A_{7/3}$ and $A_{5/3}$ have the same quantity level. Because of $A_{8/3} = 0$, the harmonic amplitude A_3 is presented in Figure 2.10(v). $A_3 \sim 3 \times 10^{-4}$ for $\Omega > 13$ and the maximum value of A_3 is about $A_3 \approx 3 \times 10^{-3}$ in the vicinity of $\Omega = 9$. Due to $A_{10/3} = 0$, the harmonic amplitude $A_{11/3}$ is plotted in Figure 2.10(vi). $A_{11/3} \sim 3 \times 10^{-5}$ for $\Omega > 13$ and the maximum value of $A_{11/3}$ is about $A_{11/3} \approx 1.4 \times 10^{-4}$ in the vicinity of $\Omega = 11$. With $A_{k/m} = 0$ for $k = 2l$, other harmonic amplitudes $A_{k/m}$ for $k = 2l + 1$ can be computed but the quantity level is very small, which will not be discussed herein.

2.3.2 *Asymmetric Period-3 Motions*

For the period-3 motions with period-6 motions to chaos, there are two types of asymmetric period-3 motions. The "P-3" in plots is used to label the period-3 motion. For the symmetric period-3 motion, $a_0^{(m)} = 0$. Constant term $a_0^{(m)}$ ($m = 3$) for the right and left sides of y-axes in analytical predictions of period-3 motions are plotted in Figure 2.11(i) and (ii), respectively. The solid and dashed curves represent the stable and unstable periodic solutions based on the 66 harmonic terms (HB66), respectively. Only $a_0^{(m)}$ for the asymmetric motion is non-zero (i.e., $a_0^{(m)} \sim 2 \times 10^{-1}$). The symmetric period-3 motion possesses $a_0^{(m)} = 0$. From the symmetric period-3 motion to the asymmetric period-3 motion, the saddle-node bifurcation occurs. For the Hopf bifurcation of asymmetric period-3 motion, the period-6 motions will take place, which is discussed later. The harmonic amplitudes $A_{k/m}$ ($k = 1, 2, \ldots, 18$ and $k = 63, 64, \ldots, 66$, $m = 3$) are presented in Figure 2.2(iii)–(xxiv), respectively. If $A_{k/m} = 0$ for $k = 2l$ ($l = 0, 1, 2, \ldots$), then periodic motion is oddly symmetric to displacement $x(t)$. In Figure 2.11, the symmetric period-3 motion is odd because $a_0^{(m)} = 0$ and $A_{k/m} = 0$ for $k = 2, 4, \ldots, 66$. The symmetric period-3 motion has two branches with three parts. The stable symmetric period-3 motion exist in ranges of $\Omega \in (2.2713, 2.28)$, $(2.7306, 2.888)$, and $(3.1523, 3.315)$. The saddle-node bifurcations of the symmetric period-3 motions are at $\Omega \in \{3.1523, 3.315, 2.7306, 2.888, 2.2713, 2.28\}$. The unstable symmetric period-3 motion exists in ranges of $\Omega \in (2.2713, 3.315)$ (lower) and $\Omega \in (2.28, 2.7306)$ and (2.888, 3.1523) (upper). For the asymmetric period-3 motions, $a_0^{(m)} \neq 0$ and $A_{k/m} \neq 0$ for $k = 2, 4, \ldots, 66$. The stable asymmetric period-3 motion is in ranges of $\Omega \in (2.28, 2.2815)$, $(2.7071, 2.7206)$, $(2.888, 2.925)$, and $(3.088, 3.1523)$. The unstable asymmetric period-3 motion is in ranges of

Table 2.1 The quantity level for harmonic amplitudes ($\delta = 0.2$, $\alpha = 1.0$, $\beta = 4.0$, $Q_0 = 100.0$)

Quantity levels	Harmonic amplitudes
$1 \sim 3$	$A_{1/3}, A_1, A_{7/3}, A_3$
$10^{-1} \sim 10^0$	$a_0^{(m)}, A_{2/3}, A_{5/3}, A_2, A_{8/3}, A_{10/3}, A_{11/3}, A_{13/3}, A_5, A_{17/3}, A_{19/3}, A_7$
$10^{-2} \sim 10^{-1}$	$A_{4/3}, A_4, A_{14/3}, A_{16/3}, A_6, A_{20/3}, A_{22/3 \sim 29/3}, A_{31/3}, A_{11}$
$10^{-3} \sim 10^{-2}$	$A_{10}, A_{32/3}, A_{34/3 \sim 15}, A_{47/3}, A_{49/3}, A_{17}$
$10^{-4} \sim 10^{-3}$	$A_{46/3}, A_{16}, A_{50/3}, A_{52/3 \sim 61/3}, A_{21}$
$10^{-5} \sim 10^{-4}$	$A_{62/3}, A_{64/3}, A_{65/3}, A_{22}$

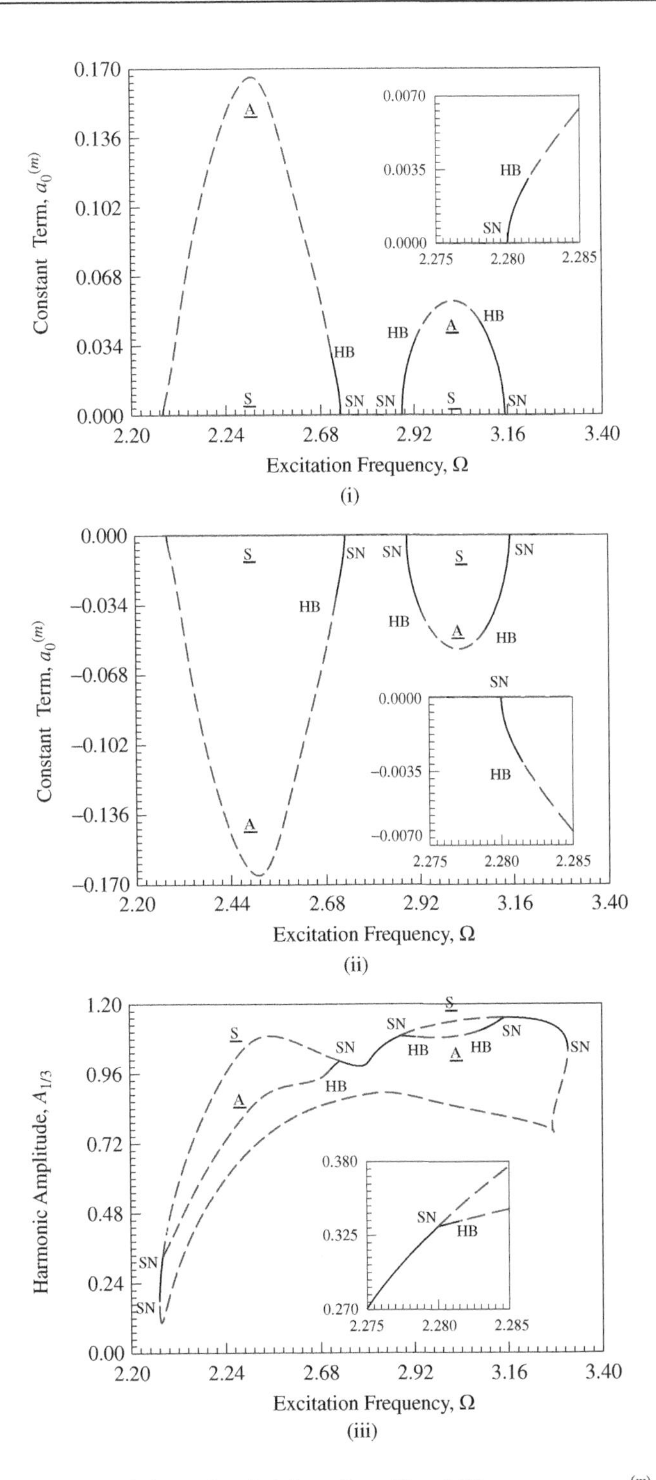

Figure 2.11 Analytical predictions of period-3 motions: (i) and (ii) constant term $a_0^{(m)}$ ($m = 3$) for right and left sides of y -axis, respectively; (iii)–(xviii) harmonic amplitudes $A_{k/m}$ ($k = 1, 2, \ldots, 12$ and $k = 63, 64, \ldots, 66$, $m = 3$). Parameters ($\delta = 0.2$, $\alpha = 1.0$, $\beta = 4.0$, $Q_0 = 100.0$)

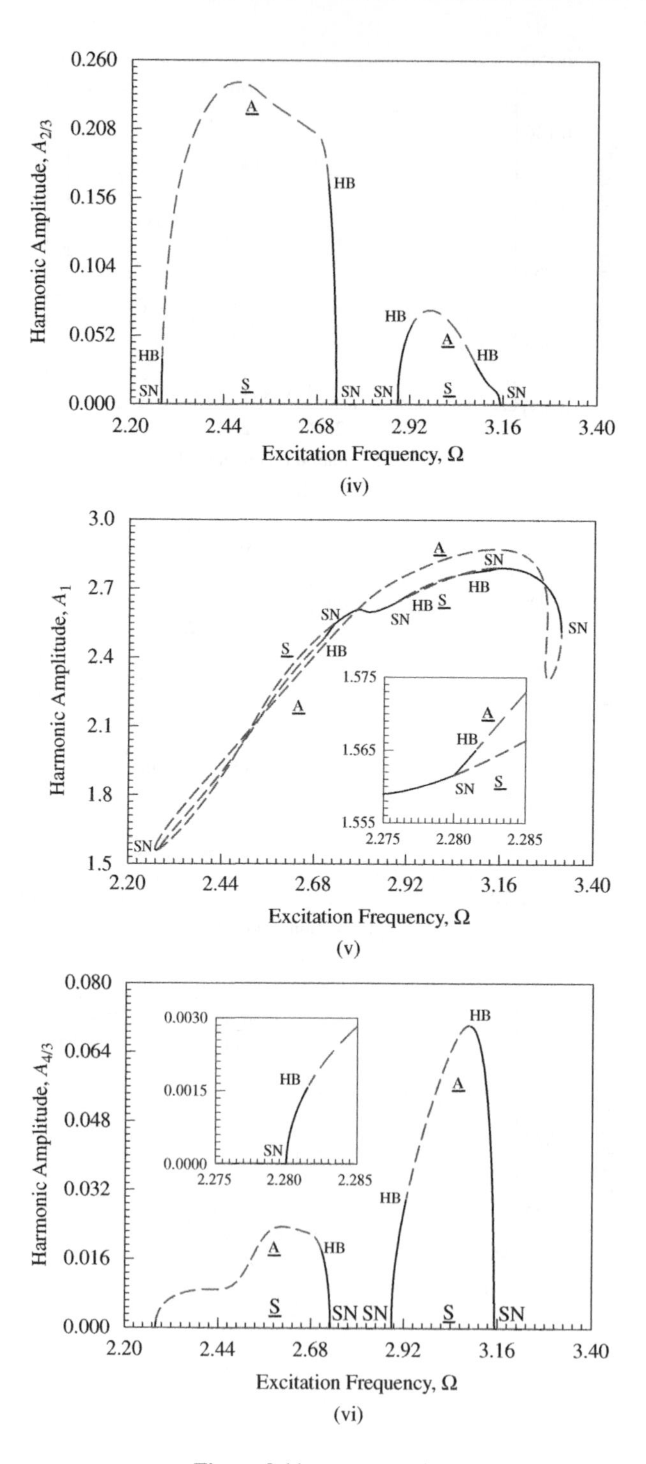

Figure 2.11 *(continued)*

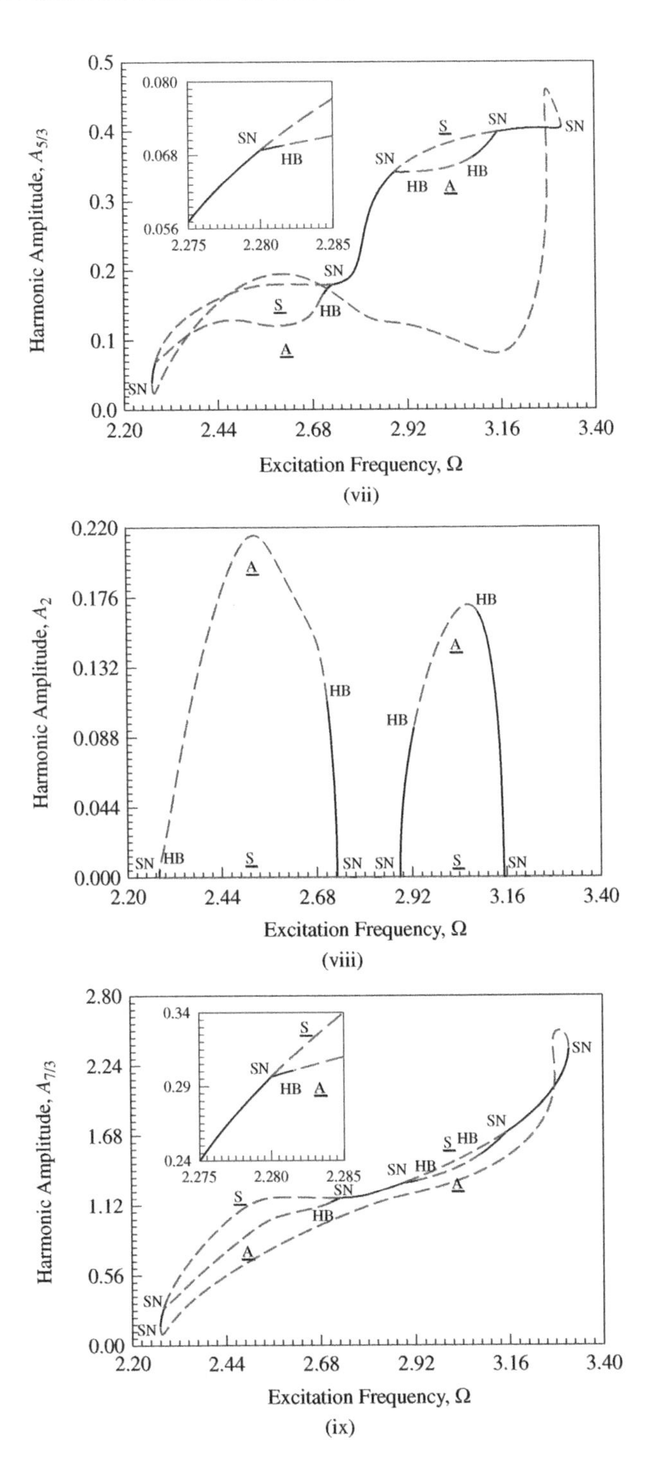

Figure 2.11 *(continued)*

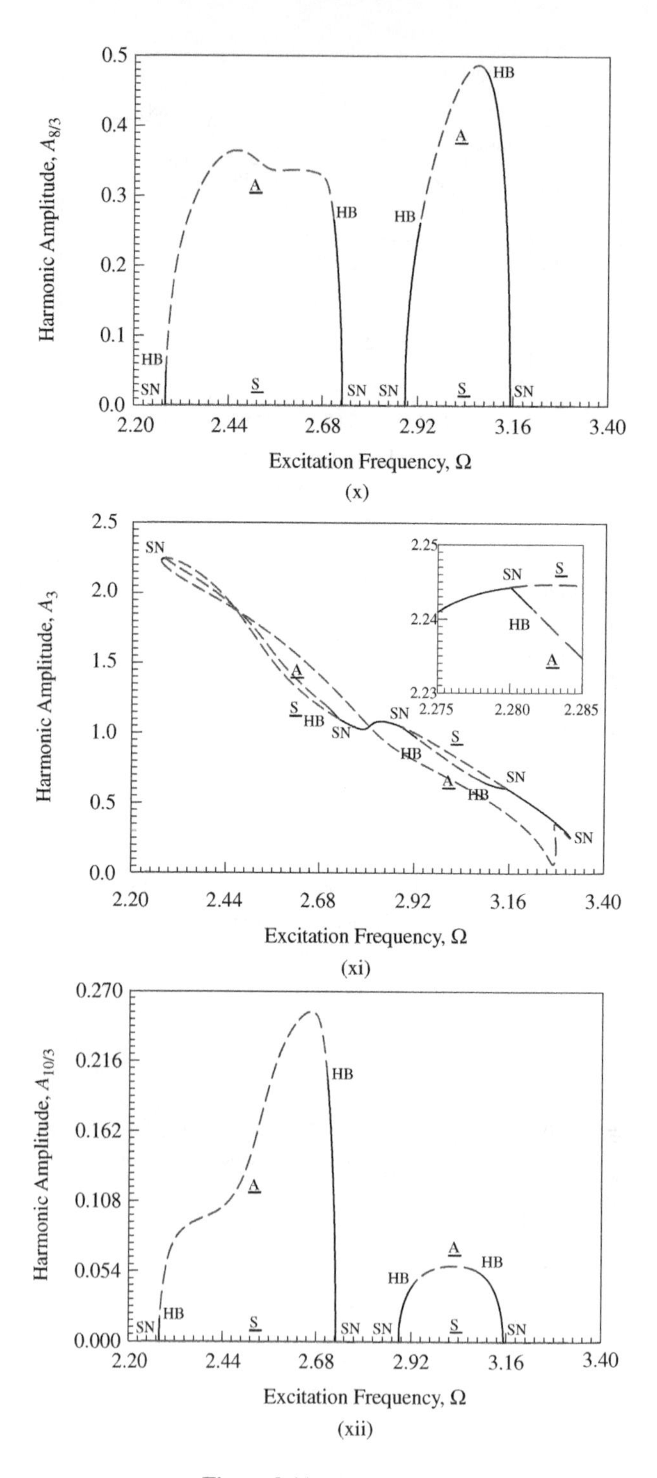

Figure 2.11 *(continued)*

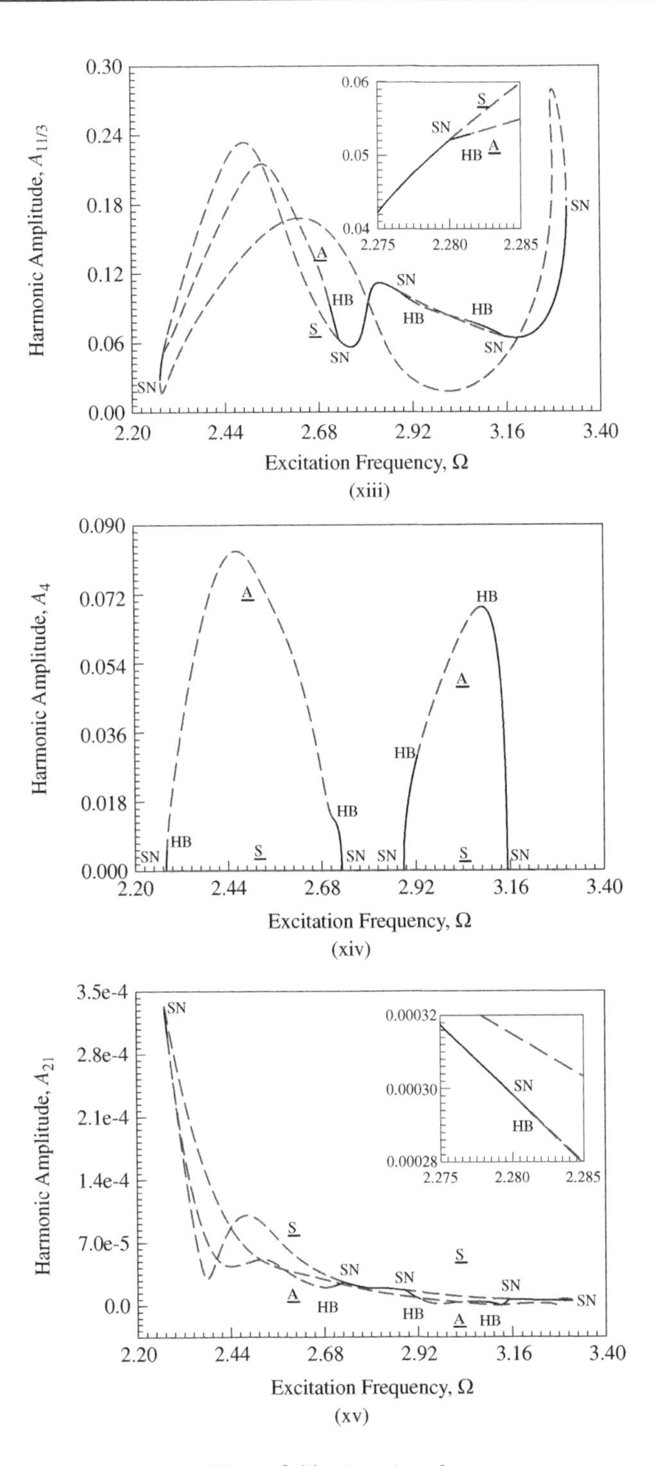

Figure 2.11 *(continued)*

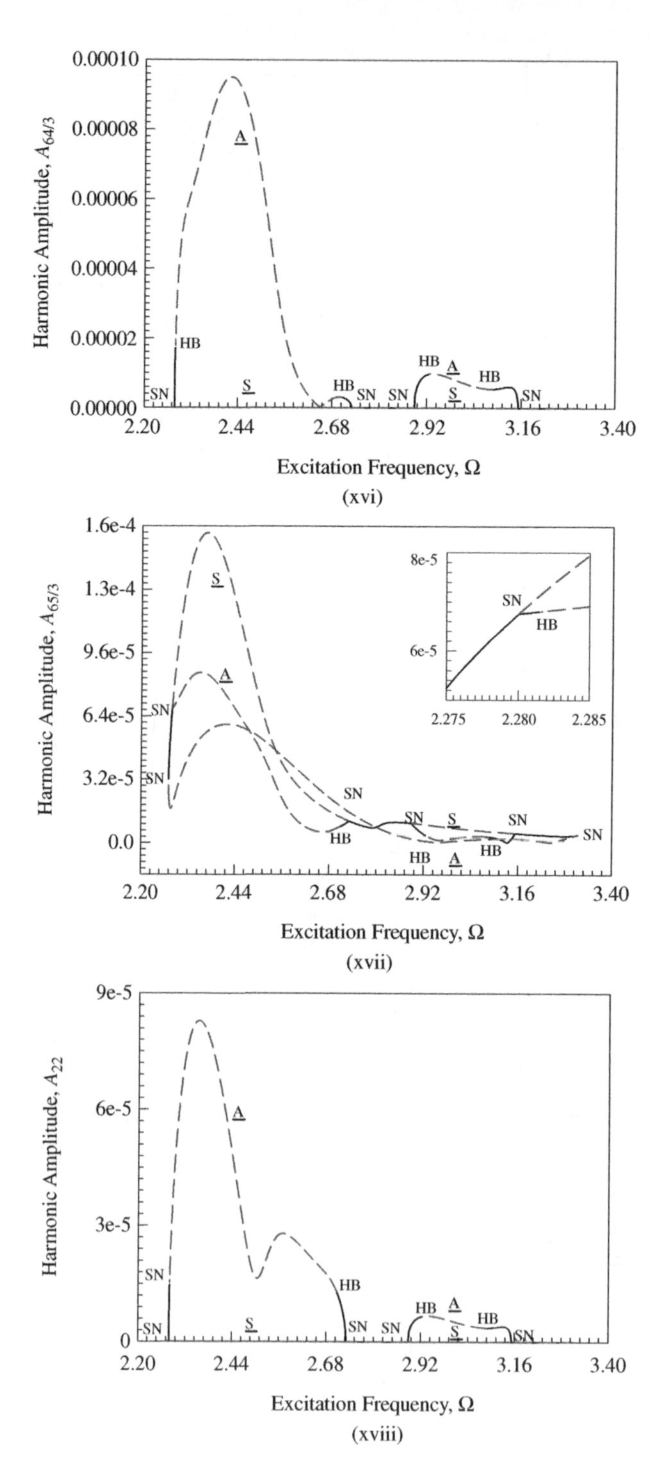

Figure 2.11 *(continued)*

$\Omega \in (2.2815, 2.7071)$ and $(2.925, 3.088)$. The Hopf bifurcations of the asymmetric period-3 motion occur at $\Omega \in \{2.2815,\ 2.7071,\ 2.925,\ 3.088\}$. The quantity level for period-3 are roughly listed in Table 2.1, which shows the effects of the harmonic terms to the period-3 motion. In Figure 2.11(iii), the harmonic amplitude $A_{1/3}$ versus excitation frequency is plotted. The value of $A_{1/3}$ is about $A_{1/3} \sim 10^0$. The lower branch of solution is for the unstable symmetric period-3 motion only, and the upper branch of the symmetric period-3 motion has the stable and unstable motions. In addition, the symmetric motion has four saddle-node bifurcations which generate two segments of asymmetric period-3 motions. In Figure 2.11(iv), the harmonic amplitude $A_{2/3}$ versus excitation frequency is plotted. The value of $A_{2/3}$ is about $A_{2/3} \sim 3 \times 10^{-1}$, which is similar to $a_0^{(m)}$ for the right hand side. In Figure 2.11(v), the harmonic amplitude A_1 versus excitation frequency is plotted. The solutions for symmetric and asymmetric motions are very close, and the quantity level of A_1 is in the range of $A_1 \in (1.0, 3.0)$, and with increasing excitation, the amplitude A_1 increases. The harmonic amplitude of $A_{4/3} \sim 8 \times 10^{-2}$ is plotted in Figure 2.11(vi), which is different from $A_{2/3} \sim 2 \times 10^{-1}$. In Figure 2.11(viii), the harmonic amplitude of $A_{5/3} \sim 5 \times 10^{-1}$ similar to the harmonic term of A_1 is plotted. The harmonic amplitude of A_2 is presented in Figure 2.11(ix) and $A_2 \sim A_{2/3}$. However, the quantity level of $A_{7/3}$ is the same as A_1 with a similar shape, as shown in Figure 2.11(ix). Suddenly, the quantity of $A_{8/3}$ increases to $A_{8/3} \sim 5 \times 10^{-1}$ is for the period-3 motion, as shown in Figure 2.11(x). The quantity level of A_3 in Figure 2.11(xi) is as large as the one of A_1, but the amplitude A_3 decreases with increasing excitation frequency. The harmonic amplitude of $A_{10/3}$ is presented in Figure 2.11(xii) and $A_{10/3}$ decreases to 3×10^{-1} from $A_3 \sim 3 \times 10^0$. In addition, the quantity level of $A_{11/3}$ is the same as $A_{10/3}$, as shown in Figure 2.11(xiii). The harmonic amplitude of A_4 decreases to 9×10^{-2} in Figure 2.11(xiv). Up to A_4 from $a_0^{(m)}$, the quantity level of harmonic amplitude is greater than 8×10^{-2}. To avoid the abundant illustrations, $A_{21} \sim 4 \times 10^{-4}$, $A_{64/3} \sim 1 \times 10^{-4}$, $A_{65/3} \sim 1.6 \times 10^{-4}$ and $A_{22} \sim 9 \times 10^{-5}$ are plotted in Figure 2.11(xv)–(xviii), respectively. Of course, the higher order harmonic terms can be considered in a similar fashion, but their effects on the period-3 motion can be ignored. For the asymmetric periodic motions with centers on the right and left sides of y-axis, the harmonic amplitudes are the same. However, the harmonic phases between the two asymmetric periodic motions possess a relation of $\varphi^L_{k/m} = \mathrm{mod}(\varphi^R_{k/m} + (k+1)\pi, 2\pi)$ $(k = 1, 2, 3 \ldots;\ m = 3)$.

2.3.3 *Period-3 to Period-6 Motions*

From the asymmetric period-3 motions, two branches of asymmetric period-3 motions exist for $\Omega \in (2.281, 2.72)$ and $\Omega \in (2.9, 3.10)$. The Hopf bifurcations of the asymmetric period-3 motions occur at $\Omega \in \{2.2815,\ 2.7071,\ 2.925,\ 3.088\}$ and the corresponding period-6 motion will appear. Compared to the period-3 motion, the range of period-6 motions is very short. Only the local view from the asymmetric period-3 to period-6 motions for $\Omega \in (2.9, 3.10)$ is plotted in Figure 2.12. The local view for the asymmetric period-3 to period-6 motions for $\Omega \in (2.281, 2.72)$ can be referred to Luo and Huang (2013b). The "P-3" and "P-6" are used to label period-3 and period-6 motions. The solid and dashed curves are for stable and unstable periodic solutions based on 132 harmonic terms (HB132). In Figure 2.12(i) and (ii), constants $a_0^{(m)}$ for the center of periodic solutions are presented for $\Omega \in (2.9, 3.10)$ on the right and left sides of y-axes. $a_0^{(m)} \sim 6 \times 10^{-2}$ $(m = 3, 6)$ for period-3 and period-6 motions are observed. The Hopf

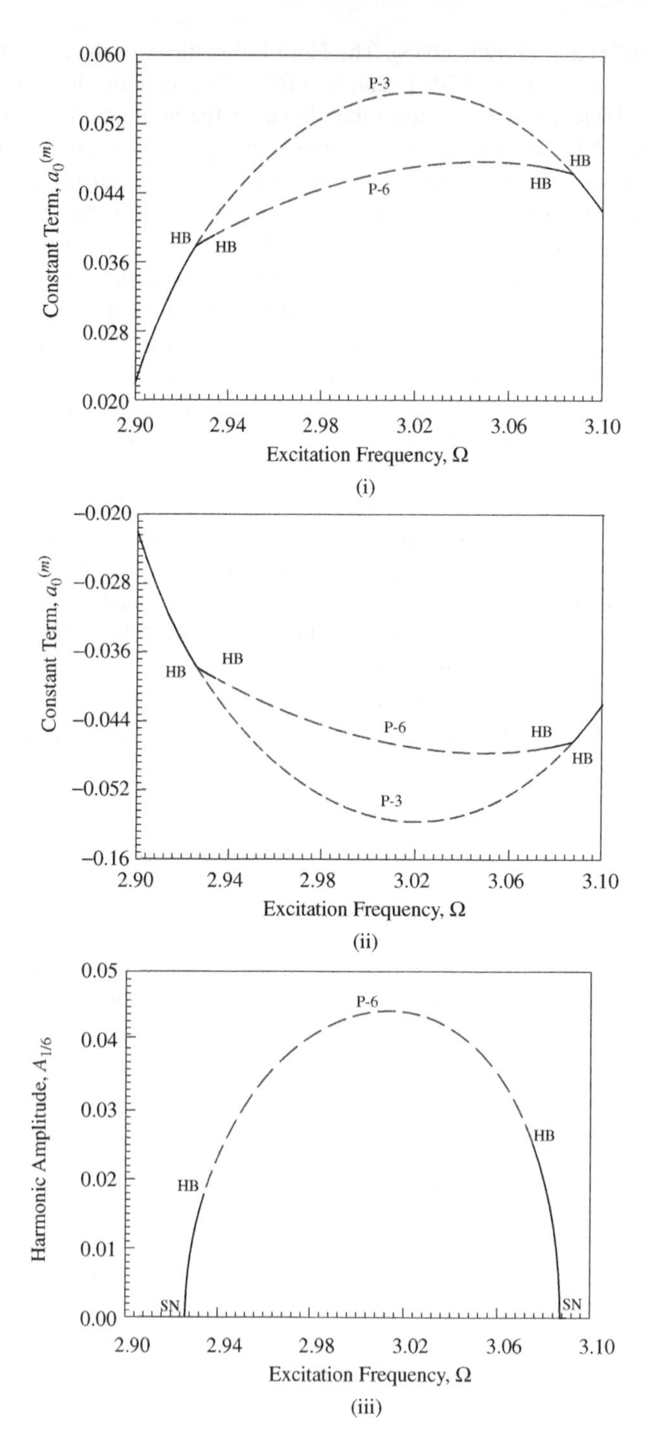

Figure 2.12 The analytical route of an asymmetric period-3 motion to chaos: (i) and (ii) constant term $a_0^{(m)}$ $(m = 3, 6)$ for right and left sides of y -axis, respectively; (iii)–(xxiv) harmonic amplitudes $A_{k/m}$ $(k = 1, 2, \ldots, 6; 8, 10, \ldots, 26; 127, 128, \ldots, 132$ and $m = 6)$. Parameters $(\delta = 0.2, \alpha = 1.0, \beta = 4.0, Q_0 = 100.0)$

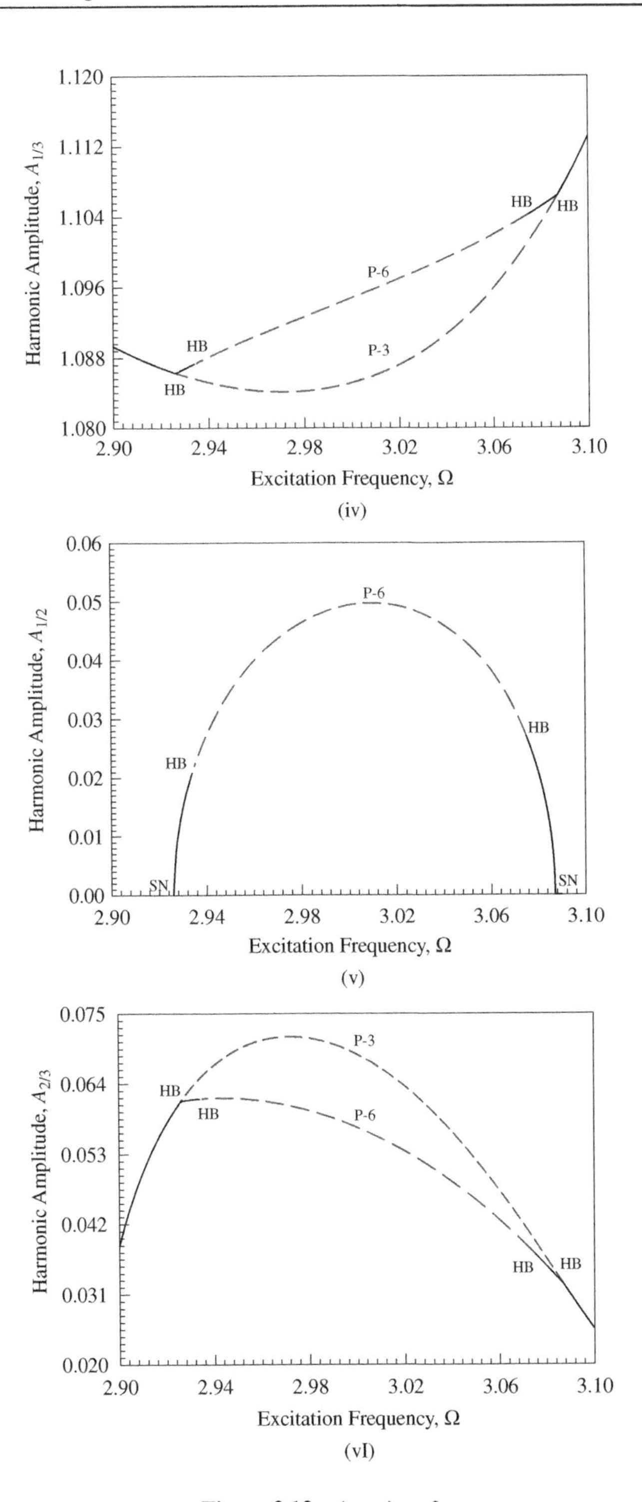

Figure 2.12 *(continued)*

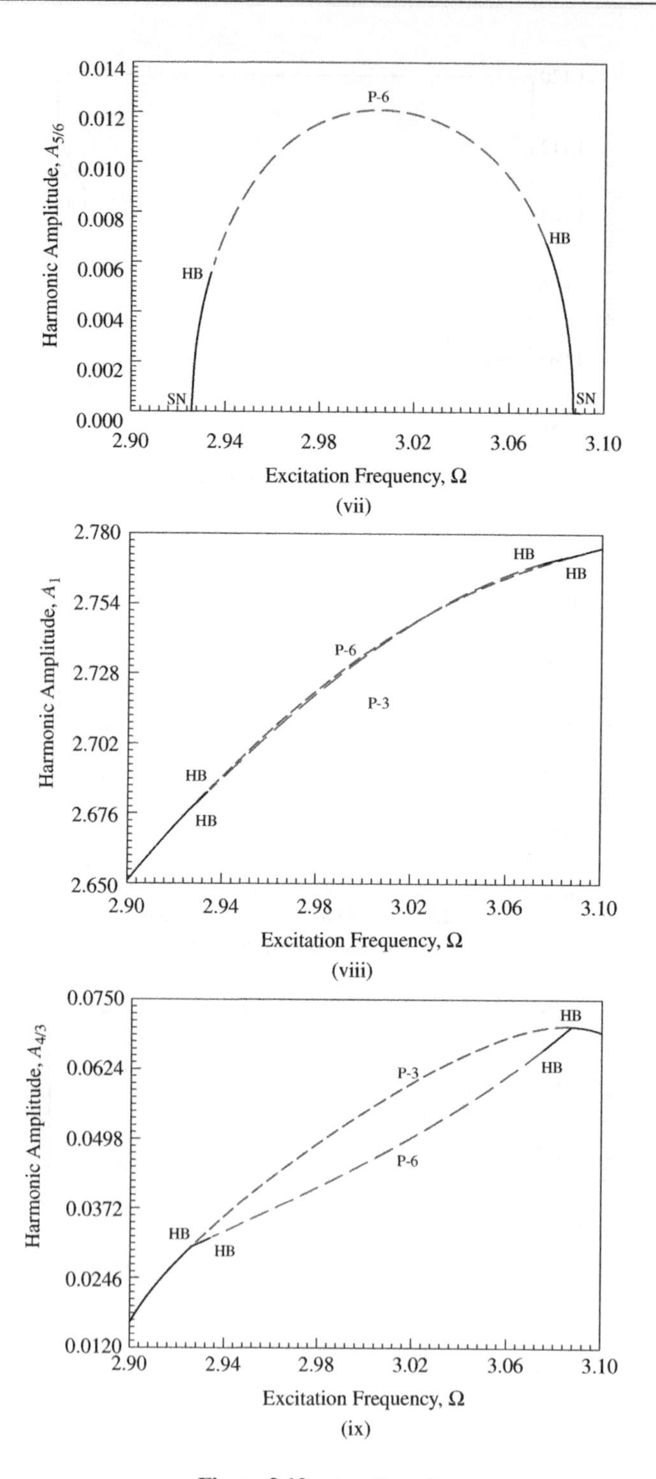

Figure 2.12 (*continued*)

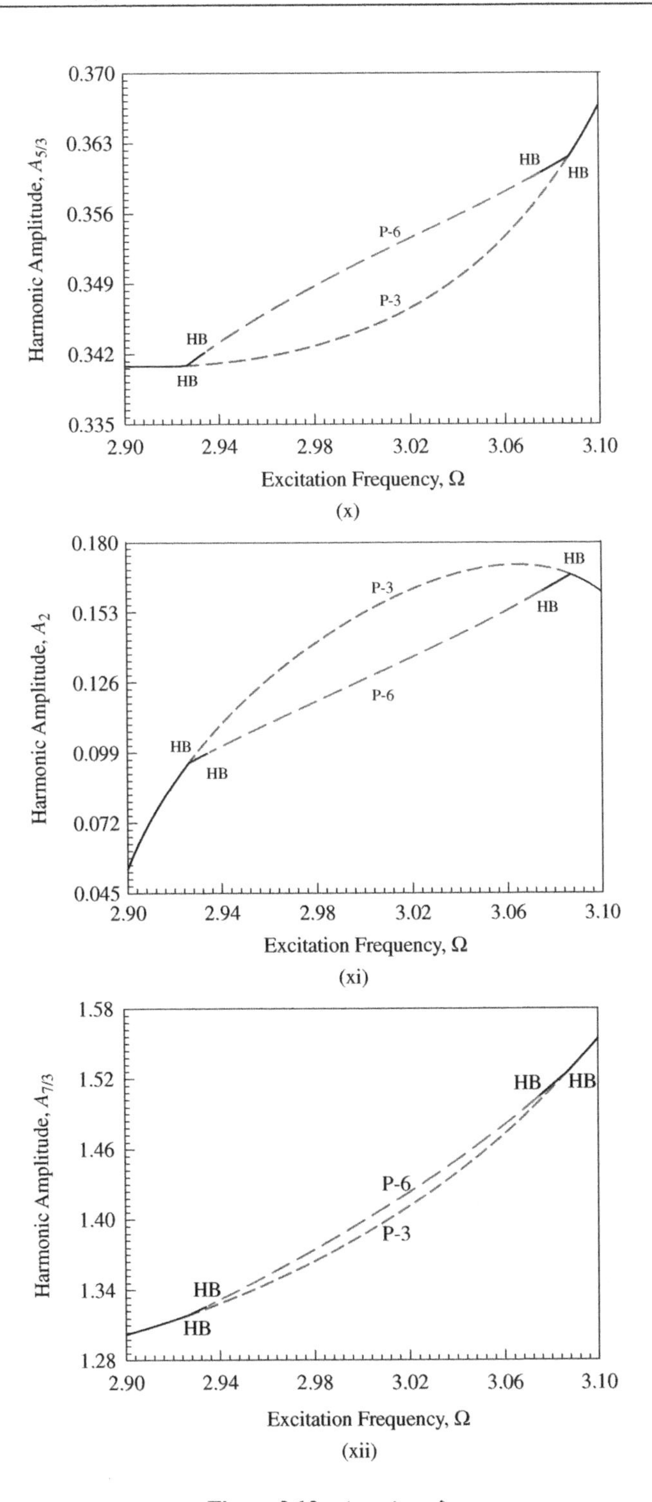

Figure 2.12 *(continued)*

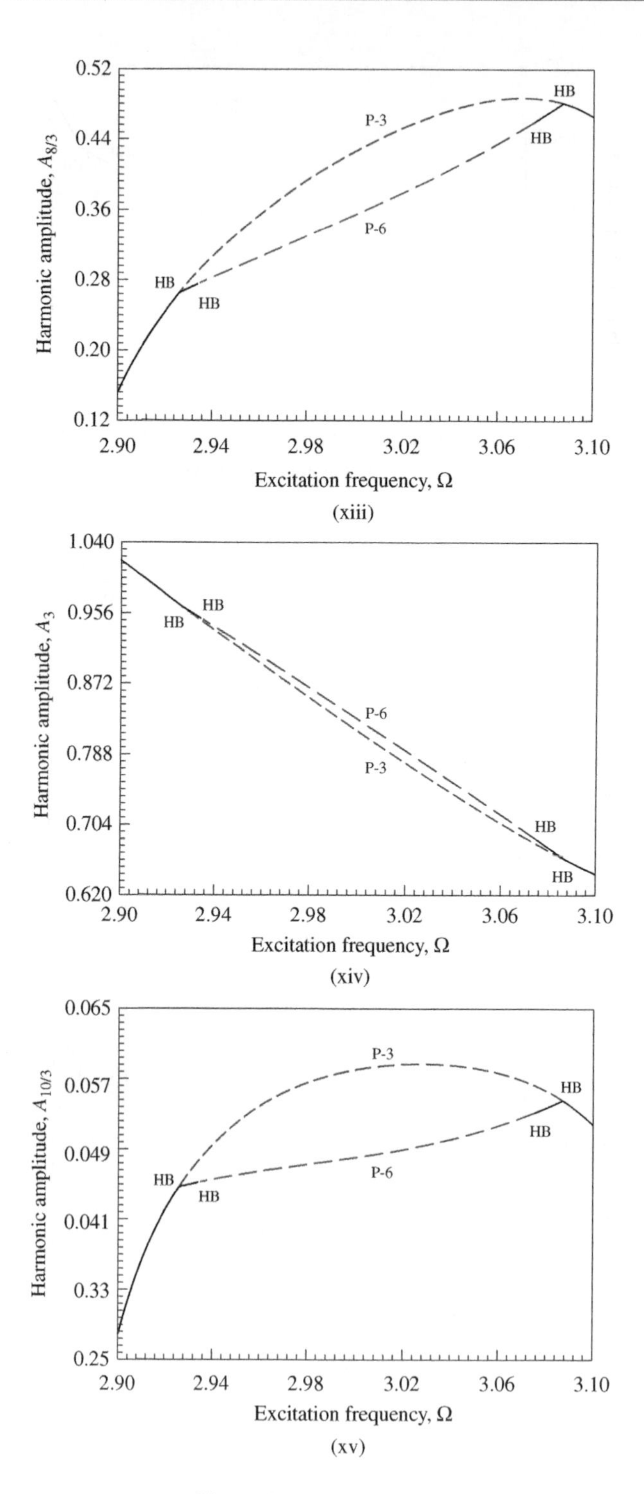

Figure 2.12 *(continued)*

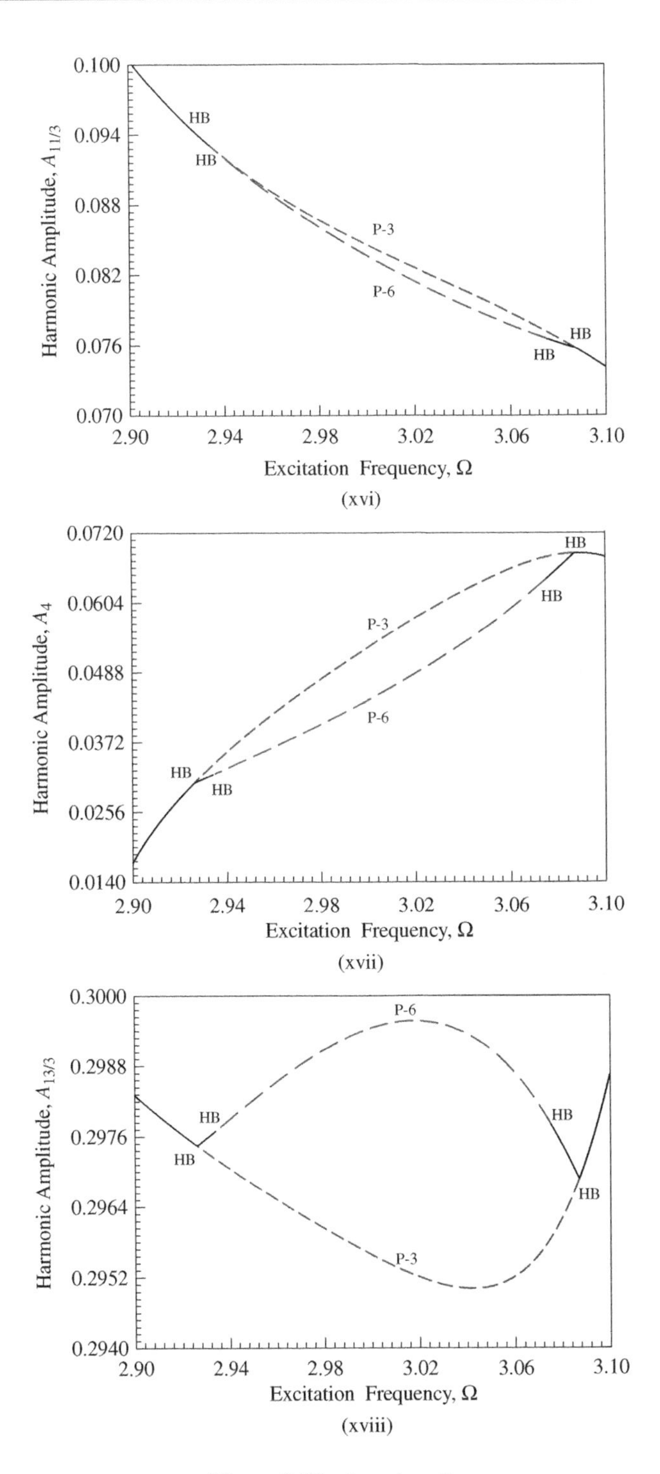

Figure 2.12 *(continued)*

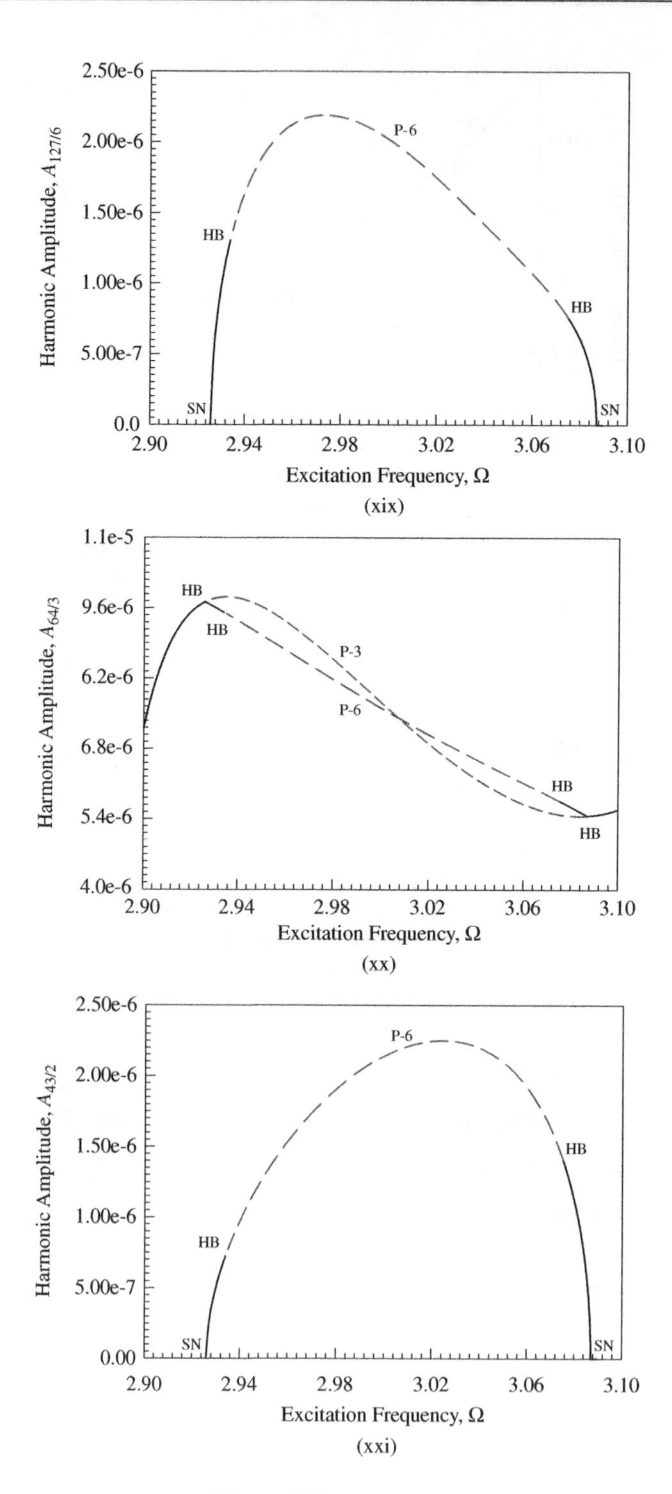

Figure 2.12 (*continued*)

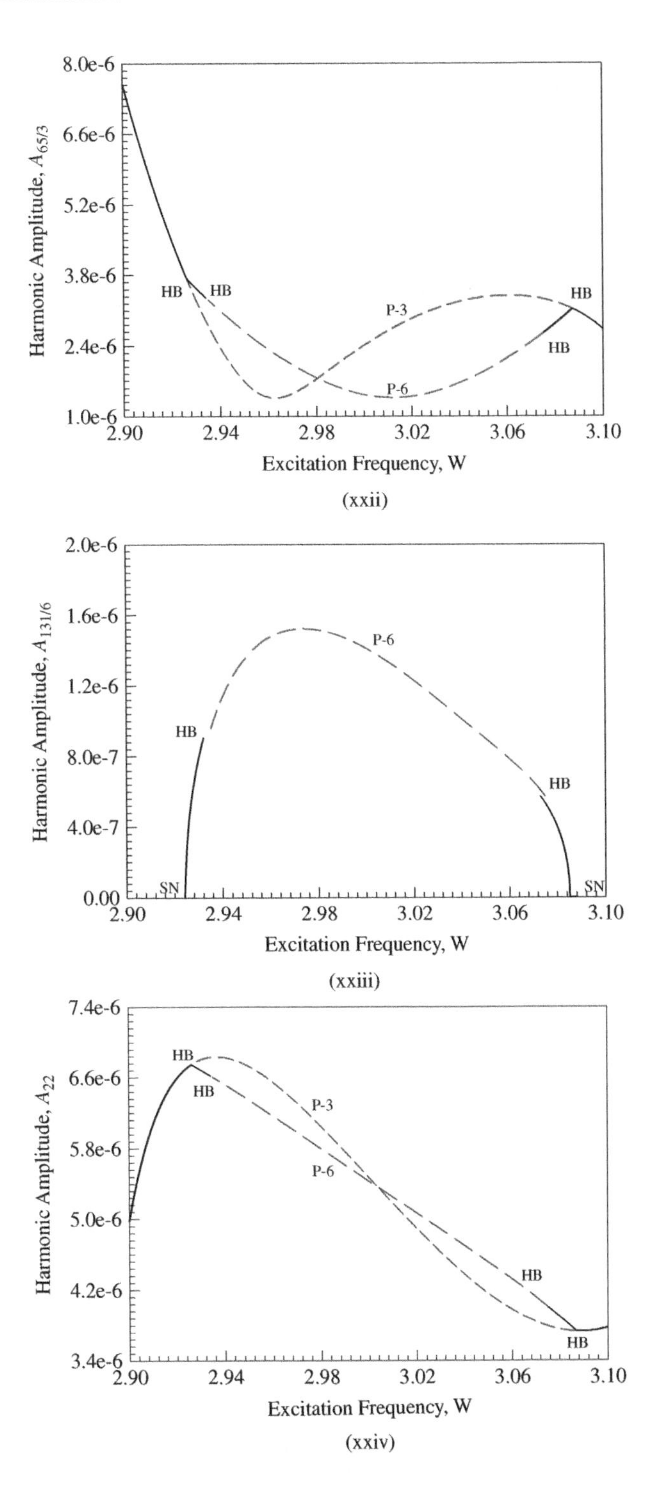

(xxii)

(xxiii)

(xxiv)

Figure 2.12 (*continued*)

bifurcations of period-3 and period-6 motions are at $\Omega \approx 2.925, 3.088$ and $\Omega \approx 2.9347, 3.0751$, respectively. In Figure 2.12(iii), harmonic amplitude $A_{1/6} \sim 5 \times 10^{-2}$ is for period-6 motion but for period-3 motion, $A_{1/6} = 0$. Since the period-6 motion appears, the saddle-nodes of the period-6 motion are at $\Omega \approx 2.925, 3.088$, which is the same as the Hopf bifurcations of period-3 motion. In Figure 2.12(iv), harmonic amplitude $A_{1/3}$ varying with excitation frequency Ω are presented, and the quantity level of $A_{1/3}$ for the period-3 and period-6 motions are very large (i.e., $A_{1/3} \sim 1$). Harmonic amplitude $A_{1/2}$ varying with excitation frequency Ω are presented in Figure 2.12(v) with $A_{1/2} \sim 5 \times 10^{-2}$ for period-6 motion. Again, for period-3 motion, $A_{1/2} = 0$. In Figure 2.12(vi), harmonic amplitude $A_{2/3}$ versus excitation frequency Ω is presented with the same quantity level of $A_{1/2}$. In Figure 2.12(vii), harmonic amplitude

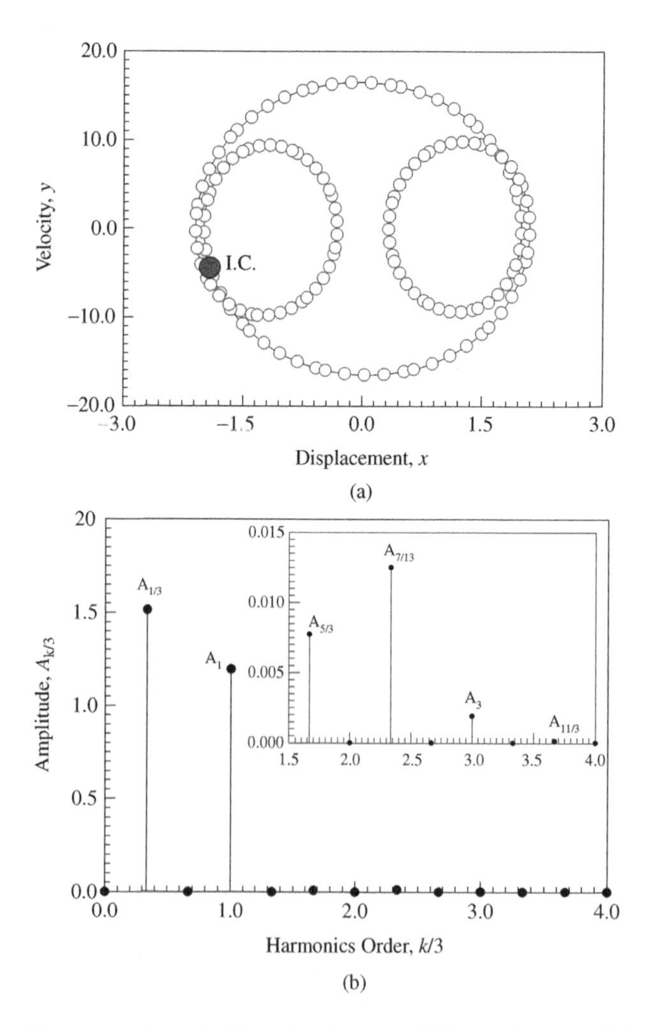

Figure 2.13 A stable symmetric period-3 motion based on 12 harmonic terms (HB12) ($\Omega = 10$): (a) trajectory and (b) amplitude spectrum ($x_0 \approx -1.913080$, $y_0 \approx -4.428830$) ($\delta = 0.2$, $\alpha = 1.0$, $\beta = 4.0$, $Q_0 = 100.0$)

$A_{5/6} \sim 10^{-2}$ is presented. Harmonic amplitude A_1 for period-3 and period-6 motions are presented in Figure 2.12(viii), and harmonic amplitudes for both periodic motions are almost identical. To save space and be compared with period-3 motion, only harmonic amplitudes of $A_{k/6} \equiv A_{l/3}$ $(k = 2l, l = 4, 5, \ldots, 13)$ are presented in Figure 2.12(ix)–(xviii). The quantity levels of harmonic amplitudes are $A_{4/3} \sim 7.5 \times 10^{-2}$, $A_{5/3} \sim 4 \times 10^{-1}$, $A_2 \sim 1.8 \times 10^{-1}$, $A_{7/3} \sim 1.6$, $A_{8/3} \sim 5 \times 10^{-1}$, $A_3 \sim 1$, $A_{10/3} \sim 6 \times 10^{-2}$, $A_{11/3} \sim 10^{-1}$, $A_4 \sim 7 \times 10^{-2}$, and $A_{13/3} \sim 3 \times 10^{-1}$. For the accuracy of period-3 and period-6 motions, the harmonic amplitudes $A_{k/6}$ $(k = 127, 128, \ldots, 132)$ are presented in Figure 2.12(xix)–(xxiv). The quantity levels of harmonic amplitudes are $A_{127/6} \sim 2.5 \times 10^{-6}$, $A_{64/3} \sim 10^{-5}$, $A_{43/2} \sim 2.5 \times 10^{-6}$, $A_{65/3} \sim 8.0 \times 10^{-6}$, $A_{131/6} \sim 2.0 \times 10^{-6}$, and $A_{22} \sim 7.0 \times 10^{-6}$. Thus, higher harmonic terms can be dropped from period-3 and period-6 motions.

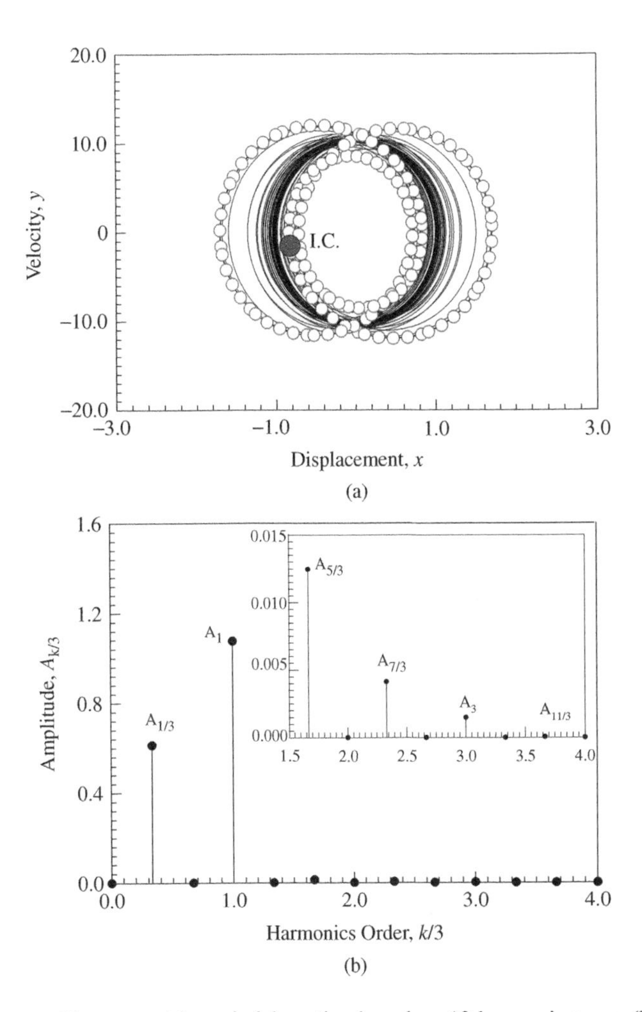

Figure 2.14 An unstable symmetric period-3 motion based on 12 harmonic terms (HB12) ($\Omega = 10$): (a) trajectory and (b) amplitude spectrum ($x_0 \approx -0.825653$, $y_0 \approx -1.550670$) ($\delta = 0.2$, $\alpha = 1.0$, $\beta = 4.0$, $Q_0 = 100.0$)

2.3.4 Numerical Illustrations

The trajectory and amplitude spectrums are illustrated in Figures 2.13 and 2.14 for independent symmetric period-3 motions. Initial conditions for numerical simulations are obtained from the approximate solutions of period-3 motion and such approximate solutions are expressed by the Fourier series expression with 12 harmonic terms (HB12). Using the parameters in Equation (2.47), trajectory in phase plane and harmonic amplitude spectrum are presented in Figure 2.13(a) and (b) with $\Omega = 10$, and the initial condition is ($x_0 \approx -1.913080$ and $y_0 \approx -4.428830$). For over 40 periods, the numerical result matches very well with the analytical solution in Figure 2.13(a). The harmonic amplitude for analytical solutions in Figure 2.13(b) are $A_{1/3} \approx 1.5$, $A_1 \approx 1.2$, $A_{5/3} \approx 7.7 \times 10^{-3}$, $A_{7/3} \approx 1.25 \times 10^{-2}$, $A_3 \approx 1.9 \times 10^{-3}$, and $A_{11/3} \approx 1.2 \times 10^{-4}$. Other harmonic terms are $a_0^{(3)} = A_{k/m} = 0$ ($k = 2l$

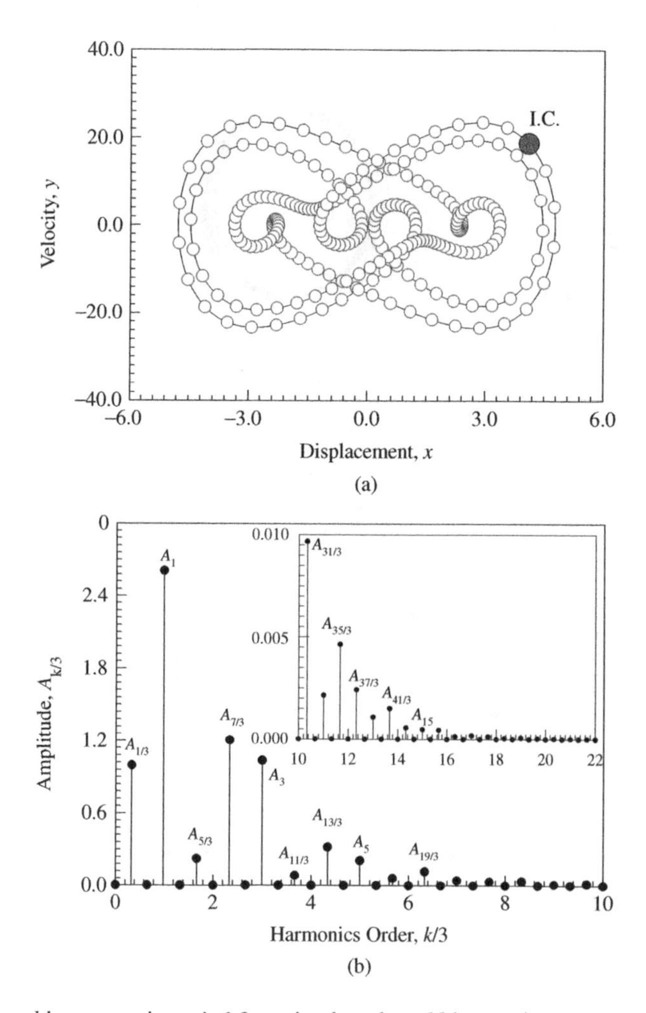

Figure 2.15 A *stable* symmetric period-3 motion based on 66 harmonic terms (HB66) for $\Omega = 2.8$: (a) trajectory and (b) amplitude spectrum ($x_0 \approx 4.088540$, $y_0 \approx 18.670000$) ($\delta = 0.2$, $\alpha = 1.0$, $\beta = 4.0$, $Q_0 = 100.0$)

and $l = 1, 2, \ldots, 6$). The symmetric period-3 motion is mainly determined by harmonic terms of $A_{1/3}$ and A_1.

For the same parameters and excitation frequency, there is an unstable symmetric period-3 motion with initial condition ($x_0 \approx 4.827070$, $y_0 \approx 1.054060$), as shown in Figure 2.14. Over 15 excitation periods, the numerical solution moves away from the analytical solution to another stable period-1 motion, which can be observed from Figure 2.14(a). This means the unstable solution is much closer to the stable-1 motion instead of the symmetric period-3 motion. For $\Omega > 13$, the unstable symmetric period-3 motion is much closer to the stable symmetric period-3 motion. The unstable symmetric period-3 motion will move to stable symmetric period-3 motion rather than period-1 motion. For the unstable period-3 motion, harmonic amplitudes in Figure 2.14(b) are $A_{1/3} \approx 6.1 \times 10^{-1}$, $A_1 \approx 1.1$, $A_{5/3} \approx 1.25 \times 10^{-2}$, $A_{7/3} \approx 4.11 \times 10^{-3}$, $A_3 \approx 1.48 \times 10^{-3}$, and $A_{11/3} \approx 5.03 \times 10^{-5}$. In addition, other harmonic terms are $a_0^{(3)} = A_{k/m} = 0$ ($k = 2l$ and $l = 1, 2, \ldots, 6$).

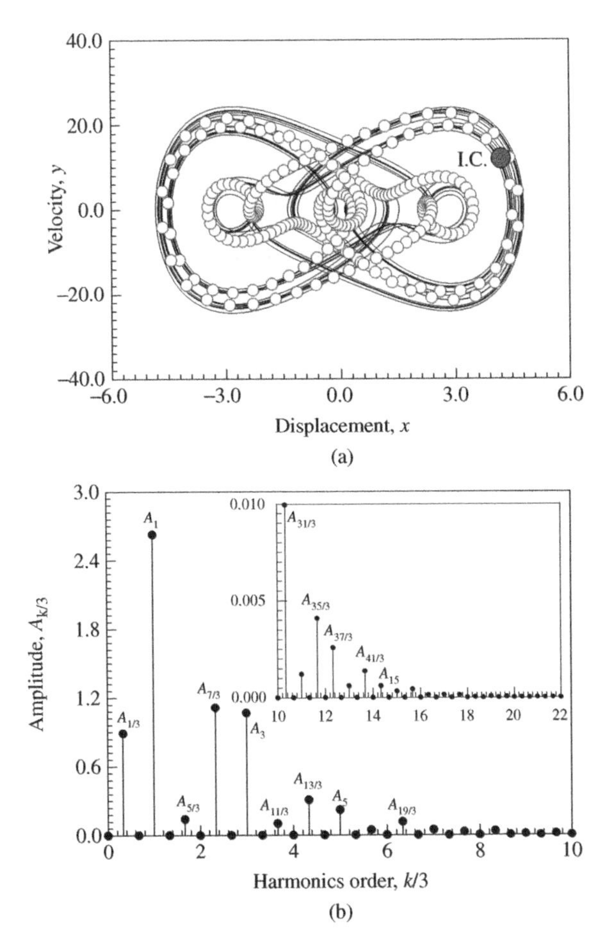

Figure 2.16 An *unstable* symmetric period-3 motion based on 66 harmonic terms (HB66) for $\Omega = 2.8$: (a) trajectory and (b) amplitude spectrum ($x_0 \approx 4.209140$, $y_0 \approx 11.907000$) ($\delta = 0.2$, $\alpha = 1.0$, $\beta = 4.0$, $Q_0 = 100.0$)

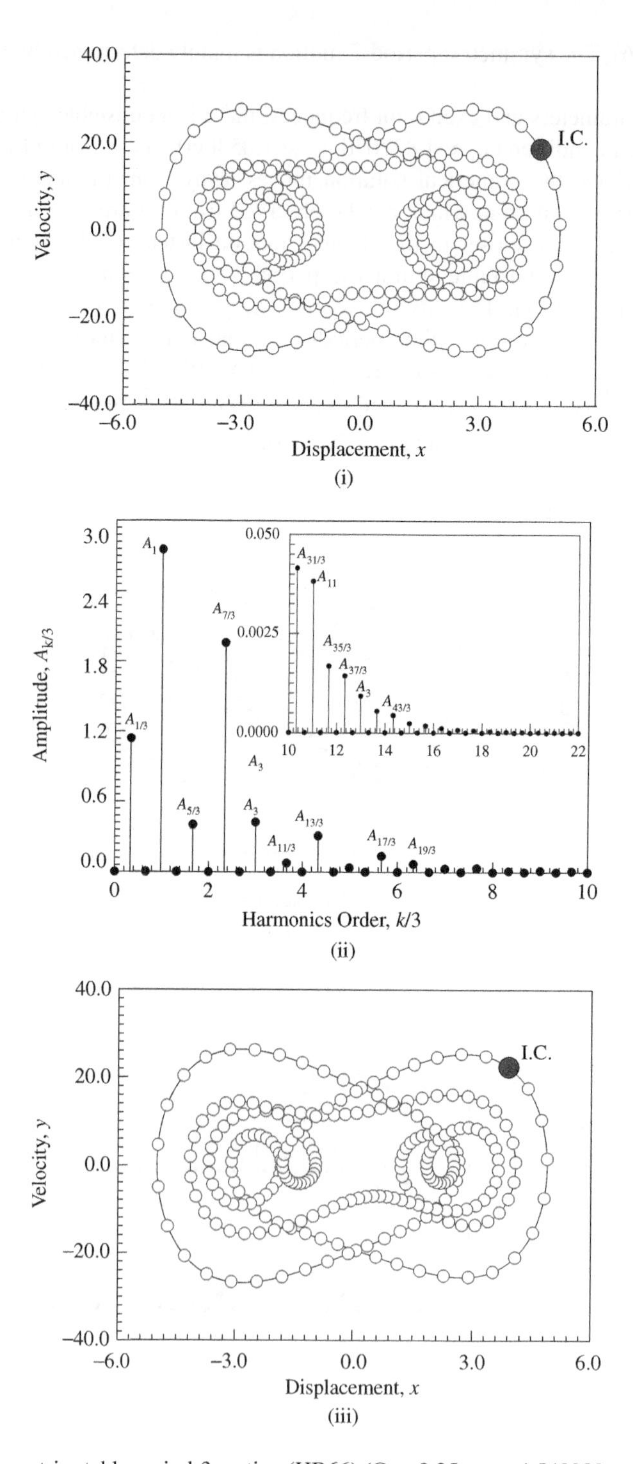

Figure 2.17 Symmetric stable period-3 motion (HB66) ($\Omega = 3.25$, $x_0 \approx 4.548080$, $y_0 \approx 8.605500$): (i) trajectory and (ii) amplitude spectrum. Asymmetric stable period-3 motion (HB66) ($\Omega = 3.088$, $x_0 \approx 3.910010$, $y_0 \approx 22.366700$): (iii) trajectory and (iv) amplitude spectrum. Asymmetric stable period-6 motion (HB132): (v) trajectory and (vi) amplitude spectrum ($\Omega = 3.08$, $x_0 \approx 4.034130$, $y_0 \approx 21.725100$) ($\delta = 0.2$, $\alpha = 1.0$, $\beta = 4.0$, $Q_0 = 100.0$)

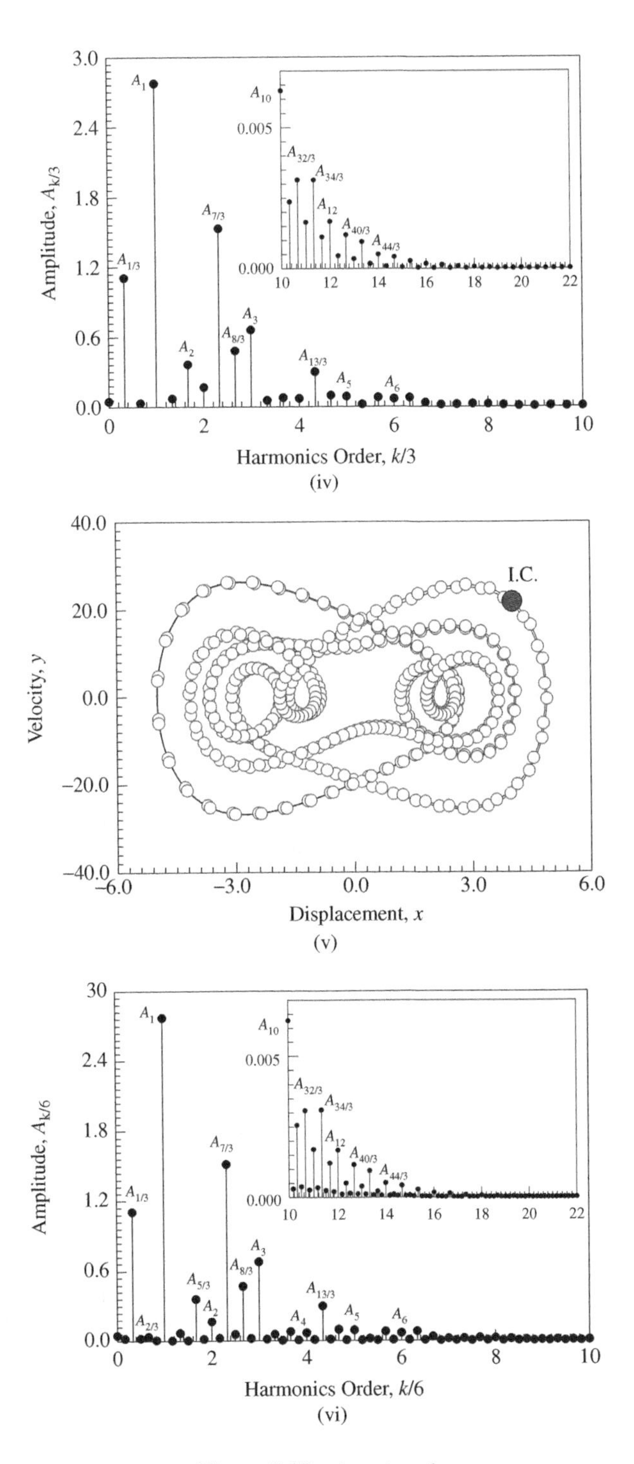

Figure 2.17 *(continued)*

Trajectories and amplitude spectrums are illustrated in Figures 2.15 and 2.16 for the approximate solutions of stable and unstable symmetric period-3 motions connected to the asymmetric period-3 motions, respectively. Once again, numerical simulations are superimposed on the analytical solution and the initial conditions are obtained from approximate solutions of period-3 motion. The analytical solution of period-3 motion is given by the Fourier series with the 66 harmonic terms (HB66). For stable period-3 motion, the initial condition are $x_0 \approx 4.088540$, and $y_0 \approx 18.670000$ for $\Omega = 2.8$ with other parameters in Equation (2.47). For 40 periods, analytical and numerical trajectories of the symmetric period-3 motion are presented in Figure 2.15(a), and both of them match very well. In addition, the harmonic amplitude spectrum is presented in Figure 2.15(b). The main harmonic amplitudes are $A_{1/3} \approx 1.0$, $A_1 \approx 2.7$, $A_{5/3} \approx 0.3$, $A_{7/3} \approx 1.2$, $A_3 \approx 1.0$, $A_{11/3} \approx 0.1$, $A_{13/3} \approx 0.3$, $A_5 \approx 0.2$, $A_{19/3} \approx 0.15$. From the zoomed area, $A_{31/3} \approx 0.009$, $A_{11} \approx 0.0025$, $A_{35/3} \approx 0.005$, $A_{37/3} \approx 0.0025$, $A_{41/3} \approx 0.002$, and $A_{15} \approx 0.001$.

For the unstable symmetric period-3 motion with the same parameters, the initial conditions are $t_0 = 0.0$, $x_0 \approx 4.209140$, and $y_0 \approx 11.907000$ for $\Omega = 2.8$. In Figure 2.16(a), the symmetries of displacement and velocity are observed. Since such a symmetric period-3 motion is unstable, after 20 periods, the unstable asymmetric period-3 motion gradually goes away to approach one of the stable periodic motions. The trajectory of the unstable period-3 motion is not clean like the stable period-3 motion because the numerical solution of unstable period-3 motion moves away from its analytical solution with increasing time. The harmonic amplitude spectrum for the unstable symmetric period-3 motion is presented in Figure 2.16(b). $A_{1/3} \approx 0.8$, $A_1 \approx 2.7$, $A_{5/3} \approx 0.2$, $A_{7/3} \approx 1.2$, $A_3 \approx 1.0$, $A_{11/3} \approx 0.1$, $A_{13/3} \approx 0.3$, $A_5 \approx 0.2$, $A_{19/3} \approx 0.15$. From the zoomed area, $A_{31/3} \approx 0.01$, $A_{11} \approx 0.002$, $A_{35/3} \approx 0.004$, $A_{37/3} \approx 0.003$, $A_{41/3} \approx 0.002$, and $A_{15} \approx 0.001$.

In Figure 2.17(i) and (ii), the stable symmetric period-3 motion at $\Omega = 3.25$ is presented with initial condition ($x_0 \approx 4.548080$ and $y_0 \approx 8.605500$). The trajectory in phase plane is different from the independent stable symmetric period-3 motions. The harmonic amplitude spectrum of the analytical solutions gives: $A_{1/3} \approx 1.2$, $A_1 \approx 2.8$, $A_{5/3} \approx 0.5$, $A_{7/3} \approx 1.9$, $A_3 \approx 0.5$, $A_{11/3} \approx 0.1$, $A_{13/3} \approx 0.3$, $A_{17/3} \approx 0.2$, $A_{19/3} \approx 0.1$. From the zoomed area, $A_{31/3} \approx 0.004$, $A_{11} \approx 0.0035$, $A_{35/3} \approx 0.002$, $A_{37/3} \approx 0.002$, $A_{13} \approx 0.0015$, and $A_{43/3} \approx 0.001$. For the asymmetric period-3 motion ($\Omega = 3.088$), its trajectory in phase plane is presented in Figure 2.17(iii). The asymmetric structure of the trajectory in phase plane is observed. The corresponding harmonic amplitude spectrum is presented in Figure 2.17(iv), and the corresponding main values are: $A_{1/3} \approx 1.1, A_1 \approx 2.8, A_{5/3} \approx 0.4, A_{7/3} \approx 1.6, A_{8/3} \approx 0.6, A_3 \approx 0.8$, $A_{13/3} \approx 0.4$. $A_5 \approx 0.1$, $A_{19/3} \approx 0.1$. From the zoomed area, $A_{10} \approx 0.006$, $A_{31/3} \approx 0.002$, $A_{32/3} \approx 0.003$, $A_{34/3} \approx 0.004$, $A_{12} \approx 0.0015$, and $A_{44/3} \approx 0.0005$. For the period-6 motion at $\Omega = 3.08$, the trajectory and harmonic amplitudes are presented with initial condition ($x_0 \approx 4.034130$, $y_0 \approx 21.725100$) in Figure 2.17(v) and (vi), respectively. The major harmonic amplitudes are $A_{1/3} \approx 1.15$, $A_1 \approx 2.7$, $A_{5/3} \approx 0.4$, $A_{7/3} \approx 1.5$, $A_{8/3} \approx 0.5$, $A_3 \approx 0.8$, $A_4 \approx 0.05$, $A_{13/3} \approx 0.3$, $A_5 \approx 0.1$, and $A_6 \approx 0.05$. From the zoomed area, $A_{10} \approx 0.0056$, $A_{31/3} \approx 0.0025$, $A_{32/3} \approx 0.003$, $A_{34/3} \approx 0.003$, $A_{12} \approx 0.002$, and $A_{40/3} \approx 0.001$. The quantitative values of the amplitudes $A_{k/6}$ ($k = 2l + 1$, $l = 1, 2, \ldots$) are very small compared to the other values $A_{l/3}$ ($l = 1, 2, \ldots$).

As in Luo and Huang (2013c,d), analytical bifurcation trees of period-m motions to chaos in a periodically forced, softening Duffing oscillator can be discussed, which will not be presented herein. The bifurcation trees of period-m motions to chaos in double-well Duffing oscillators can also be referred to Luo and Huang (2012c,d).

3

Self-Excited Nonlinear Oscillators

In this chapter, analytical solutions for period-m motions in periodically forced, self-excited oscillators are presented in the Fourier series form with finite harmonic terms, and the stability and bifurcation of the corresponding period-m motions are completed. The period-m motions in the periodically forced, van der Pol oscillator will be discussed first as an example, and the limit cycles for the van der Pol oscillator without any excitation will be discussed as well. The period-m motions are in independent periodic solution windows embedded in quasi-periodic and chaotic motions. The period-m motions for the van der Pol-Duffing oscillator will be presented, and the bifurcation tree of period-m motion will be discussed. For a better understanding of complex period-m motions in such a van der Pol-Duffing oscillator, trajectories and amplitude spectrums are illustrated numerically.

3.1 van del Pol Oscillators

In this section, the van del Pol oscillator will be discussed. The appropriate analytical solutions of period-m motions embedded in quasi-periodic and chaotic motions will be presented with finite harmonic terms in the Fourier series solution based on the prescribed accuracy of harmonic amplitudes. A limited cycle of the van der Pol oscillator without any periodic excitations will be discussed first, and under periodic excitations, the period-m motions in the van der Pol oscillator will be presented.

3.1.1 Analytical Solutions

Consider a generalized van der Pol oscillator

$$\ddot{x} + (-\alpha_1 + \alpha_2 x^2)\dot{x} + \alpha_3 x = Q_0 \cos \Omega t \tag{3.1}$$

where α_i $(i = 1, 2, 3)$ are system coefficients for the van der Pol oscillator. Q_0 and Ω are excitation amplitude and frequency, respectively. In Luo (2012a), the standard form of Equation (3.1) can be written as

$$\ddot{x} = F(x, \dot{x}, t) \tag{3.2}$$

Analytical Routes to Chaos in Nonlinear Engineering, First Edition. Albert C. J. Luo.

where

$$F(\dot{x}, x, t) = -\dot{x}(-\alpha_1 + \alpha_2 x^2) - \alpha_3 x + Q_0 \cos \Omega t. \tag{3.3}$$

The analytical solution of period-1 motion for the above equation is

$$x^{(m)*} = a_0^{(m)}(t) + \sum_{k=1}^{N} b_{k/m}(t) \cos\left(\frac{k}{m}\theta\right) + c_{k/m}(t) \sin\left(\frac{k}{m}\theta\right) \tag{3.4}$$

where $a_0^{(m)}(t)$, $b_{k/m}(t)$, and $c_{k/m}(t)$ vary with time and $\theta = \Omega t$. The first and second order of derivatives of $x^*(t)$ are

$$\begin{aligned}
\dot{x}^{(m)*} &= \dot{a}_0^{(m)} + \sum_{k=1}^{N} \left(\dot{b}_{k/m} + \frac{k}{m}\Omega c_{k/m}\right) \cos\left(\frac{k}{m}\theta\right) \\
&\quad + \left(\dot{c}_{k/m} - \frac{k}{m}\Omega b_{k/m}\right) \sin\left(\frac{k}{m}\theta\right),
\end{aligned} \tag{3.5}$$

$$\begin{aligned}
\ddot{x}^{(m)*} &= \ddot{a}_0^{(m)} + \sum_{k=1}^{N} \left[\ddot{b}_{k/m} + 2\left(\frac{k}{m}\Omega\right)\dot{c}_{k/m} - \left(\frac{k}{m}\Omega\right)^2 b_{k/m}\right] \cos\left(\frac{k}{m}\theta\right) \\
&\quad + \left[\ddot{c}_{k/m} - 2\left(\frac{k}{m}\Omega\right)\dot{b}_{k/m} - \left(\frac{k}{m}\Omega\right)^2 c_{k/m}\right] \sin\left(\frac{k}{m}\theta\right).
\end{aligned} \tag{3.6}$$

Substitution of Equations (3.4)–(3.6) into Equation (3.1) and application of the virtual work principle for a basis of constant, $\cos(k\theta/m)$ and $\sin(k\theta/m)$ $(k = 1, 2, \ldots)$ as a set of virtual displacements gives

$$\begin{aligned}
&\ddot{a}_0^{(m)} = F_0^{(m)}(a_0^{(m)}, \mathbf{b}^{(m)}, \mathbf{c}^{(m)}, \dot{a}_0^{(m)}, \dot{\mathbf{b}}^{(m)}, \dot{\mathbf{c}}^{(m)}), \\
&\ddot{b}_{k/m} + 2\frac{k\Omega}{m}\dot{c}_{k/m} - \left(\frac{k\Omega}{m}\right)^2 b_{k/m} = F_{1k}^{(m)}(a_0^{(m)}, \mathbf{b}^{(m)}, \mathbf{c}^{(m)}, \dot{a}_0^{(m)}, \dot{\mathbf{b}}^{(m)}, \dot{\mathbf{c}}^{(m)}), \\
&\ddot{c}_{k/m} - 2\frac{k\Omega}{m}\dot{b}_{k/m} - \left(\frac{k\Omega}{m}\right)^2 c_{k/m} = F_{2k}^{(m)}(a_0^{(m)}, \mathbf{b}^{(m)}, \mathbf{c}^{(m)}, \dot{a}_0^{(m)}, \dot{\mathbf{b}}^{(m)}, \dot{\mathbf{c}}^{(m)}) \\
&\text{for } k = 1, 2, \ldots, N
\end{aligned} \tag{3.7}$$

where

$$\begin{aligned}
&F_0^{(m)}(a_0^{(m)}, \mathbf{b}^{(m)}, \mathbf{c}^{(m)}, \dot{a}_0^{(m)}, \dot{\mathbf{b}}^{(m)}, \dot{\mathbf{c}}^{(m)}) \\
&\quad = \frac{1}{mT}\int_0^{mT} F(x^{(m)*}, \dot{x}^{(m)*}, t)dt \\
&\quad = \alpha_1 \dot{a}_0^{(m)} - \alpha_3 a_0^{(m)} - \alpha_2 f_0^{(m)}, \\
&F_{1k}^{(m)}(a_0^{(m)}, \mathbf{b}^{(m)}, \mathbf{c}^{(m)}, \dot{a}_0^{(m)}, \dot{\mathbf{b}}^{(m)}, \dot{\mathbf{c}}^{(m)}) \\
&\quad = \frac{2}{mT}\int_0^{mT} F(x^{(m)*}, \dot{x}^{(m)*}, t) \cos\left(\frac{k}{m}\Omega t\right) dt \\
&\quad = \alpha_1 \left(\dot{b}_{k/m} + \frac{k}{m}\Omega c_{k/m}\right) - \alpha_3 b_{k/m} + Q_0 \delta_k^m - \alpha_2 f_{1k/m},
\end{aligned}$$

$$\begin{aligned}
&F_{2k}^{(m)}(a_0^{(m)}, \mathbf{b}^{(m)}, \mathbf{c}^{(m)}, \dot{a}_0^{(m)}, \dot{\mathbf{b}}^{(m)}, \dot{\mathbf{c}}^{(m)}) \\
&= \frac{2}{mT}\int_0^{mT} F(x^{(m)*}, \dot{x}^{(m)*}, t)\sin\left(\frac{k}{m}\Omega t\right) dt \\
&= \alpha_1\left(\dot{c}_{k/m} - \frac{k}{m}\Omega b_{k/m}\right) - \alpha_3 c_{k/m} - \alpha_2 f_{2k/m}
\end{aligned} \tag{3.8}$$

and

$$f_0^{(m)} = \dot{a}_0^{(m)}(a_0^{(m)})^2 + \sum_{n=1}^{6}\sum_{l=1}^{N}\sum_{j=1}^{N}\sum_{i=1}^{N} f_n^{(m,0)}(i,j,l) \tag{3.9}$$

with

$$\begin{aligned}
f_1^{(m,0)}(i,j,l) &= \frac{1}{2N}\dot{a}_0^{(m)}(b_{i/m}b_{j/m} + c_{i/m}c_{j/m})\delta_{i-j}^0, \\
f_2^{(m,0)}(i,j,l) &= \frac{1}{N}a_0^{(m)}\left[b_{i/m}\left(\dot{b}_{l/m} + \frac{l}{m}\Omega c_{l/m}\right) + c_{i/m}\left(\dot{c}_{l/m} - \frac{l}{m}\Omega b_{l/m}\right)\right]\delta_{l-i}^0, \\
f_3^{(m,0)}(i,j,l) &= \frac{1}{4}(b_{i/m}b_{j/m} - c_{i/m}c_{j/m})\left(\dot{b}_{l/m} + \frac{l}{m}\Omega c_{k/m}\right)\delta_{l-i-j}^0, \\
f_4^{(m,0)}(i,j,l) &= \frac{1}{4}(b_{j/m}c_{i/m} + b_{i/m}c_{j/m})\left(\dot{c}_{l/m} - \frac{l}{m}\Omega b_{l/m}\right)\delta_{l-i-j}^0, \\
f_5^{(m,0)}(i,j,l) &= \frac{1}{4}(b_{i/m}b_{j/m} + c_{i/m}c_{j/m})\left(\dot{b}_{l/m} + \frac{l}{m}\Omega c_{l/m}\right)(\delta_{l-i+j}^0 + \delta_{l+i-j}^0), \\
f_6^{(m,0)}(i,j,l) &= \frac{1}{4}(b_{j/m}c_{i/m} - b_{i/m}c_{j/m})\left(\dot{c}_{l/m} - \frac{l}{m}\Omega b_{l/m}\right)(\delta_{l-i+j}^0 + \delta_{l+i-j}^0);
\end{aligned} \tag{3.10}$$

and

$$f_{1k/m} = \sum_{n=1}^{8}\sum_{l=1}^{N}\sum_{j=1}^{N}\sum_{i=1}^{N} f_n^{(m,1)}(i,j,l,k) \tag{3.11}$$

with

$$\begin{aligned}
f_1^{(m,1)}(i,j,l,k) &= \frac{1}{N^2}\left[2\dot{a}_0^{(m)}a_0^{(m)}b_{i/m}\delta_i^k + (a_0^{(m)})^2(\dot{b}_{l/m} + \frac{k}{m}\Omega c_{l/m})\delta_l^k\right], \\
f_2^{(m,1)}(i,j,l,k) &= \frac{1}{2N}\dot{a}_0^{(m)}\left[(b_{i/m}b_{j/m} + c_{i/m}c_{j/m})\delta_{i-j}^k\right. \\
&\quad \left. +(b_{i/m}b_{j/m} - c_{i/m}c_{j/m})\delta_{i+j}^k\right], \\
f_3^{(m,1)}(i,j,l,k) &= \frac{1}{N}a_0^{(m)}b_{i/m}\left(\dot{b}_{l/m} + \frac{l}{m}\Omega c_{l/m}\right)(\delta_{|l-i|}^k + \delta_{l+i}^k), \\
f_4^{(m,1)}(i,j,l,k) &= \frac{1}{N}a_0^{(m)}c_{i/m}\left(\dot{c}_{l/m} - \frac{l}{m}\Omega b_{l/m}\right)(\delta_{|l-i|}^k - \delta_{l+i}^k), \\
f_5^{(m,1)}(i,j,l,k) &= \frac{1}{4}(b_{i/m}b_{j/m} - c_{i/m}c_{j/m})\left(\dot{b}_{l/m} + \frac{l}{m}\Omega c_{l/m}\right)(\delta_{|l-i-j|}^k + \delta_{l+i+j}^k), \\
f_6^{(m,1)}(i,j,l,k) &= \frac{1}{4}(b_{j/m}c_{i/m} + b_{i/m}c_{j/m})\left(\dot{c}_{l/m} - \frac{l}{m}\Omega b_{l/m}\right)(\delta_{|l-i-j|}^k - \delta_{l+i+j}^k), \\
f_7^{(m,1)}(i,j,l,k) &= \frac{1}{4}(b_{i/m}b_{j/m} + c_{i/m}c_{j/m})\left(\dot{b}_{l/m} + \frac{l}{m}\Omega c_{l/m}\right)(\delta_{|l-i-j|}^k + \delta_{l+i+j}^k),
\end{aligned}$$

$$f_8^{(m,1)}(i,j,l,k) = \frac{1}{4}(b_{j/m}c_{i/m} - b_{i/m}c_{j/m})\left(\dot{c}_{l/m} - \frac{l}{m}\Omega b_{l/m}\right)(\delta^k_{|l-i-j|} - \delta^k_{l+i+j}) \tag{3.12}$$

and

$$f_{2k/m} = \sum_{n=1}^{9}\sum_{l=1}^{N}\sum_{j=1}^{N}\sum_{i=1}^{N} f_n^{(m,2)}(i,j,l,k) \tag{3.13}$$

with

$$\begin{aligned}
f_1^{(m,2)}(i,j,l,k) &= \frac{1}{N^2}\left[2\dot{a}_0^{(m)}a_0^{(m)}c_{i/m}\delta^k_i + (a_0^{(m)})^2\left(\dot{c}_{l/m} - \frac{l}{m}\Omega b_{l/m}\right)\delta^k_l\right],\\
f_2^{(m,2)}(i,j,l,k) &= \frac{1}{2N}\dot{a}_0^{(m)}(b_{j/m}c_{i/m} - b_{i/m}c_{j/m})\mathrm{sgn}(i-j)\delta^k_{|i-j|},\\
f_3^{(m,2)}(i,j,l,k) &= \frac{1}{2N}\dot{a}_0^{(m)}(b_{j/m}c_{i/m} + b_{i/m}c_{j/m})\delta^k_{i+j},\\
f_4^{(m,2)}(i,j,l,k) &= \frac{1}{N}a_0^{(m)}b_{i/m}\left(\dot{c}_{l/m} - \frac{l}{m}\Omega b_{l/m}\right)[\delta^k_{l+i} + \mathrm{sgn}(l-i)\delta^k_{|l-i|}],\\
f_5^{(m,2)}(i,j,l,k) &= \frac{1}{N}a_0^{(m)}c_{i/m}\left(\dot{b}_{l/m} + \frac{l}{m}\Omega c_{l/m}\right)[\delta^k_{l+i} - \mathrm{sgn}(l-i)\delta^k_{|l-i|}],\\
f_6^{(m,2)}(i,j,l,k) &= \frac{1}{4}(b_{i/m}c_{j/m} + b_{j/m}c_{i/m})\left(\dot{b}_{l/m} + \frac{l}{m}\Omega c_{l/m}\right)\\
&\quad\times[\delta^k_{l+i+j} - \mathrm{sgn}(l-i-j)\delta^k_{|l-i-j|}],\\
f_7^{(m,2)}(i,j,l,k) &= \frac{1}{4}(b_{i/m}b_{j/m} - c_{i/m}c_{j/m})\left(\dot{c}_{k/m} - \frac{l}{m}\Omega b_{l/m}\right)\\
&\quad\times[\delta^k_{l+i+j} + \mathrm{sgn}(l-i-j)\delta^k_{|l-i-j|}],\\
f_8^{(m,2)}(i,j,l,k) &= \frac{1}{4}(b_{i/m}b_{j/m} + c_{i/m}c_{j/m})\left(\dot{c}_{l/m} - \frac{l}{m}\Omega b_{l/m}\right)\\
&\quad\times[\mathrm{sgn}(l-i+j)\delta^k_{|l-i+j|} + \mathrm{sgn}(l+i-j)\delta^k_{|l+i-j|}],\\
f_9^{(m,2)}(i,j,l,k) &= \frac{1}{4}(b_{i/m}c_{j/m} - b_{j/m}c_{i/m})\left(\dot{b}_{l/m} + \frac{l}{m}\Omega c_{l/m}\right)\\
&\quad\times[\mathrm{sgn}(l-i+j)\delta^k_{|l-i+j|} - \mathrm{sgn}(l+i-j)\delta^k_{|l+i-j|}].
\end{aligned} \tag{3.14}$$

Define

$$\begin{aligned}
\mathbf{z}^{(m)} &\triangleq (a_0^{(m)}, \mathbf{b}^{(m)}, \mathbf{c}^{(m)})^{\mathrm{T}}\\
&= (a_0^{(m)}, b_{1/m}, \dots, b_{N/m}, c_{1/m}, \dots, c_{N/m})^{\mathrm{T}}\\
&\equiv (z_0^{(m)}, z_1^{(m)}, \dots, z_{2N}^{(m)})^{\mathrm{T}},\\
\mathbf{z}_1 &\triangleq \dot{\mathbf{z}} = (\dot{a}_0^{(m)}, \dot{\mathbf{b}}^{(m)}, \dot{\mathbf{c}}^{(m)})^{\mathrm{T}}\\
&= (\dot{a}_0^{(m)}, \dot{b}_{1/m}, \dots, \dot{b}_{N/m}, \dot{c}_{1/m}, \dots, \dot{c}_{N/m})^{\mathrm{T}}\\
&\equiv (\dot{z}_0^{(m)}, \dot{z}_1^{(m)}, \dots, \dot{z}_{2N}^{(m)})^{\mathrm{T}},
\end{aligned} \tag{3.15}$$

where

$$\mathbf{b}^{(m)} = (b_{1/m}, \dots, b_{N/m})^{\mathrm{T}},$$
$$\mathbf{c}^{(m)} = (c_{1/m}, \dots, c_{N/m})^{\mathrm{T}}. \tag{3.16}$$

Equation (3.7) can be expressed in the form of

$$\dot{\mathbf{z}}^{(m)} = \mathbf{z}_1^{(m)} \text{ and } \dot{\mathbf{z}}_1^{(m)} = \mathbf{g}^{(m)}(\mathbf{z}^{(m)}, \mathbf{z}_1^{(m)}) \tag{3.17}$$

where

$$\mathbf{g}^{(m)}(\mathbf{z}^{(m)}, \mathbf{z}_1^{(m)}) = \begin{pmatrix} F_0^{(m)}(\mathbf{z}^{(m)}, \mathbf{z}_1^{(m)}) \\ \mathbf{F}_1^{(m)}(\mathbf{z}^{(m)}, \mathbf{z}_1^{(m)}) - 2\mathbf{k}_1 \frac{\Omega}{m}\dot{\mathbf{c}}^{(m)} + \mathbf{k}_2\left(\frac{\Omega}{m}\right)^2 \mathbf{b}^{(m)} \\ \mathbf{F}_2^{(m)}(\mathbf{z}^{(m)}, \mathbf{z}_1^{(m)}) + 2\mathbf{k}_1 \frac{\Omega}{m}\dot{\mathbf{b}}^{(m)} + \mathbf{k}_2\left(\frac{\Omega}{m}\right)^2 \mathbf{c}^{(m)} \end{pmatrix} \tag{3.18}$$

and

$$\mathbf{k}_1 = diag(1, 2, \dots, N),$$
$$\mathbf{k}_2 = diag(1, 2^2, \dots, N^2),$$
$$\mathbf{F}_1^{(m)} = (F_{11}^{(m)}, F_{12}^{(m)}, \dots, F_{1N}^{(m)})^{\mathrm{T}},$$
$$\mathbf{F}_2^{(m)} = (F_{21}^{(m)}, F_{22}^{(m)}, \dots, F_{2N}^{(m)})^{\mathrm{T}}$$
$$\text{for } N = 1, 2, \dots, \infty. \tag{3.19}$$

Introducing

$$\mathbf{y}^{(m)} \equiv (\mathbf{z}^{(m)}, \mathbf{z}_1^{(m)}) \text{ and } \mathbf{f}^{(m)} = (\mathbf{z}_1^{(m)}, \mathbf{g}^{(m)})^{\mathrm{T}}, \tag{3.20}$$

Equation (3.17) becomes

$$\dot{\mathbf{y}}^{(m)} = \mathbf{f}^{(m)}(\mathbf{y}^{(m)}). \tag{3.21}$$

The steady-state solutions for periodic motions of the van del Pol oscillator in Equation (3.1) can be obtained by setting $\dot{\mathbf{y}}^{(m)} = \mathbf{0}$, that is,

$$F_0^{(m)}(\mathbf{z}^{(m)}, \mathbf{0}) = 0,$$
$$\mathbf{F}_1^{(m)}(\mathbf{z}^{(m)}, \mathbf{0}) - \mathbf{k}_2\left(\frac{\Omega}{m}\right)^2 \mathbf{b}^{(m)} = \mathbf{0},$$
$$\mathbf{F}_2^{(m)}(\mathbf{z}^{(m)}, \mathbf{0}) - \mathbf{k}_2\left(\frac{\Omega}{m}\right)^2 \mathbf{c}^{(m)} = \mathbf{0}. \tag{3.22}$$

The $(2N+1)$ nonlinear equations in Equation (3.22) are solved by the Newton–Raphson method. The linearized equation at equilibrium point $\mathbf{y}^* = (\mathbf{z}^*, \mathbf{0})^{\mathrm{T}}$ is given by

$$\Delta\dot{\mathbf{y}}^{(m)} = D\mathbf{f}(\mathbf{y}^{(m)*})\Delta\mathbf{y}^{(m)} \tag{3.23}$$

where

$$D\mathbf{f}(\mathbf{y}^{(m)*}) = \partial\mathbf{f}(\mathbf{y}^{(m)})/\partial\mathbf{y}^{(m)}|_{\mathbf{y}^{(m)*}}. \tag{3.24}$$

The corresponding eigenvalues are determined by

$$|D\mathbf{f}(\mathbf{y}^{(m)*}) - \lambda \mathbf{I}_{2(2N+1)\times 2(2N+1)}| = 0. \tag{3.25}$$

where

$$D\mathbf{f}(\mathbf{y}^{(m)*}) = \begin{bmatrix} \mathbf{0}_{(2N+1)\times(2N+1)} & \mathbf{I}_{(2N+1)\times(2N+1)} \\ \mathbf{G}_{(2N+1)\times(2N+1)} & \mathbf{H}_{(2N+1)\times(2N+1)} \end{bmatrix} \tag{3.26}$$

and

$$\mathbf{G} = \frac{\partial \mathbf{g}^{(m)}}{\partial \mathbf{z}^{(m)}} = (\mathbf{G}^{(0)}, \mathbf{G}^{(c)}, \mathbf{G}^{(s)})^{\mathrm{T}} \tag{3.27}$$

$$\begin{aligned} \mathbf{G}^{(0)} &= (G_0^{(0)}, G_1^{(0)}, \ldots, G_{2N}^{(0)}), \\ \mathbf{G}^{(c)} &= (\mathbf{G}_1^{(c)}, \mathbf{G}_2^{(c)}, \ldots, \mathbf{G}_N^{(c)})^{\mathrm{T}}, \\ \mathbf{G}^{(s)} &= (\mathbf{G}_1^{(s)}, \mathbf{G}_2^{(s)}, \ldots, \mathbf{G}_N^{(s)})^{\mathrm{T}} \end{aligned} \tag{3.28}$$

for $N = 1, 2, \ldots \infty$ with

$$\begin{aligned} \mathbf{G}_k^{(c)} &= (G_{k0}^{(c)}, G_{k1}^{(c)}, \ldots, G_{k(2N)}^{(c)}), \\ \mathbf{G}_k^{(s)} &= (G_{k0}^{(s)}, G_{k1}^{(s)}, \ldots, G_{k(2N)}^{(s)}) \end{aligned} \tag{3.29}$$

for $k = 1, 2, \ldots N$. The corresponding components are

$$\begin{aligned} G_r^{(0)} &= -\alpha_3 \delta_0^r - \alpha_2 g_r^{(0)}, \\ G_{kr}^{(c)} &= \left(\frac{k\Omega}{m}\right)^2 \delta_k^r + \alpha_1 \left(\frac{k}{m}\Omega\right) \delta_{k+N}^r - \alpha_3 \delta_k^r - \alpha_2 g_{kr}^{(c)}, \\ G_{kr}^{(s)} &= \left(\frac{k\Omega}{m}\right)^2 \delta_{k+N}^r - \alpha_1 \left(\frac{k}{m}\Omega\right) \delta_k^r - \alpha_3 \delta_{k+N}^r - \alpha_2 g_{kr}^{(s)}, \end{aligned} \tag{3.30}$$

where

$$g_r^{(0)} = g_0^{(0)}(r) + \sum_{n=1}^{16} \sum_{l=1}^{N} \sum_{j=1}^{N} \sum_{i=1}^{N} g_n^{(0)}(i,j,l,r) \tag{3.31}$$

with

$$\begin{aligned} g_0^{(0)}(r) &= 2\dot{a}_0^{(m)} a_0^{(m)} \delta_r^0, \\ g_1^{(0)}(i,j,l,r) &= \frac{1}{N} b_{i/m} \left(\dot{b}_{k/m} + \frac{l}{m}\Omega c_{l/m}\right) \delta_{l-i}^0 \delta_0^r, \\ g_2^{(0)}(i,j,l,r) &= \frac{1}{N} c_{i/m} \left(\dot{c}_{l/m} - \frac{l}{m}\Omega b_{l/m}\right) \delta_{l-i}^0 \delta_0^r, \\ g_3^{(0)}(i,j,l,r) &= \frac{1}{N} \left[\dot{a}_0^{(m)} b_{i/m} \delta_{i-j}^0 + a_0^{(m)} \left(\dot{b}_{k/m} + \frac{l}{m}\Omega c_{l/m}\right) \delta_{l-i}^0\right] \delta_i^r, \\ g_4^{(0)}(i,j,l,r) &= -\frac{1}{N} a_0^{(m)} \frac{k}{m}\Omega c_{i/m} \delta_{l-i}^0 \delta_l^r, \\ g_5^{(0)}(i,j,l,r) &= \frac{1}{2} b_{j/m} \left(\dot{b}_{l/m} + \frac{l}{m}\Omega c_{l/m}\right) \delta_{l-i-j}^0 \delta_i^r, \end{aligned}$$

$$
\begin{aligned}
g_6^{(0)}(i,j,l,r) &= \frac{1}{2}c_{j/m}\left(\dot{c}_{l/m} - \frac{l}{m}\Omega b_{l/m}\right)\delta^0_{l-i-j}\delta^r_i,\\
g_7^{(0)}(i,j,l,r) &= -\frac{1}{4}\frac{l}{m}\Omega(b_{j/m}c_{i/m} + b_{i/m}c_{j/m})\delta^0_{l-i-j}\delta^r_l,\\
g_8^{(0)}(i,j,l,r) &= \frac{1}{2}b_{j/m}\left(\dot{b}_{l/m} + \frac{l}{m}\Omega c_{l/m}\right)(\delta^0_{l-i+j} + \delta^0_{l+i-j})\delta^r_i,\\
g_9^{(0)}(i,j,l,r) &= -\frac{1}{4}\frac{l}{m}\Omega(b_{j/m}c_{i/m} - b_{i/m}c_{j/m})(\delta^0_{l-i+j} + \delta^0_{l+i-j})\delta^r_l,\\
g_{10}^{(0)}(i,j,l,r) &= \frac{1}{N}\left[\dot{a}_0^{(m)}c_{j/m}\delta^0_{i-j}\delta^r_{j+N} + \frac{l}{m}\Omega a_0^{(m)}b_{i/m}\delta^0_{l-i}\delta^r_{l+N}\right],\\
g_{11}^{(0)}(i,j,l,r) &= \frac{1}{N}a_0^{(m)}\left(\dot{c}_{l/m} - \frac{l}{m}\Omega b_{l/m}\right)\delta^0_{l-i}\delta^r_{i+N},\\
g_{12}^{(0)}(i,j,l,r) &= -\frac{1}{2}c_{j/m}\left(\dot{b}_{l/m} + \frac{l}{m}\Omega c_{l/m}\right)\delta^0_{l-i-j}\delta^r_{i+N},\\
g_{13}^{(0)}(i,j,l,r) &= \frac{1}{4}\frac{l}{m}\Omega(b_{i/m}b_{j/m} - c_{i/m}c_{j/m})\delta^0_{l-i-j}\delta^r_{l+N},\\
g_{14}^{(0)}(i,j,l,r) &= \frac{1}{2}b_{j/m}\left(\dot{c}_{l/m} - \frac{l}{m}\Omega b_{l/m}\right)\delta^0_{l-i-j}\delta^r_{i+N},\\
g_{15}^{(0)}(i,j,l,r) &= \frac{1}{2}c_{j/m}\left(\dot{b}_{l/m} + \frac{l}{m}\Omega c_{l/m}\right)(\delta^0_{l-i+j} + \delta^0_{l+i-j})\delta^r_{i+N},\\
g_{16}^{(0)}(i,j,l,r) &= \frac{1}{4}\frac{l}{m}\Omega(b_{i/m}b_{j/m} + c_{i/m}c_{j/m})(\delta^0_{l-i+j} + \delta^0_{l+i-j})\delta^r_{l+N},
\end{aligned}
\tag{3.32}
$$

and

$$
g_{kr}^{(c)} = \sum_{n=1}^{18}\sum_{l=1}^{N}\sum_{j=1}^{N}\sum_{i=1}^{N} g_n^{(1)}(i,j,l,k,r) \tag{3.33}
$$

with

$$
\begin{aligned}
g_1^{(1)}(i,j,l,k,r) &= \frac{2}{N^2}\left[\dot{a}_0^{(m)}b_{i/m}\delta^k_i + 2a_0^{(m)}\left(\dot{b}_{l/m} + \frac{l}{m}\Omega c_{l/m}\right)\delta^k_l\right]\delta^r_0,\\
g_2^{(1)}(i,j,l,k,r) &= \frac{1}{N}b_{i/m}(\dot{b}_{l/m} + \frac{l}{m}\Omega c_{l/m})(\delta^k_{|l-i|} + \delta^k_{l+i})\delta^r_0,\\
g_3^{(1)}(i,j,l,k,r) &= \frac{1}{N}c_{i/m}\left(\dot{c}_{k/m} - \frac{l}{m}\Omega b_{l/m}\right)(\delta^k_{|l-i|} - \delta^k_{l+i})\delta^r_0,\\
g_4^{(1)}(i,j,l,k,r) &= \frac{2}{N^2}\dot{a}_0^{(m)}a_0^{(m)}\delta^k_i\delta^r_i + \frac{1}{N}\dot{a}_0^{(m)}b_{j/m}\delta^r_i(\delta^k_{|i-j|} + \delta^k_{i+j})\delta^r_i,\\
g_5^{(1)}(i,j,l,k,r) &= \frac{1}{N}a_0^{(m)}\left(\dot{b}_{l/m} + \frac{l}{m}\Omega c_{l/m}\right)(\delta^k_{|l-i|} + \delta^k_{l+i})\delta^r_i,\\
g_6^{(1)}(i,j,l,k,r) &= -\frac{1}{N}\frac{l}{m}\Omega a_0^{(m)}c_{i/m}(\delta^k_{|l-i|} - \delta^k_{l+i})\delta^r_k,\\
g_7^{(1)}(i,j,l,k,r) &= \frac{1}{2}b_{j/m}\left(\dot{b}_{l/m} + \frac{l}{m}\Omega c_{l/m}\right)(\delta^k_{|l-i-j|} + \delta^k_{l+i+j})\delta^r_i,\\
g_8^{(1)}(i,j,l,k,r) &= \frac{1}{2}c_{j/m}\left(\dot{c}_{l/m} - \frac{l}{m}\Omega b_{l/m}\right)(\delta^k_{|l-i-j|} - \delta^k_{l+i+j})\delta^r_i,
\end{aligned}
$$

$$
\begin{aligned}
g_9^{(1)}(i,j,l,k,r) &= -\frac{1}{4}\frac{l}{m}\Omega(b_{j/m}c_{i/m} + b_{i/m}c_{j/m})(\delta^k_{|l-i-j|} - \delta^k_{l+i+j})\delta^r_l, \\
g_{10}^{(1)}(i,j,l,k,r) &= \frac{1}{2}b_{j/m}\left(\dot{b}_{l/m} + \frac{l}{m}\Omega c_{l/m}\right)(\delta^k_{|l-i+j|} + \delta^k_{|l+i-j|})\delta^r_i, \\
g_{11}^{(1)}(i,j,l,k,r) &= -\frac{1}{4}\frac{l}{m}\Omega(b_{j/m}c_{i/m} - b_{i/m}c_{j/m})(\delta^k_{|l-i+j|} - \delta^k_{|l+i-j|})\delta^r_l, \\
g_{12}^{(1)}(i,j,l,k,r) &= \frac{1}{N^2}\frac{l}{m}\Omega(a_0^{(m)})^2\delta^k_l\delta^r_{l+N} + \frac{1}{N}\dot{a}_0^{(m)}c_{j/m}(\delta^k_{|i-j|} - \delta^k_{i+j})\delta^r_{i+N}, \\
g_{13}^{(1)}(i,j,l,k,r) &= \frac{1}{N}\frac{k}{m}\Omega a_0^{(m)}b_{i/m}(\delta^k_{|l-i|} + \delta^k_{l+i})\delta^r_{l+N}, \\
g_{14}^{(1)}(i,j,l,k,r) &= -\frac{1}{2}c_{j/m}\left(\dot{b}_{l/m} + \frac{l}{m}\Omega c_{l/m}\right)(\delta^k_{|l-i-j|} + \delta^k_{l+i+j})\delta^r_{i+N}, \\
g_{15}^{(1)}(i,j,l,k,r) &= \frac{1}{4}\frac{l}{m}\Omega(b_{i/m}b_{j/m} - c_{i/m}c_{j/m})(\delta^k_{|l-i-j|} + \delta^k_{l+i+j})\delta^r_{l+N}, \\
g_{16}^{(1)}(i,j,l,k,r) &= \frac{1}{2}b_{j/m}\left(\dot{c}_{l/m} - \frac{l}{m}\Omega b_{l/m}\right)(\delta^k_{|l-i-j|} - \delta^k_{l+i+j})\delta^r_{i+N}, \\
g_{17}^{(1)}(i,j,l,k,r) &= \frac{1}{2}c_{j/m}\left(\dot{b}_{l/m} + \frac{l}{m}\Omega c_{l/m}\right)(\delta^k_{|l-i+j|} + \delta^k_{|l+i-j|})\delta^r_{i+N}, \\
g_{18}^{(1)}(i,j,l,k,r) &= \frac{1}{4}\frac{l}{m}\Omega(b_{i/m}b_{j/m} + c_{i/m}c_{j/m})(\delta^k_{|l-i+j|} + \delta^k_{|l+i-j|})\delta^r_{l+N};
\end{aligned}
\tag{3.34}
$$

and

$$
g_{kr}^{(s)} = \sum_{n=1}^{21}\sum_{l=1}^{N}\sum_{j=1}^{N}\sum_{i=1}^{N} g_n^{(2)}(i,j,l,k,r) \tag{3.35}
$$

with

$$
\begin{aligned}
g_1^{(2)}(i,j,l,k,r) &= \frac{2}{N^2}\left[\dot{a}_0^{(m)}c_{i/m}\delta^k_i + a_0^{(m)}\left(\dot{c}_{l/m} - \frac{l}{m}\Omega b_{l/m}\right)\delta^k_l\right]\delta^r_0, \\
g_2^{(2)}(i,j,l,k,r) &= \frac{1}{N}b_{i/m}\left(\dot{c}_{l/m} - \frac{l}{m}\Omega b_{l/m}\right)(\delta^k_{l+i} + \text{sgn}(l-i)\delta^k_{|l-i|})\delta^r_0, \\
g_3^{(2)}(i,j,l,k,r) &= \frac{1}{N}c_{i/m}\left(\dot{b}_{l/m} + \frac{l}{m}\Omega c_{l/m}\right)(\delta^k_{l+i} - \text{sgn}(l-i)\delta^k_{|l-i|})\delta^r_0, \\
g_4^{(2)}(i,j,l,k,r) &= -\frac{1}{N^2}\frac{l}{m}\Omega(a_0^{(m)})^2\delta^k_l\delta^r_l + \frac{1}{2N}\dot{a}_0^{(m)}c_{i/m} \\
&\quad \times(\text{sgn}(i-j)\delta^k_{|i-j|} + \delta^k_{i+j})\delta^r_j, \\
g_5^{(2)}(i,j,l,k,r) &= \frac{1}{2N}\dot{a}_0^{(m)}c_{j/m}(\delta^k_{i+j} - \text{sgn}(i-j)\delta^k_{|i-j|})\delta^r_i, \\
g_6^{(2)}(i,j,l,k,r) &= \frac{1}{N}a_0^{(m)}\left(\dot{c}_{l/m} - \frac{l}{m}\Omega b_{l/m}\right)(\delta^k_{l+i} + \text{sgn}(l-i)\delta^k_{|l-i|})\delta^r_i, \\
g_7^{(2)}(i,j,l,k,r) &= -\frac{1}{N}\frac{l}{m}\Omega a_0^{(m)}b_{i/m}[\delta^k_{l+i} + \text{sgn}(l-i)\delta^k_{|l-i|}]\delta^r_l, \\
g_8^{(2)}(i,j,l,k,r) &= \frac{1}{2}c_{j/m}\left(\dot{b}_{l/m} + \frac{l}{m}\Omega c_{l/m}\right)(\delta^k_{l+i+j} - \text{sgn}(l-i-j)\delta^k_{|l-i-j|})\delta^r_i,
\end{aligned}
$$

$$
\begin{aligned}
g_9^{(2)}(i,j,l,k,r) &= \frac{1}{2}b_{j/m}\left(\dot{c}_{l/m} - \frac{l}{m}\Omega b_{l/m}\right)[\delta^k_{l+i+j} + \mathrm{sgn}(l-i-j)\delta^k_{|l-i-j|}]\delta^r_i,\\
g_{10}^{(2)}(i,j,l,k,r) &= -\frac{1}{4}\frac{l}{m}\Omega(b_{i/m}b_{j/m} - c_{i/m}c_{j/m})[\delta^k_{l+i+j}\\
&\quad + \mathrm{sgn}(l-i-j)\delta^k_{|l-i-j|}]\delta^r_l,\\
g_{11}^{(2)}(i,j,l,k,r) &= \frac{1}{4}b_{j/m}\left(\dot{c}_{l/m} - \frac{l}{m}\Omega b_{l/m}\right)[\mathrm{sgn}(l-i+j)\delta^k_{|l-i+j|}\\
&\quad + \mathrm{sgn}(l+i-j)\delta^k_{|l+i-j|}]\delta^r_i,\\
g_{12}^{(2)}(i,j,l,k,r) &= -\frac{1}{4}\frac{l}{m}\Omega(b_{i/m}b_{j/m} + c_{i/m}c_{j/m})[\mathrm{sgn}(l-i+j)\delta^k_{|l-i+j|}\\
&\quad + \mathrm{sgn}(l+i-j)\delta^k_{|l+i-j|}]\delta^r_l,\\
g_{13}^{(2)}(i,j,l,k,r) &= \frac{2}{N^2}\dot{a}_0^{(m)}a_0^{(m)}\delta^k_i\delta^r_{i+N} + \frac{1}{2N}\dot{a}_0^{(m)}b_{j/m}\\
&\quad \times[\mathrm{sgn}(i-j)\delta^k_{|i-j|} + \delta^k_{i+j}]\delta^r_{i+N},\\
g_{14}^{(2)}(i,j,l,k,r) &= \frac{1}{2N}\dot{a}_0^{(m)}b_{i/m}[\delta^k_{i+j} - \mathrm{sgn}(i-j)\delta^k_{|i-j|}]\delta^r_{j+N},\\
g_{15}^{(2)}(i,j,l,k,r) &= \frac{1}{N}a_0^{(m)}\left(\dot{b}_{l/m} + \frac{l}{m}\Omega c_{l/m}\right)[\delta^k_{l+i} - \mathrm{sgn}(l-i)\delta^k_{|l-i|}]\delta^r_{i+N},\\
g_{16}^{(2)}(i,j,l,k,r) &= \frac{1}{N}\frac{l}{m}\Omega a_0^{(m)}c_{i/m}[\delta^k_{l+i} - \mathrm{sgn}(l-i)\delta^k_{|l-i|}]\delta^r_{l+N},\\
g_{17}^{(2)}(i,j,l,k,r) &= \frac{1}{2}b_{j/m}\left(\dot{b}_{l/m} + \frac{l}{m}\Omega c_{l/m}\right)\\
&\quad \times[\delta^k_{l+i+j} - \mathrm{sgn}(l-i-j)\delta^k_{|l-i-j|}]\delta^r_{i+N},\\
g_{18}^{(2)}(i,j,l,k,r) &= \frac{1}{4}\frac{l}{m}\Omega(b_{i/m}c_{j/m} + b_{j/m}c_{i/m})\\
&\quad \times[\delta^k_{l+i+j} - \mathrm{sgn}(l-i-j)\delta^k_{|l-i-j|}]\delta^r_{l+N},\\
g_{19}^{(2)}(i,j,l,k,r) &= -\frac{1}{2}c_{j/m}\left(\dot{c}_{l/m} - \frac{l}{m}\Omega b_{l/m}\right)\\
&\quad \times[\delta^k_{l+i+j} + \mathrm{sgn}(l-i-j)\delta^k_{|l-i-j|}]\delta^r_{i+N},\\
g_{20}^{(2)}(i,j,l,k,r) &= \frac{1}{2}c_{j/m}\left(\dot{c}_{l/m} - \frac{l}{m}\Omega b_{l/m}\right)[\mathrm{sgn}(l-i+j)\delta^k_{|l-i+j|}\\
&\quad + \mathrm{sgn}(l+i-j)\delta^k_{|l+i-j|}]\delta^r_{i+N},\\
g_{21}^{(2)}(i,j,l,k,r) &= \frac{1}{4}\frac{l}{m}\Omega(b_{i/m}c_{j/m} - b_{j/m}c_{i/m})[\mathrm{sgn}(l-i+j)\delta^k_{|l-i+j|}\\
&\quad - \mathrm{sgn}(l+i-j)\delta^k_{|l+i-j|}]\delta^r_{l+N}
\end{aligned}
\tag{3.36}
$$

for $r = 0, 1, \ldots, 2N$.

$$
\mathbf{H} = \frac{\partial \mathbf{g}^{(m)}}{\partial \mathbf{z}_1^{(m)}} = (\mathbf{H}^{(0)}, \mathbf{H}^{(c)}, \mathbf{H}^{(s)})^{\mathrm{T}}
\tag{3.37}
$$

where

$$\begin{aligned}\mathbf{H}^{(0)} &= (H_0^{(0)}, H_1^{(0)}, \ldots, H_{2N}^{(0)}),\\ \mathbf{H}^{(c)} &= (\mathbf{H}_1^{(c)}, \mathbf{H}_2^{(c)}, \ldots, \mathbf{H}_N^{(c)})^{\mathrm{T}},\\ \mathbf{H}^{(s)} &= (\mathbf{H}_1^{(s)}, \mathbf{H}_2^{(s)}, \ldots, \mathbf{H}_N^{(s)})^{\mathrm{T}}\end{aligned} \tag{3.38}$$

for $N = 1, 2, \ldots \infty$, with

$$\begin{aligned}\mathbf{H}_k^{(c)} &= (H_{k0}^{(c)}, H_{k1}^{(c)}, \ldots, H_{k(2N)}^{(c)}),\\ \mathbf{H}_k^{(s)} &= (H_{k0}^{(s)}, H_{k1}^{(s)}, \ldots, H_{k(2N)}^{(s)})\end{aligned} \tag{3.39}$$

for $k = 1, 2, \ldots N$. The corresponding components are

$$\begin{aligned}H_r^{(0)} &= \alpha_1\delta_0^r - \alpha_2 h_r^{(0)}\\ H_{kr}^{(c)} &= -2\left(\frac{k\Omega}{m}\right)\delta_{k+N}^r + \alpha_1\delta_k^r - \alpha_2 h_{kr}^{(c)},\\ H_{kr}^{(s)} &= 2\left(\frac{k\Omega}{m}\right)\delta_k^r + \alpha_1\delta_{k+N}^r - \alpha_2 h_{kr}^{(s)}\end{aligned} \tag{3.40}$$

for $r = 0, 1, \ldots, 2N$.

$$h_r^{(0)} = (a_0^{(m)})^2\delta_0^r + \sum_{n=1}^{6}\sum_{l=1}^{N}\sum_{j=1}^{N}\sum_{i=1}^{N} h_n^{(0)}(i,j,l,r) \tag{3.41}$$

with

$$\begin{aligned}h_1^{(0)}(i,j,l,r) &= \frac{1}{2N} b_{i/m} b_{j/m}\delta_{i-j}^0\delta_0^r,\\ h_2^{(0)}(i,j,l,r) &= \frac{1}{2N}(c_{i/m}c_{j/m}\delta_{i-j}^0\delta_0^r + 2a_0^{(m)} b_{i/m}\delta_{l-i}^0\delta_l^r),\\ h_3^{(0)}(i,j,l,r) &= \frac{1}{4}(b_{i/m}b_{j/m} - c_{i/m}c_{j/m})\delta_{l-i-j}^0\delta_l^r,\\ h_4^{(0)}(i,j,l,r) &= \frac{1}{4}(b_{i/m}b_{j/m} + c_{i/m}c_{j/m})(\delta_{l-i+j}^0 + \delta_{l+i-j}^0)\delta_l^r,\\ h_5^{(0)}(i,j,l,r) &= \frac{1}{N}a_0^{(m)} c_{i/m}\delta_{l-i}^0\delta_{l+N}^r + \frac{1}{4}(b_{j/m}c_{i/m} + b_{i/m}c_{j/m})\delta_{l-i-j}^0\delta_{l+N}^r,\\ h_6^{(0)}(i,j,l,r) &= \frac{1}{4}(b_{j/m}c_{i/m} - b_{i/m}c_{j/m})(\delta_{l-i+j}^0 + \delta_{l+i-j}^0)\delta_{l+N}^r;\end{aligned} \tag{3.42}$$

and

$$h_{kr}^{(c)} = \sum_{n=1}^{9}\sum_{l=1}^{N}\sum_{j=1}^{N}\sum_{i=1}^{N} h_n^{(1)}(i,j,l,k,r) \tag{3.43}$$

with

$$\begin{aligned}h_1^{(1)}(i,j,l,k,r) &= 2\frac{1}{N^2}\dot{a}_0^{(m)} b_{i/m}\delta_l^k\delta_0^r,\\ h_2^{(1)}(i,j,l,k,r) &= \frac{1}{2N} b_{i/m} b_{j/m}(\delta_{|l-i|}^k + \delta_{l+i}^k)\delta_0^r,\end{aligned}$$

$$
\begin{aligned}
h_3^{(1)}(i,j,l,k,r) &= \frac{1}{2N} c_{i/m} c_{j/m} (\delta_{|l-i|}^k - \delta_{l+i}^k) \delta_0^r, \\
h_4^{(1)}(i,j,l,k,r) &= \frac{1}{N^2} (a_0^{(m)})^2 \delta_l^k \delta_l^r + \frac{1}{N} a_0^{(m)} b_{i/m} (\delta_{|l-i|}^k + \delta_{l+i}^k) \delta_l^r, \\
h_5^{(1)}(i,j,l,k,r) &= \frac{1}{4} (b_{i/m} b_{j/m} - c_{i/m} c_{j/m}) (\delta_{|l-i-j|}^k + \delta_{l+i+j}^k) \delta_l^r, \\
h_6^{(1)}(i,j,l,k,r) &= \frac{1}{4} (b_{i/m} b_{j/m} + c_{i/m} c_{j/m}) (\delta_{|l-i+j|}^k + \delta_{|l+i-j|}^k) \delta_l^r, \\
h_7^{(1)}(i,j,l,k,r) &= \frac{1}{N} a_0^{(m)} c_{i/m} (\delta_{|l-i|}^k - \delta_{l+i}^k) \delta_{l+N}^r, \\
h_8^{(1)}(i,j,l,k,r) &= \frac{1}{4} (b_{j/m} c_{i/m} + b_{i/m} c_{j/m}) (\delta_{|l-i-j|}^k - \delta_{l+i+j}^k) \delta_{l+N}^r, \\
h_9^{(1)}(i,j,l,k,r) &= \frac{1}{4} (b_{j/m} c_{i/m} - b_{i/m} c_{j/m}) (\delta_{|l-i+j|}^k - \delta_{|l+i-j|}^k) \delta_{l+N}^r;
\end{aligned} \tag{3.44}
$$

and

$$
h_{kr}^{(s)} = \sum_{n=1}^{6} \sum_{l=1}^{N} \sum_{j=1}^{N} \sum_{i=1}^{N} h_n^{(2)}(i,j,l,k,r) \tag{3.45}
$$

with

$$
\begin{aligned}
h_1^{(2)}(i,j,l,k,r) &= \frac{2}{N^2} a_0^{(m)} c_{i/m} \delta_i^k \delta_0^r, \\
h_2^{(2)}(i,j,l,k,r) &= \frac{1}{2N} b_{j/m} c_{i/m} [\operatorname{sgn}(i-j) \delta_{|i-j|}^k + \delta_{i+j}^k] \delta_0^r, \\
h_3^{(2)}(i,j,l,k,r) &= \frac{1}{2N} b_{i/m} c_{j/m} [\delta_{i+j}^k - \operatorname{sgn}(i-j) \delta_{|i-j|}^k] \delta_0^r, \\
h_4^{(2)}(i,j,l,k,r) &= \frac{1}{N} a_0^{(m)} c_{i/m} [\delta_{l+i}^k - \operatorname{sgn}(l-i) \delta_{|l-i|}^k] \delta_l^r, \\
h_5^{(2)}(i,j,l,k,r) &= \frac{1}{4} (b_{i/m} c_{j/m} + b_{j/m} c_{i/m}) [\delta_{l+i+j}^k - \operatorname{sgn}(l-i-j) \delta_{|l-i-j|}^k] \delta_l^r, \\
h_6^{(2)}(i,j,l,k,r) &= \frac{1}{4} (b_{i/m} c_{j/m} - b_{j/m} c_{i/m}) [\operatorname{sgn}(l-i+j) \delta_{|l-i+j|}^k, \\
&\quad - \operatorname{sgn}(l+i-j) \delta_{|l+i-j|}^k] \delta_l^r.
\end{aligned} \tag{3.46}
$$

From Luo (2012a), the eigenvalues of $D\mathbf{f}(\mathbf{y}^{(m)*})$ are classified as

$$
(n_1, n_2, n_3 | n_4, n_5, n_6). \tag{3.47}
$$

The corresponding boundary between the stable and unstable solution is given by the saddle-node bifurcation (SN) and Hopf bifurcation (HB).

3.1.2 Frequency-Amplitude Characteristics

As in Luo and Lakeh (2013a), the curves of harmonic amplitudes varying with excitation frequency Ω are illustrated. The harmonic amplitude and phase are defined by

$$
A_{k/m} \equiv \sqrt{b_{k/m}^2 + c_{k/m}^2} \text{ and } \varphi_{k/m} = \arctan \frac{c_{k/m}}{b_{k/m}}. \tag{3.48}
$$

The corresponding solution in Equation (3.1) becomes

$$x^*(t) = a_0^{(m)} + \sum_{k=1}^{N} A_{k/m} \cos\left(\frac{k}{m}\Omega t + \varphi_{k/m}\right). \tag{3.49}$$

Consider system parameters as

$$\alpha_1 = 0.5, \alpha_2 = 5.0, \alpha_3 = 10.0. \tag{3.50}$$

For the van der Pol oscillator, without any periodic entrainments ($Q_0 = 0$), a limit cycle exists for a set of specific parameters. For the limit cycle, its frequency is very important, which is strongly dependent on α_1 and α_3 but weakly dependent on α_2. Thus the frequencies of the limit cycle versus parameter α_2 are presented in Figure 3.1 for $\alpha_3 = 5, 10, 20$. The frequency of the limit cycle decreases with increasing parameter α_2. For $\alpha_3 = 5$, frequency $\omega \approx 2.236340$ at $\alpha_1 = 0.1$ and $\omega \approx 0.804484$ at $\alpha_1 = 20$ are obtained. The limit cycle frequency cannot be expressed by $\omega = \sqrt{\alpha_3} = 2.2360679\ldots$. Similarly, for $\alpha_3 = 10$, $\omega \approx 3.162610$ for $\alpha_1 = 0.1$ and $\omega \approx 1.467790$ for $\alpha_1 = 20$ are obtained. For $\alpha_3 = 20$, $\omega \approx 4.473070$ for $\alpha_1 = 0.1$ and $\omega \approx 2.587110$ for $\alpha_1 = 20$ are obtained.

The parameters of the oscillator are $\alpha_1 = 0.5$, $\alpha_2 = 5$, and $\alpha_3 = 10$. With varying excitation frequency, harmonic amplitude and the corresponding stability and bifurcation analysis of the periodic solution will be presented herein. To consider excitation amplitude effects on period-1 motion, the excitation amplitudes $Q_0 = 0.5, 1.0, 2.0, 4.0, 8.0$ are selected, and the harmonic amplitudes versus excitation frequency are presented in Figure 3.2, which are based on 22 harmonic terms to compute. The solid and dashed curves are for stable and unstable periodic motion. Acronym "HB" represents the Hopf bifurcation. For such excitation amplitudes, a stable portion and two unstable portions for such frequency-amplitude curves are obtained. With increasing excitation amplitude, the range for stable period-1 motion increases, and becomes small with decreasing excitation amplitude. Such a range becomes one point for $\Omega \approx 3.1574$

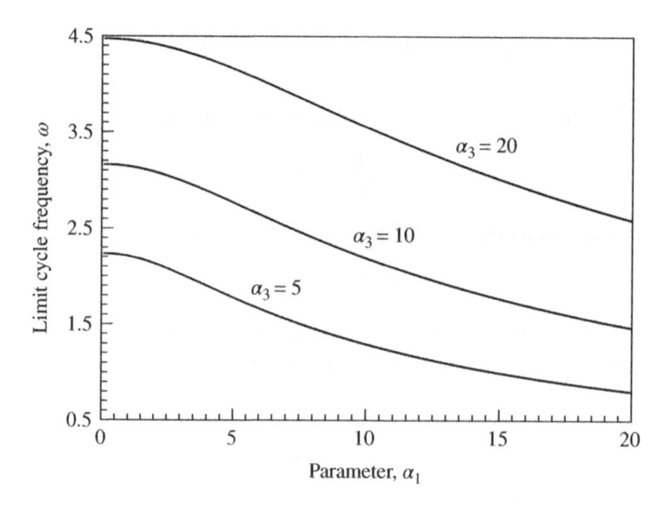

Figure 3.1 Limit cycle frequency of the van der Pol oscillator without any entrainments. ($\alpha_2 = 5$, and $Q_0 = 0$)

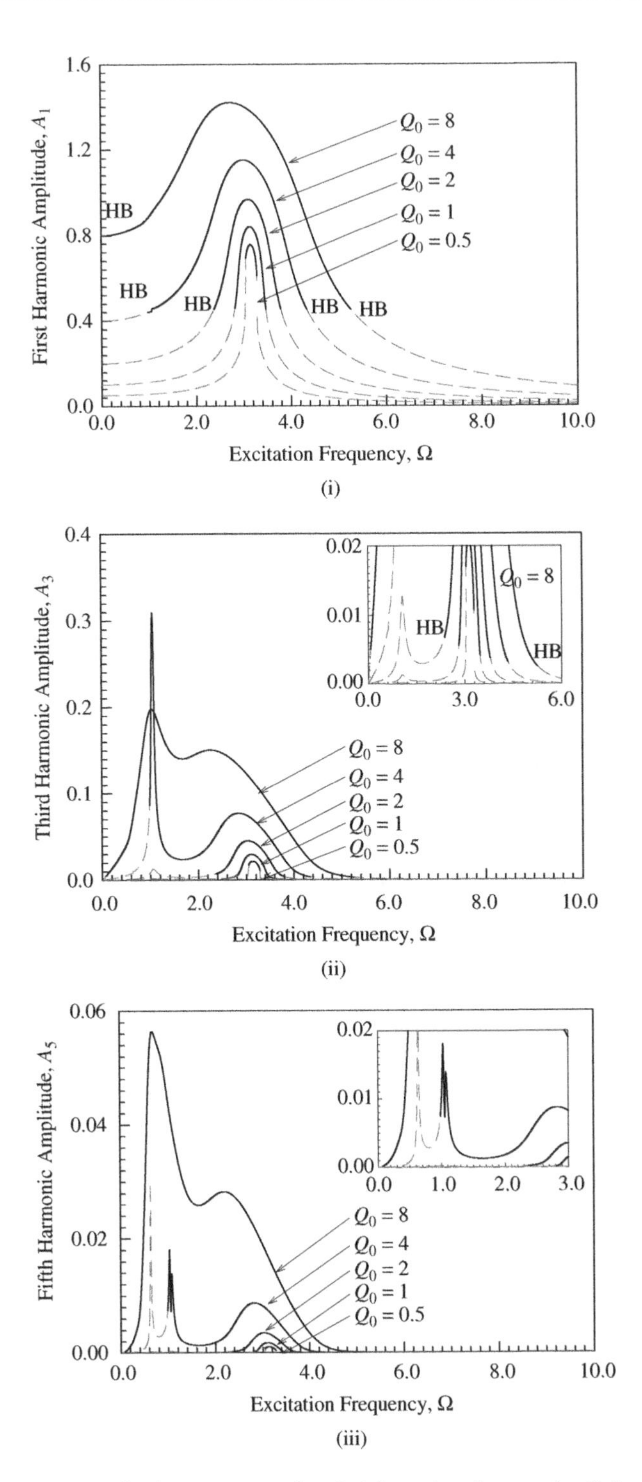

Figure 3.2 Frequency-amplitude responses of period-1 motion for van der Pol oscillator from the 22 harmonic terms (HB22): (i)–(vi) $A_1 - A_{11}$ ($\alpha_1 = 0.5$, $\alpha_2 = 5$, $\alpha_3 = 10$, $Q_0 = 0.5, 1.0, 2.0, 4.0, 8.0$). Acronym "HB" represents Hopf bifurcation

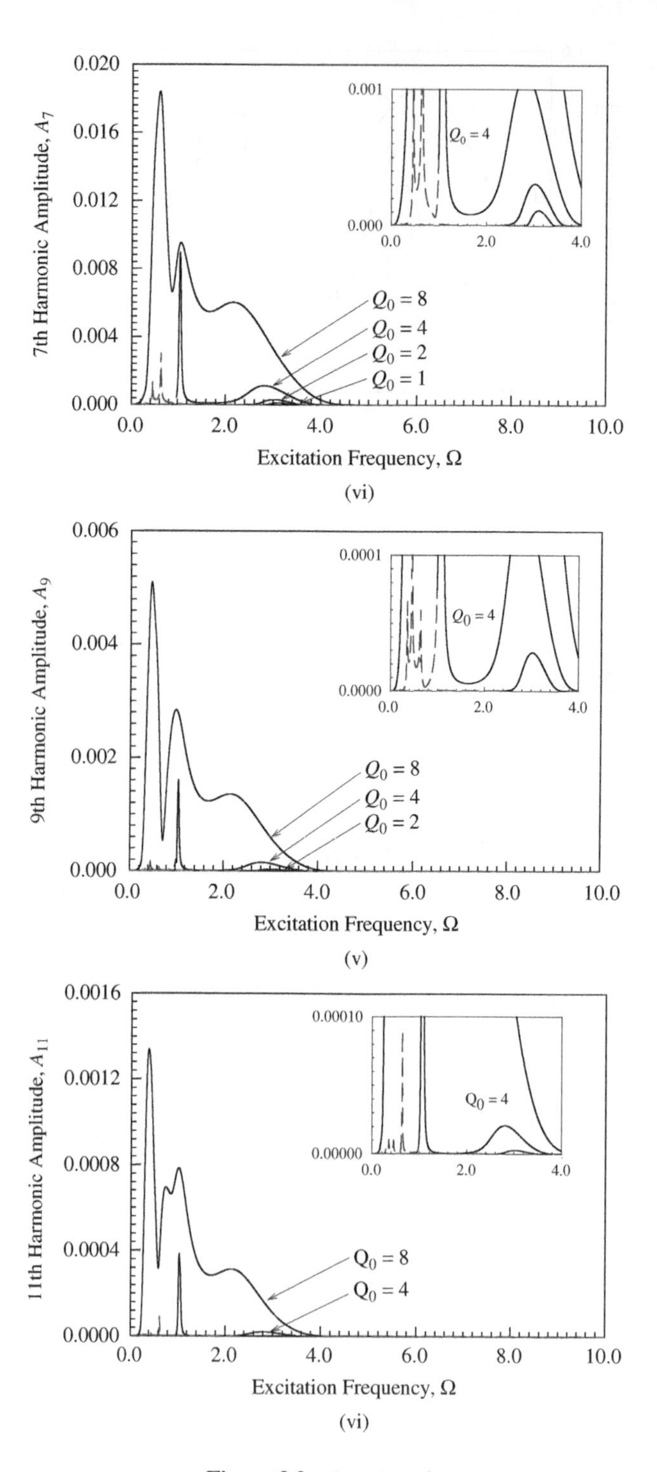

Figure 3.2 *(continued)*

at $Q_0 = 0$, which is a stable limit cycle. With increasing excitation amplitude Q_0, the harmonic terms should increase to get enough accurate period-1 motions. For $Q_0 = 4$, the 18 harmonic terms (HB18) will be included in the approximate solution to give $A_{17} \sim 10^{-6}$. But for $Q_0 = 8$, the 22 harmonic terms (HB22) will be taken into account to achieve $A_{21} \sim 10^{-6}$. For $Q_0 = 0.5$, only eight harmonic terms (HB8) can achieve $A_7 \sim 10^{-6}$. In Figure 3.2, only the first 11 harmonic terms are plotted due to limitation of the number of pages.

From Figure 3.2(i)–(vi), the decay rates of harmonic amplitudes with increasing harmonic orders are presented clearly. For excitation amplitude $Q_0 = 8.0$, we have $A_1 \sim 10^0$, $A_3 \sim 2 \times 10^{-1}$, $A_5 \sim 6 \times 10^{-2}$, $A_7 \sim 2 \times 10^{-2}$, $A_9 \sim 6 \times 10^{-3}$, $A_{11} \sim 2 \times 10^{-3}$, $A_{13} \sim 4 \times 10^{-4}$, $A_{15} \sim 10^{-4}$, $A_{17} \sim 3 \times 10^{-5}$, $A_{19} \sim 1.0 \times 10^{-5}$, and $A_{21} \sim 4 \times 10^{-6}$. $a_0 = 0$ and $A_{2m} = 0$ $(m = 1, 2, \ldots, 11)$. The harmonic amplitudes for other Q_0 can be discussed in a similar fashion. From the stability and bifurcation analysis, the parameter map (Ω, Q_0) is based on eight harmonic terms (HB8) for $\alpha_1 = 0.5$, $\alpha_2 = 5$, and $\alpha_3 = 10$, as shown in Figure 3.3. The shaded and white areas are for stable and unstable period-1 motions, respectively. For more a accurate parameter map, the higher order harmonic terms should be included.

Consider another set of system parameters as

$$\alpha_1 = 5.0, \alpha_2 = 5.0, \alpha_3 = 1.0, Q_0 = 5.0 \tag{3.51}$$

The frequency-amplitude curves for period-1, period-3, and period-5 motions based on 100 harmonic terms accordingly are presented in Figure 3.4. The acronyms "HB" and "SN" are used to represent the Hopf bifurcation and saddle-node bifurcation, respectively. Solid and dashed curves represent stable and unstable period-*m* motions. For the period-1 motion, the Hopf bifurcation is observed. After the Hopf bifurcation, the quasi-periodic motions exist and further the chaotic motions can be developed. For period-3 and period-5 motions, the saddle-node bifurcations are observed. After the saddle-node bifurcation, the period-3 and

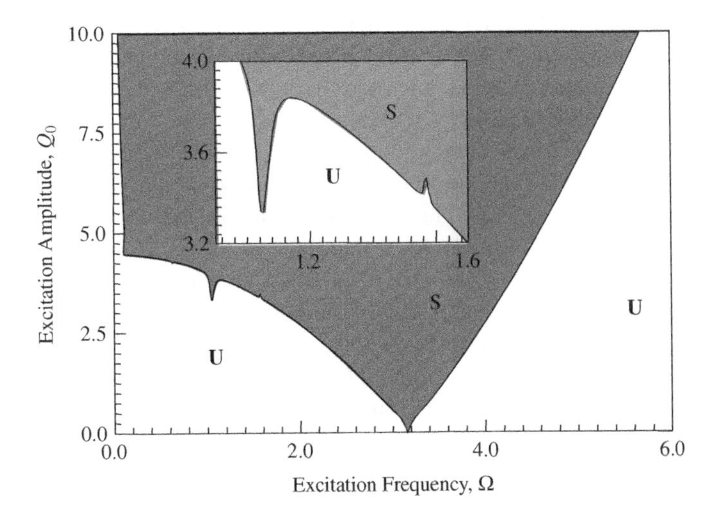

Figure 3.3 Parameter map for the period-1 motion of the periodically forced van der Pol oscillator. The gray and white areas are for stable and unstable period-1 motions, respectively ($\alpha_1 = 0.5$, $\alpha_2 = 5$, $\alpha_3 = 10$). The stability boundary is based on the Hopf bifurcation

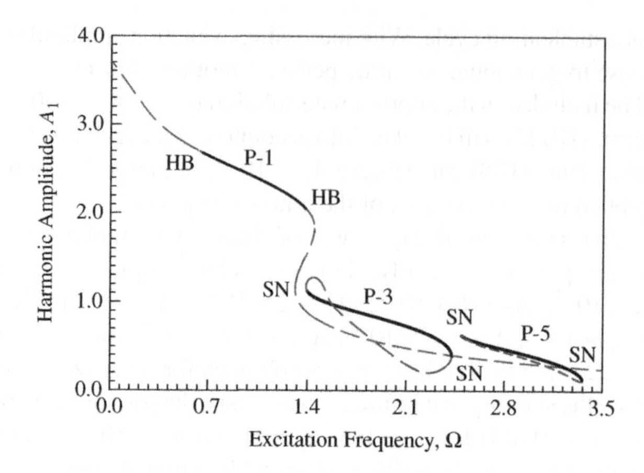

Figure 3.4 Frequency-amplitude curves of harmonic term based on 100 harmonic terms for period-1, period-3, and period-5 motions in the van der Pol oscillator ($\alpha_1 = 5.0, \alpha_2 = 5.0, \alpha_3 = 1.0, Q_0 = 5.0$)

period-5 motions disappear. Such periodic motions are embedded in chaotic motions or are switched to the other periodic motions. In Figure 3.4, it is observed that three periodic motions possess gaps. In such gaps, the quasi-periodic and chaotic motions will exist. For 100 harmonic terms, curves of harmonic frequency-amplitudes can be similarly presented. The curves shapes may be different, but with increasing harmonics order, the harmonic amplitudes will decrease.

For a better understanding of period-m motions, the comprehensive discussion of frequency-amplitude responses should be completed herein. In Figure 3.5, the harmonic amplitudes A_{2m-1} $(m = 1, 2, \ldots, 10)$, A_{55} and A_{89} are presented. For $\Omega \in (0, \infty)$, the effects of harmonic terms are $A_1 < 4.0, A_3 < 1.0, A_5 < 5 \times 10^{-1}, A_7 < 4 \times 10^{-1}, A_9 < 3 \times 10^{-1}$. For harmonic amplitude $A_{89} < 9 \times 10^{-3}$, the stable period-1 motion has $A_{89} \in (10^{-10}, 10^{-5})$. The Hopf bifurcations of period-1 motion occurs at $\Omega \approx 0.665$ and $\Omega = 1.385$. After the two Hopf bifurcations, the quasi-periodic motions of the periodically excited, van der Pol oscillator are observed. For small excitation frequency, more harmonic terms should be involved in the finite Fourier series solution of period-1 motion.

For period-3 motion, the harmonic amplitudes $A_{(2m-1)/3}$ $(m = 1, 2, \ldots, 11)$ and A_{59} are presented for $\Omega \in (1.3904, 2.371)$ in Figure 3.6. The stable period-3 motion is on the upper portion of the closed loop in $\Omega \in (1.3990, 2.371)$. The harmonic amplitudes for the stable and unstable period-3 motion are $A_{1/3} \in (1.0, 2.2)$, $A_1 \in (0.2, 1.3)$, $A_{5/3} \in (0.15, 0.7)$, $A_{7/3} \in (0.05, 0.3)$, $A_3 \in (0.03, 0.21)$. To save pages , $A_{59} \in (10^{-7}, 2 \times 10^{-3})$. For small excitation frequency, more harmonic terms should be involved in the finite Fourier series solution of period-3 motion. The two saddle-node bifurcations occur at $\Omega \approx 1.3990$ and $\Omega \approx 2.371$. Once the period-3 motion disappears, the chaotic motions will be around. In other words, the period-3 motion is embedded in chaotic motions of the periodically forced van der Pol oscillator.

For period-5 motion, the harmonic amplitudes $A_{(2m-1)/5}$ $(m = 1, 2, \ldots, 13)$ and A_{99} are presented for $\Omega \in (2.4953, 3.3540)$ in Figure 3.7. The stable period-5 motion is on the upper portion of the closed loop in $\Omega \in (2.4953, 3.3540)$. The harmonic amplitudes for stable and unstable period-5 motions are $A_{1/5} \in (1.8, 2.2)$, $A_{3/5} \in (0.3, 0.7)$, $A_1 \in (0.15, 0.7)$, $A_{7/5} \in (0.1, 0.7)$, $A_{9/5} \in (0.07, 0.17)$, $A_{11/5} \in (0.04, 0.13)$, $A_{13/5} \in (0.03, 0.11)$, $A_3 \in (0.025, 0.08)$. To save page number, $A_{99} \in (10^{-8}, 3 \times 10^{-5})$. For small

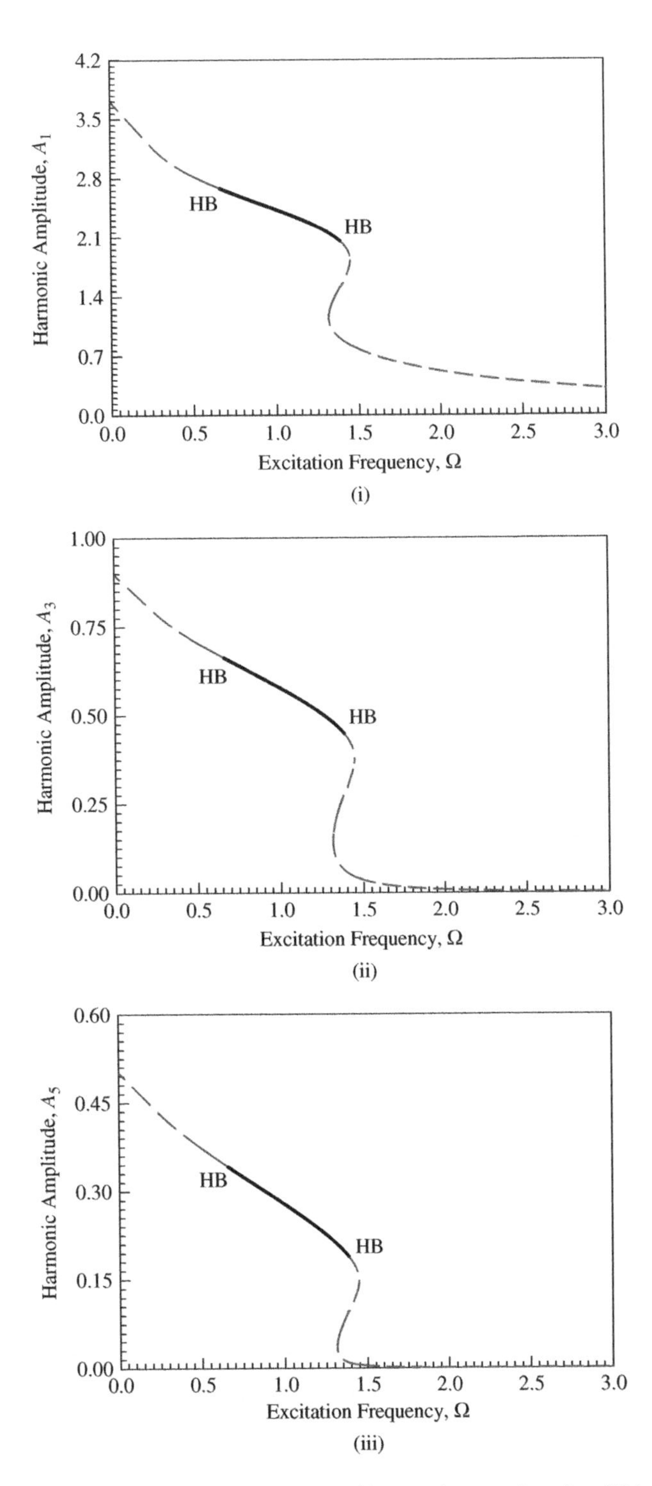

Figure 3.5 Frequency-amplitude backbone curves of harmonic terms based on 90 harmonic terms for period-1 motion in the van der Pol oscillator: (i)–(v) A_{2m-1} ($m = 1, 2, \ldots, 5$) and (vi) A_{89} ($\alpha_1 = 5.0, \alpha_2 = 5.0, \alpha_3 = 1.0, Q_0 = 5.0$)

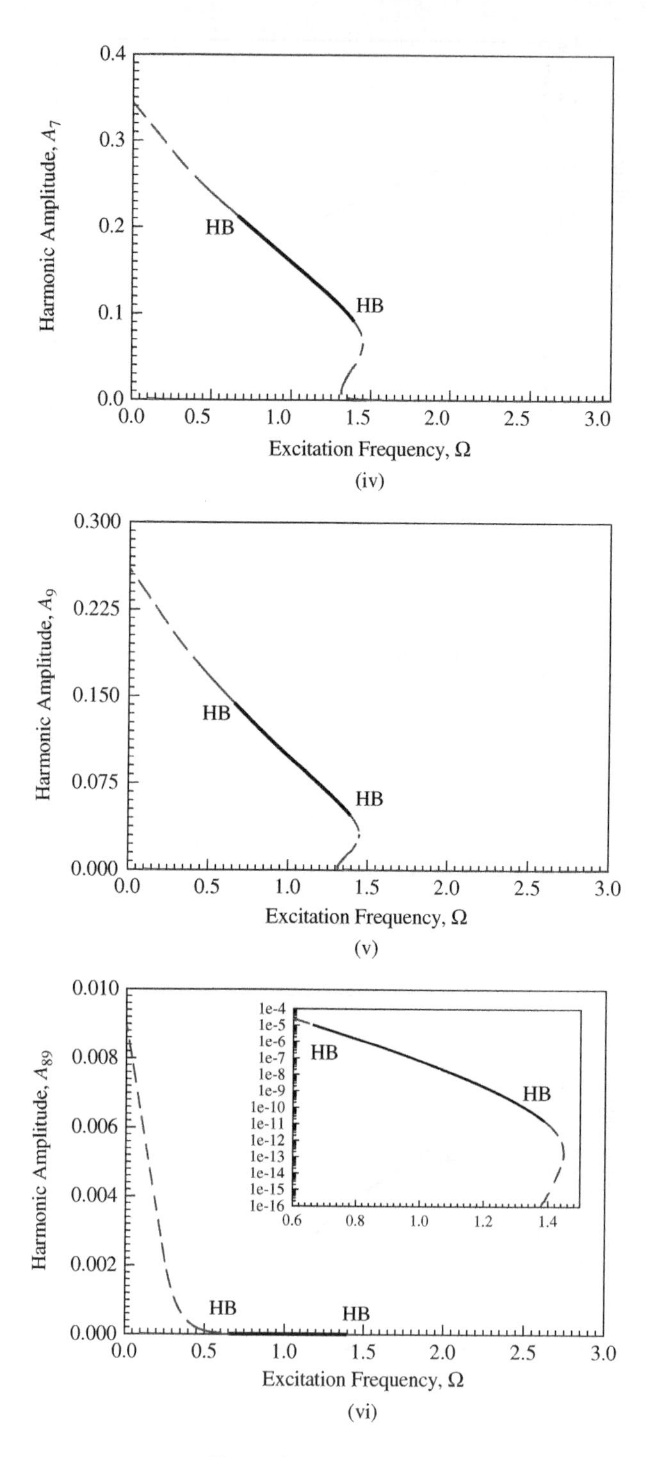

Figure 3.5 *(continued)*

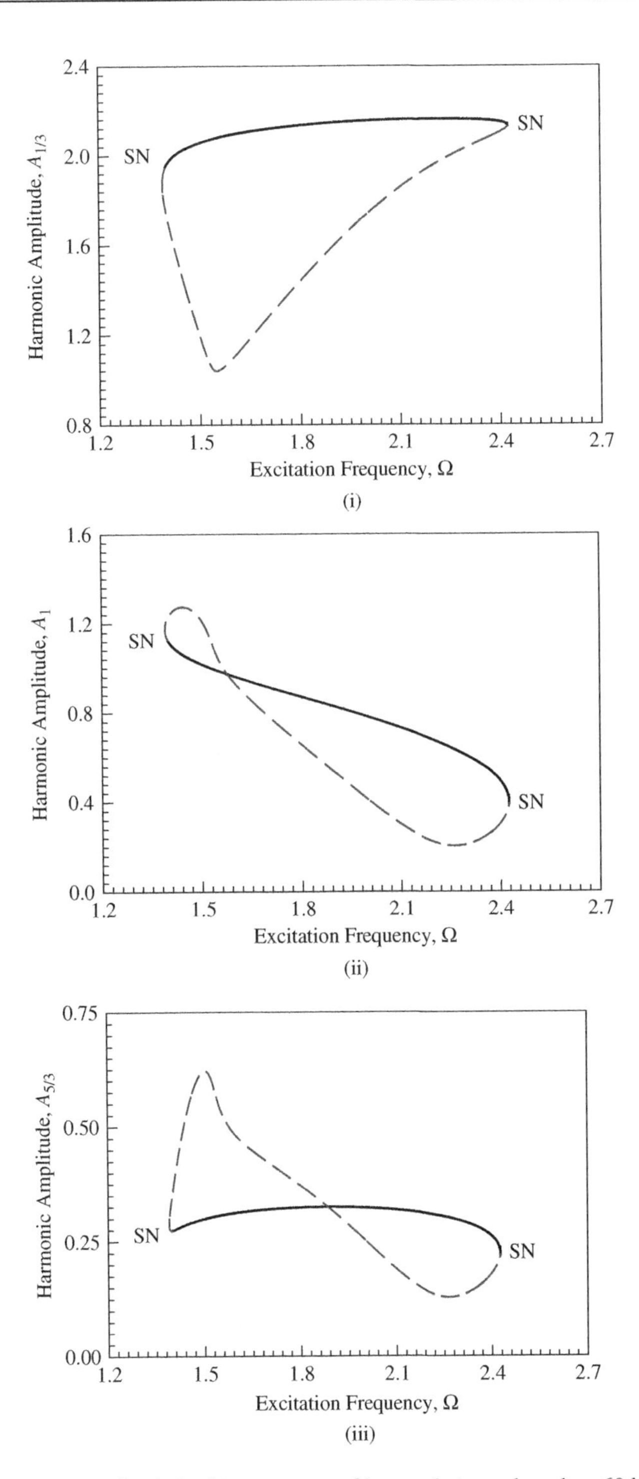

Figure 3.6 Frequency-amplitude backbone curves of harmonic terms based on 60 harmonic terms for period-3 motion in the van der Pol oscillator: (i)–(v) $A_{(2m-1)/3}$ ($m = 1, 2, \ldots, 5$) and (vi) $A_{59/3}$ ($\alpha_1 = 5.0$, $\alpha_2 = 5.0$, $\alpha_3 = 1.0$, $Q_0 = 5.0$)

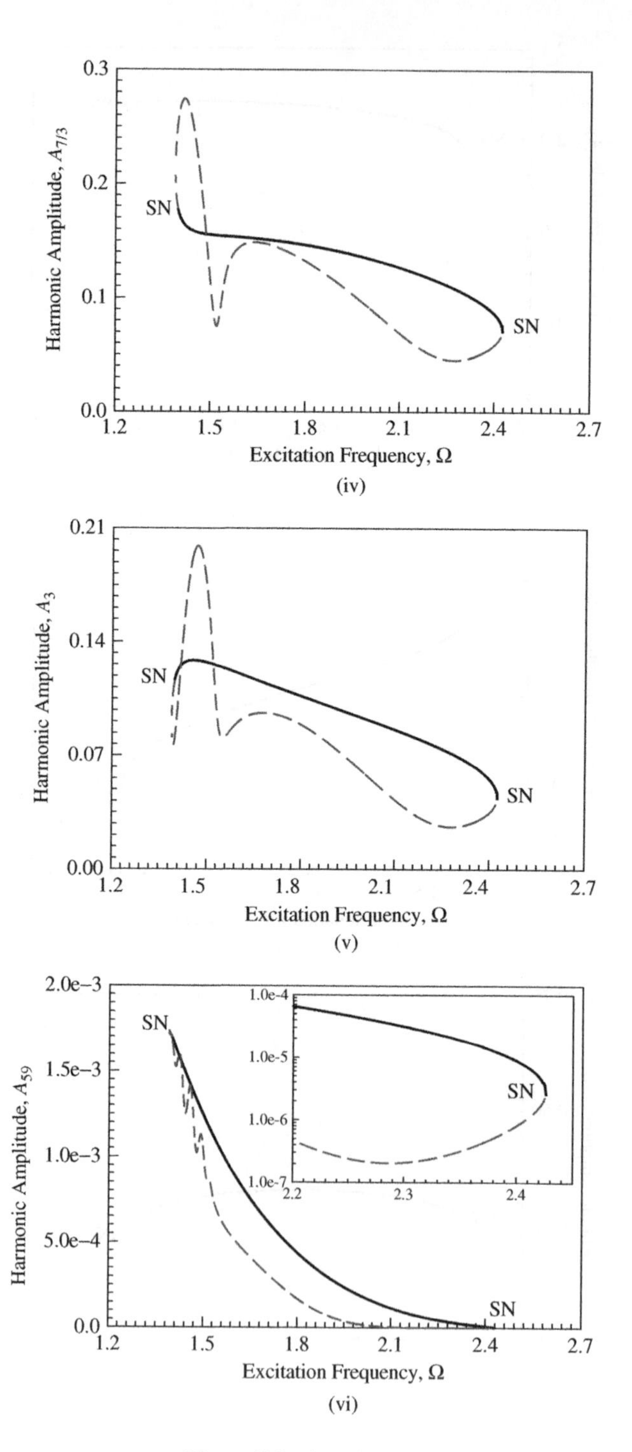

Figure 3.6 (*continued*)

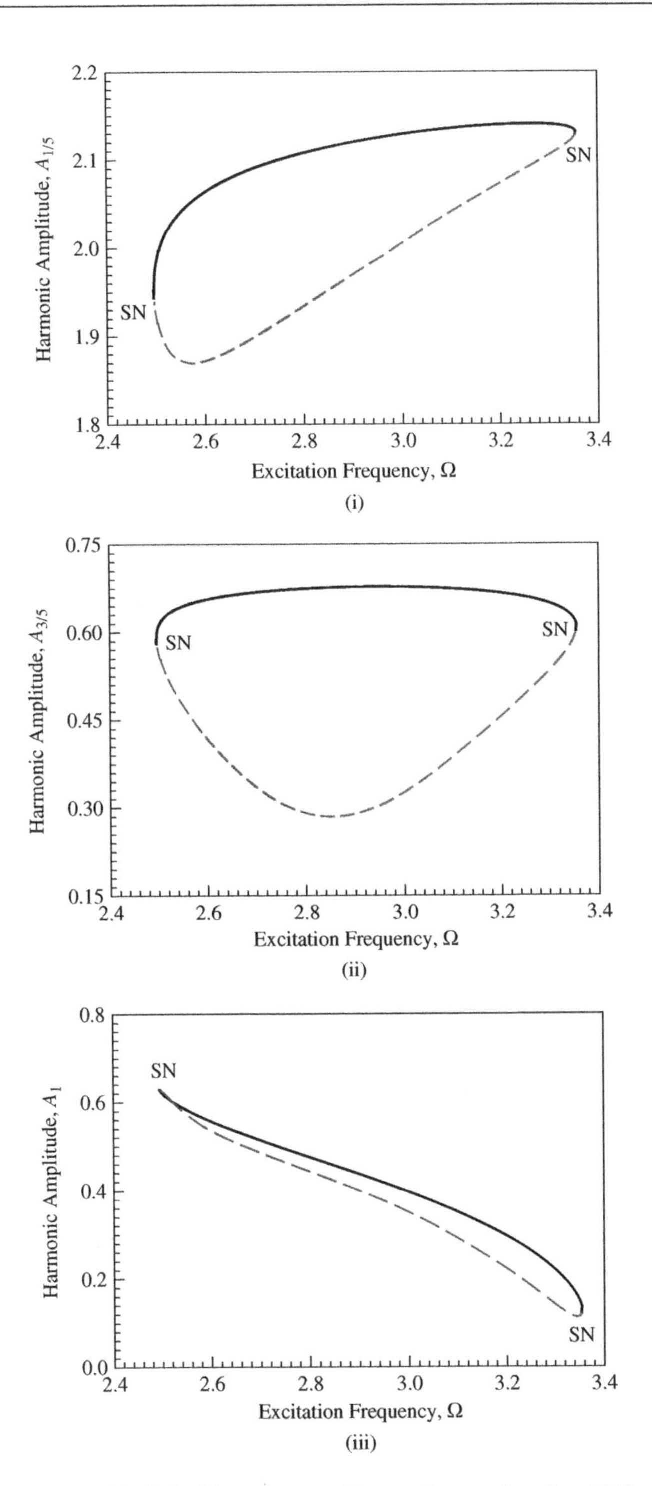

Figure 3.7 Frequency-amplitude backbone curves of harmonic terms based on 100 harmonic terms for period-5 motion in the van der Pol oscillator: (i)–(v) $A_{(2m-1)/5}$ ($m = 1, 2, \ldots, 7$) and (vi) $A_{99/5}$ ($\alpha_1 = 5.0$, $\alpha_2 = 5.0$, $\alpha_3 = 1.0$, $Q_0 = 5.0$)

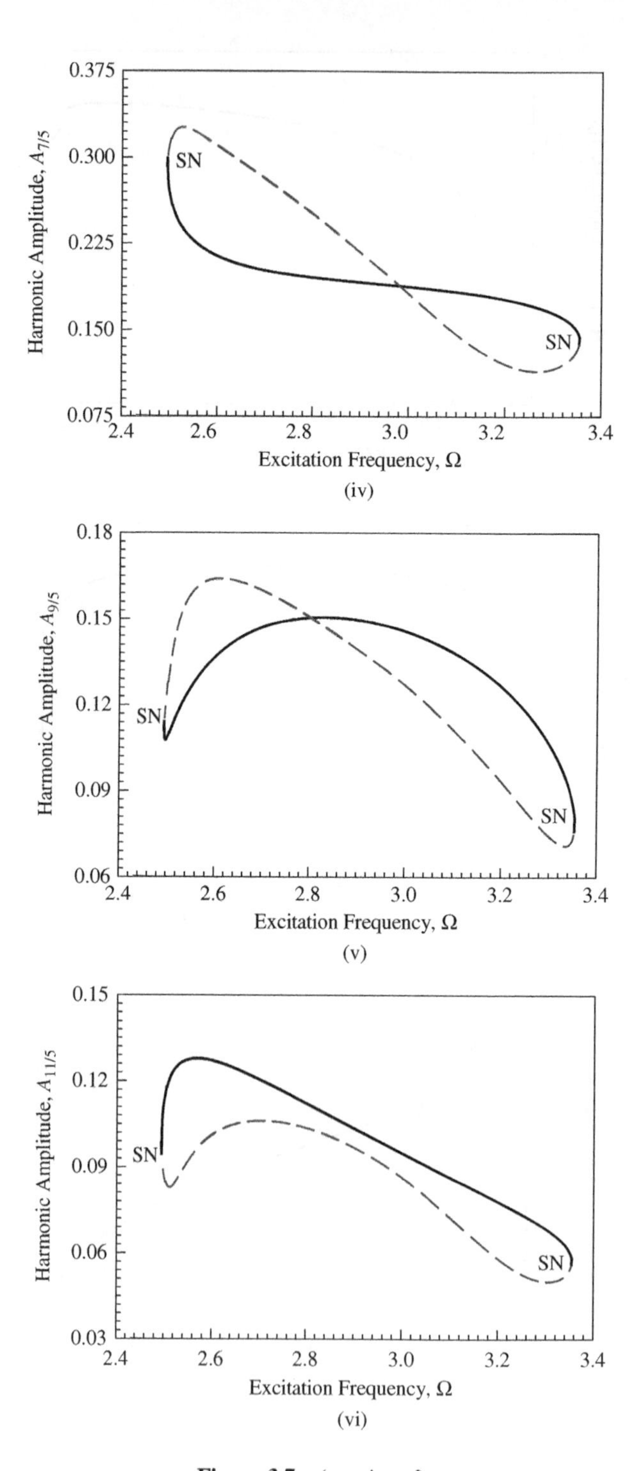

Figure 3.7 (*continued*)

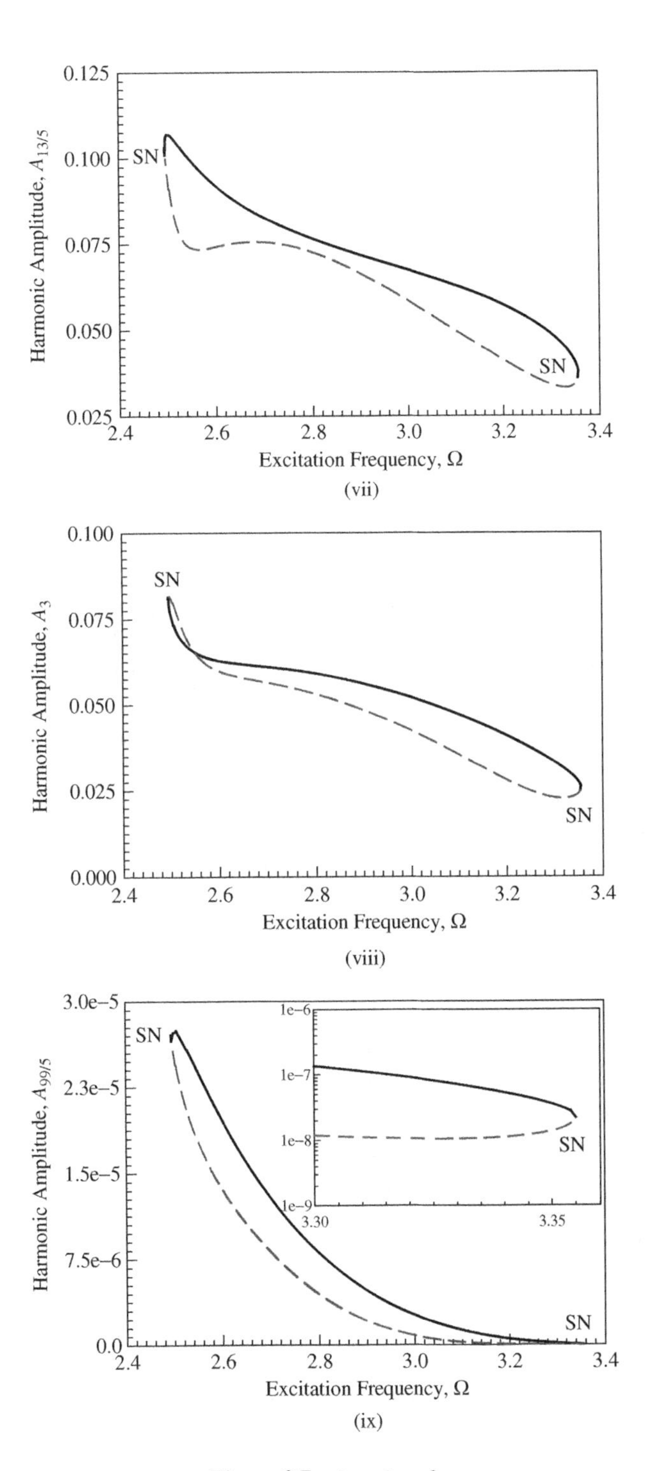

Figure 3.7 *(continued)*

excitation frequency, more harmonic terms should be involved in the finite Fourier series solution of period-5 motion. The two saddle-node bifurcations of period-5 motions occur at $\Omega \approx 2.4953$ and $\Omega \approx 3.3540$. Once the period-5 motion disappears, the chaotic motions will also be around. In other words, the period-5 motion is embedded in chaotic motions of the periodically forced van der Pol oscillator.

3.1.3 *Numerical Illustrations*

In this section, numerical simulations are carried out by the symplectic scheme. The initial conditions for numerical simulation are computed from the approximate solutions with 22 harmonic terms (HB22). The numerical results are depicted by solid curves, but the analytical solutions based on the 22 harmonic terms (HB22) are given by red circular symbols.

The trajectory, displacement, and velocity responses, and analytic harmonic amplitude spectrum of stable limit cycle are presented in Figure 3.8(a)–(d) with $\Omega \approx 3.1574$ and $Q_0 = 0$ plus parameters ($\alpha_1 = 0.5$, $\alpha_2 = 5.0$, $\alpha_3 = 10.0$). Analytical solutions of limit cycle are based on 22 harmonic term solution (HB22). The initial condition is $x_0 \approx 0.0844137$, and $y_0 \approx 2.02906$. The analytical and numerical solutions of limit cycle for such van der Pol oscillator match very well, as shown in Figure 3.8(a)–(c). In Figure 3.8(d), the main harmonic amplitudes are $A_1 \approx 0.632689$, $A_3 \approx 0.012491$, $A_5 \sim 4.1 \times 10^{-4}$, $A_7 \sim 1.5 \times 10^{-5}$, $A_9 \sim 5.9 \times 10^{-7}$, and $A_{11} \sim 2.3 \times 10^{-8}$, and so on with $A_{2l} = 0$ ($l = 0, 1, 2, \ldots$). For a rough approximate solution, one can use two harmonic terms. To obtain the solution accuracy less than 10^{-8}, 11 harmonic terms should be included in the Fourier series expression in Equation (3.4). Through illustrations of displacement and velocity time histories, we did not observe any transient motion before the stable limit cycle is formed. Exactly speaking, such transient motion can be ignored with the accuracy tolerance of $\varepsilon < 10^{-8}$.

For $Q_0 \neq 0$, the $\Omega = 1.048$ and $Q_0 = 4.0$ are used for the numerical illustration of stable period-1 motion. The corresponding initial condition is $x_0 \approx 0.147019$, and $y_0 \approx 0.270182$, which is computed from the analytical solution with 22 harmonic terms (HB22). For 40 periods, the trajectories, displacement, and velocity responses, and analytical harmonic amplitude of stable period-1 motion are presented in Figure 3.9(a)–(d). This period-1 motion with two knots in phase plane is observed and the trajectory is of skew symmetry, as shown in Figure 3.9(a)–(c). In Figure 3.9(d), the main harmonic amplitudes are $A_1 \approx 0.447715$, $A_3 \approx 0.309094$, $A_5 \sim 8.5 \times 10^{-3}$, $A_7 \sim 8.9 \times 10^{-3}$, $A_9 \sim 1.2 \times 10^{-3}$, $A_{11} \sim 3.7 \times 10^{-3}$, $A_{13} \sim 1.1 \times 10^{-4}$, $A_{15} \sim 1.6 \times 10^{-5}$, $A_{17} \sim 7.3 \times 10^{-6}$, $A_{19} \sim 1.2 \times 10^{-6}$, $A_{21} \sim 4.2 \times 10^{-7}$, and so on with $A_{2l} = 0$ ($l = 0, 1, 2, \ldots$). From the harmonic amplitude spectrum in Figures 3.8(d) and 3.9(d), both of them have different pattern. To determine periodic solutions in the same oscillator, one cannot use the same expression to determine the periodic motions. For this case, one should used 22 harmonic terms to achieve the solution precision less than 10^{-8}.

The trajectories and harmonic amplitudes of period-1, period-3, and period-5 motions are presented in Figures 3.10–3.12 to verify analytical periodic motions in such an oscillator. The displacement, velocity, trajectories, and harmonic amplitude spectrums of period-1, period-3, and period-5 motions are presented for $\Omega = 1.385, 2.3, 3.2$. The input data for numerical simulations is in Table 3.1. In Figure 3.10(a)–(d), 90 harmonic terms (HB90) are used in the analytical solution of period-1 motion for $\Omega = 1.385$. The analytical and numerical results of the period-1 motion match very well. The main amplitudes are $A_1 \approx 2.057854$, $A_3 \approx 0.448938$,

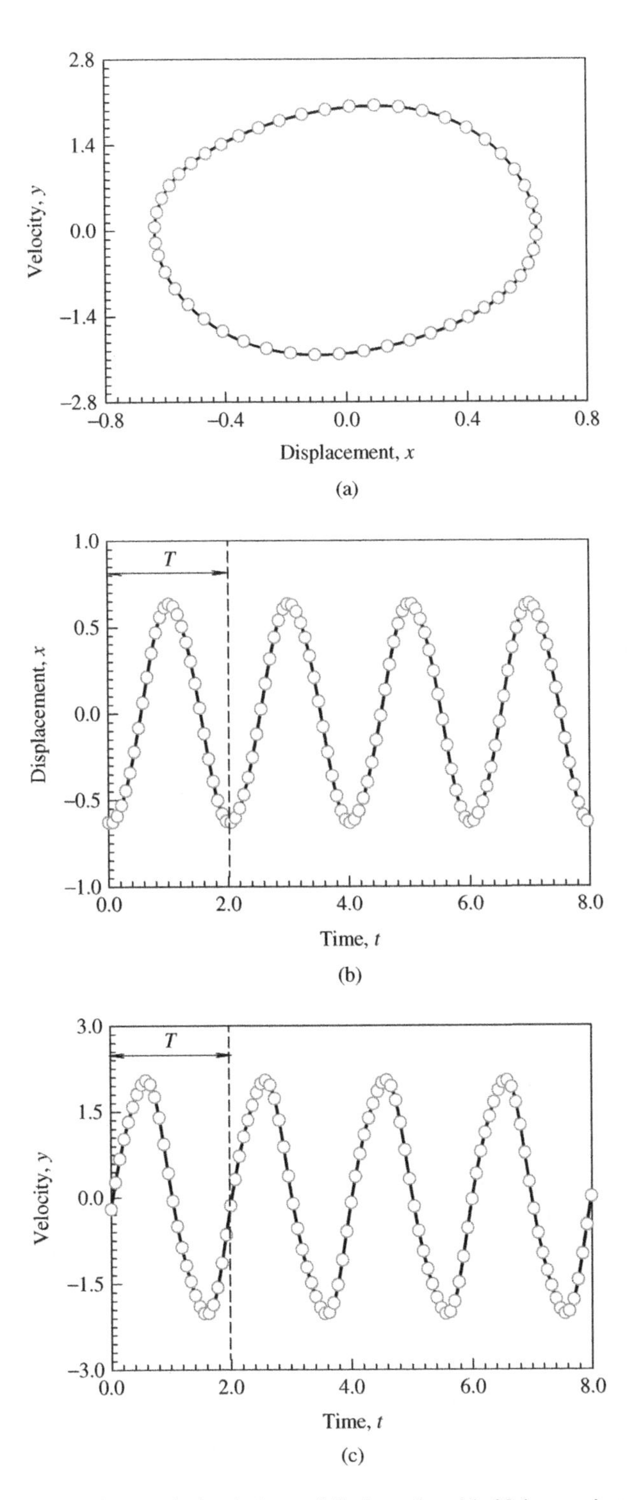

Figure 3.8 Analytical and numerical solutions of limit cycle with 22 harmonic terms (HB22) with $\Omega \approx 3.1574$ and $Q_0 = 0$: (a) phase plane, (b) displacement, (c) velocity, and (d) analytical amplitude spectrum. Solid and symbol curves are for numerical and analytical solutions. The initial condition is $x_0 = 0.0844137$ and $y_0 = 2.02906$ ($\alpha_1 = 0.5$, $\alpha_2 = 5$, $\alpha_3 = 10$)

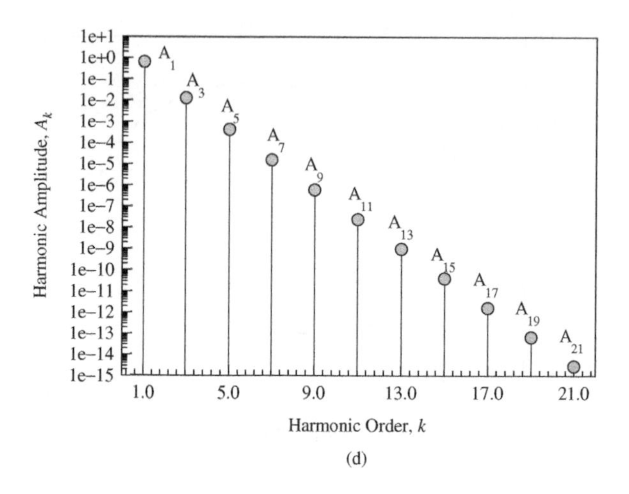

Figure 3.8 *(continued)*

Table 3.1 Input data for numerical illustrations ($\alpha_1 = 5.0$, $\alpha_2 = 5.0$, $\alpha_3 = 1.0$, $Q_0 = 5.0$)

	Ω	Initial condition $(x_0, \dot{x}_0)$	
Figure 3.10(a)–(d)	1.385	$(-1.781350, 0.579786)$	HB90 (P-1)
Figure 3.11(a)–(d)	2.3	$(1.707910, 0.330133)$	HB60 (P-3)
Figure 3.12(a)–(d)	3.2	$(-0.245784, 0.098412)$	HB100 (P-5)

$A_5 \approx 0.189691$, $A_7 \approx 0.092938$, $A_9 \approx 0.048419$, $A_{11} \approx 0.026053$, $A_{13} \approx 0.014295$, $A_{2l-1} \sim 10^{-3}$ $(l = 8 \sim 11)$, 10^{-4} $(l = 12 \sim 15)$, 10^{-5} $(l = 16 \sim 20)$, …, 10^{-11} $(l = 43 \sim 45)$. Due to over-damping, the pattern of period-1 motion is different from the one in Figure 3.9. More harmonic terms are needed for the analytical solution.

The trajectory for this period-1 motion has one cycle, and the higher order harmonic terms have significant contributions on the periodic motion. For the period-3 motion, 60 harmonic terms are used for approximate analytical solutions of period-3 motion. In Figure 3.11(a) and (b) the time-histories of displacement and velocity are presented. The analytical and numerical results match very well. In Figure 3.11(c), the trajectory of period-3 motion for $\Omega = 2.3$ is illustrated and the period-3 motion possesses a large cycle with two small cycles, different from the period-1 motion. In Figure 3.11(d), the harmonic amplitudes are presented with $A_{1/3} \approx 2.162036$, $A_1 \approx 0.657342$, $A_{5/3} \approx 0.306563$, $A_{7/3} \approx 0.170757$, $A_3 \approx 0.113623$, $A_{11/3} \approx 0.076060$, $A_{13/3} \approx 0.052751$, $A_5 \approx 0.037229$, $A_{17/3} \approx 0.026567$, $A_{19/3} \approx 0.019141$, $A_7 \approx 0.013880$, $A_{23/3} \approx 0.010120$, $A_{(2l-1)/3} \sim 10^{-3}$ $(l = 12 \sim 18)$, 10^{-4} $(l = 19 \sim 27)$, and 10^{-5} $(m = 28 \sim 30)$. The pattern of harmonic distributions is different from period-1 motion.

For period-5 motion, the displacement and velocity responses are presented in Figure 3.12(a) and (b) for $\Omega = 3.2$, and the analytical solution is given by 100 harmonic terms (HB100). Compared to period-1 and period-3 motions, the period-5 motion have more waving in the periodic motion. In Figure 3.12(c), the trajectory of a period-5 motion in phase plane is presented, and the period-5 motion has a big cycle with four small cycles at the

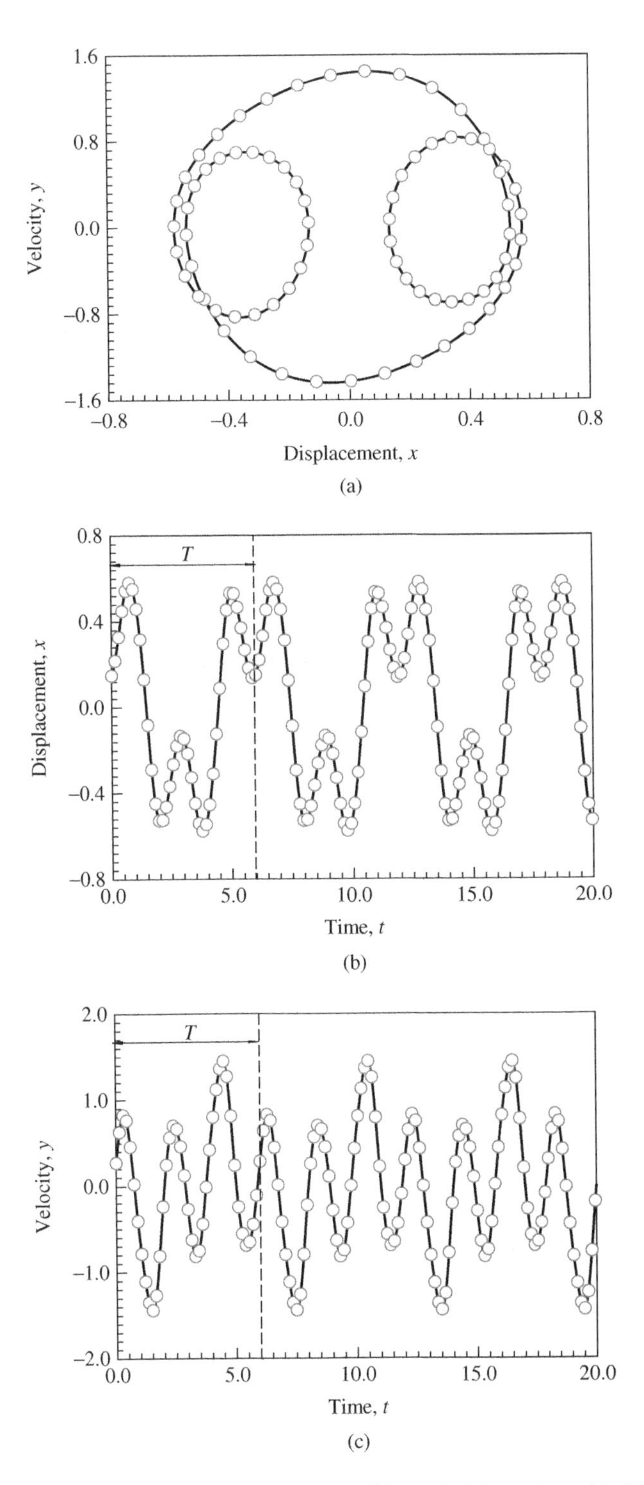

Figure 3.9 Analytical and numerical solutions of stable period-1 motion with 22 harmonic terms (HB22) at $\Omega = 1.048$ and $Q_0 = 4.0$: (a) phase plane, (b) displacement, (c) velocity, and (d) harmonic amplitude spectrum. Solid and symbol curves are for numerical and analytical solutions. The initial condition is $x_0 = 0.391628$ and $y_0 = -0.556119$ ($\alpha_1 = 0.5$, $\alpha_2 = 5$, $\alpha_3 = 10$)

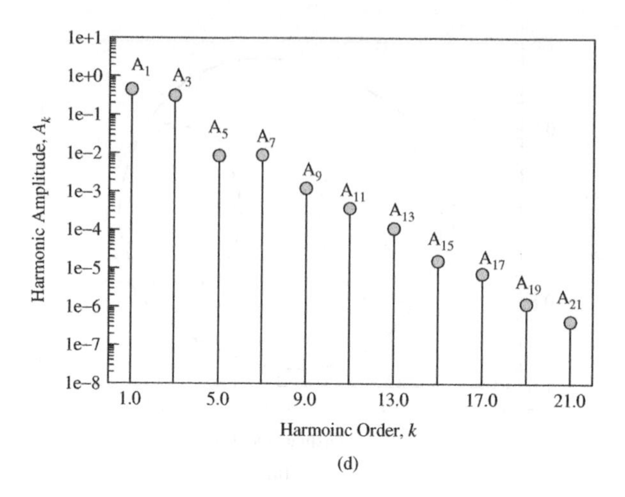

Figure 3.9 *(continued)*

two ends symmetrically. In Figure 3.12(d), the harmonic amplitudes are $A_{1/5} \approx 2.139456$, $A_{3/5} \approx 0.666350$, $A_1 \approx 0.296105$, $A_{7/5} \approx 0.939406$, $A_{9/5} \approx 0.126531$, $A_{11/5} \approx 0.077993$, $A_{13/5} \approx 0.057344$, $A_3 \approx 0.041025$, $A_{17/5} \approx 0.029733$, $A_{19/5} \approx 0.022050$, $A_{21/5} \approx 0.016265$, $A_{23/5} \approx 0.012120$, $A_{(2l-1)/5} \sim 10^{-3}$ $(l = 13 \sim 20)$, 10^{-4} $(l = 21 \sim 28)$, 10^{-5} $(l = 20 \sim 38)$, 10^{-6} $(l = 39 \sim 47)$, 10^{-7} $(l = 48 \sim 50)$.

3.2 van del Pol-Duffing Oscillators

In this section, the van del Pol-Duffing oscillator will be discussed. The appropriate analytical solutions of period-m motions will be presented with finite harmonic terms in the Fourier series solutions based on the prescribed accuracy of harmonic amplitudes. The bifurcation trees of period-m motions to chaos in the van der Pol-Duffing oscillator will be presented, and numerical and analytical solutions of period-m motions will be illustrated.

3.2.1 *Finite Fourier Series Solutions*

Consider a van der Pol-Duffing oscillator

$$\ddot{x} + (-\alpha_1 + \alpha_2 x^2)\dot{x} + \alpha_3 x + \alpha_4 x^3 = Q_0 \cos \Omega t \tag{3.52}$$

where α_i $(i = 1, 2, \ldots, 4)$ are system coefficients for the van der Pol-Duffing oscillator. Q_0 and Ω are excitation amplitude and frequency, respectively. In Luo (2012a), the standard form of Equation (3.52) can be written as

$$\ddot{x} = F(x, \dot{x}, t) \tag{3.53}$$

where

$$F(\dot{x}, x, t) = -\dot{x}(-\alpha_1 + \alpha_2 x^2) - \alpha_3 x - \alpha_4 x^3 + Q_0 \cos \Omega t. \tag{3.54}$$

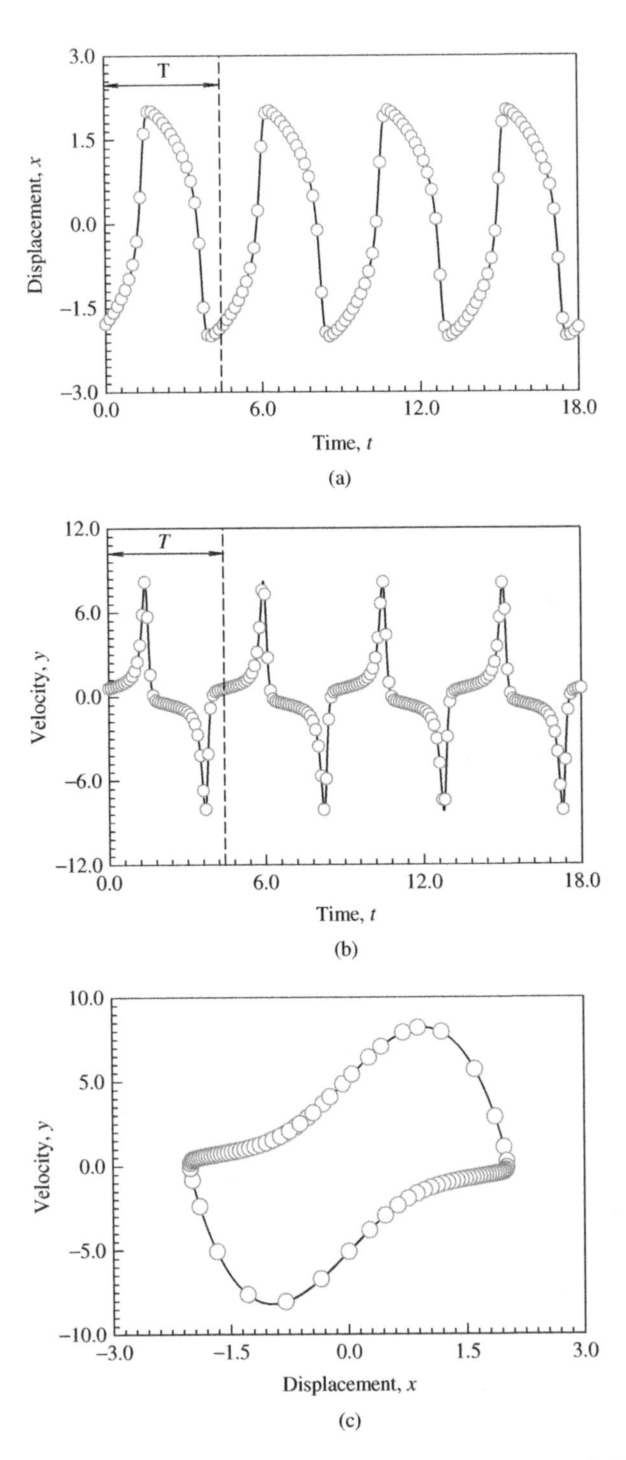

Figure 3.10 Period-1 motion: (a) displacement, (b) velocity, (c) trajectory, and (d) amplitude. Initial condition $(x_0, \dot{x}_0) = (-1.781350, 0.579786)$, $(\alpha_1 = 5.0, \alpha_2 = 5.0, \alpha_3 = 1.0, Q_0 = 5.0)$

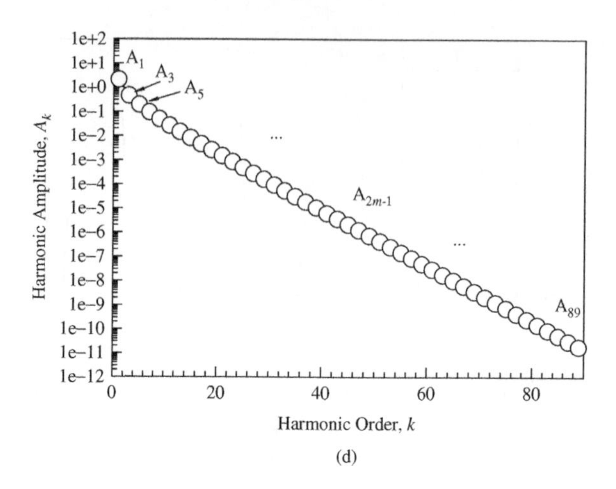

(d)

Figure 3.10 *(continued)*

The analytical solution of period-m motion for the above equation is

$$x^{(m)*} = a_0^{(m)}(t) + \sum_{k=1}^{N} b_{k/m}(t) \cos\left(\frac{k}{m}\theta\right) + c_{k/m}(t) \sin\left(\frac{k}{m}\theta\right) \tag{3.55}$$

where $a_0^{(m)}(t)$, $b_{k/m}(t)$ and $c_{k/m}(t)$ vary with time and $\theta = \Omega t$. The first and second order of derivatives of $x^*(t)$ are

$$\begin{aligned} \dot{x}^{(m)*} &= \dot{a}_0^{(m)} + \sum_{k=1}^{N} \left[\left(\dot{b}_{k/m} + \frac{k}{m}\Omega c_{k/m}\right) \cos\left(\frac{k}{m}\theta\right)\right. \\ &\quad \left. + \left(\dot{c}_{k/m} - \frac{k}{m}\Omega b_{k/m}\right) \sin\left(\frac{k}{m}\theta\right)\right], \end{aligned} \tag{3.56}$$

$$\begin{aligned} \ddot{x}^{(m)*} &= \ddot{a}_0^{(m)} + \sum_{k=1}^{N} \left[\ddot{b}_{k/m} + 2\left(\frac{k}{m}\Omega\right) \dot{c}_{k/m} - \left(\frac{k}{m}\Omega\right)^2 b_{k/m}\right] \cos\left(\frac{k}{m}\theta\right) \\ &\quad + \left[\ddot{c}_{k/m} - 2\left(\frac{k}{m}\Omega\right) \dot{b}_{k/m} - \left(\frac{k}{m}\Omega\right)^2 c_{k/m}\right] \sin\left(\frac{k}{m}\theta\right)\Bigg]. \end{aligned} \tag{3.57}$$

Substitution of Equations (3.55)–(3.57) into Equation (3.52) and application of the virtual work principle for a basis of constant, $\cos(k\theta/m)$ and $\sin(k\theta/m)$ ($k = 1, 2, \ldots$) as a set of virtual displacements gives

$$\begin{aligned} &\ddot{a}_0^{(m)} = F_0^{(m)}(a_0^{(m)}, \mathbf{b}^{(m)}, \mathbf{c}^{(m)}, \dot{a}_0^{(m)}, \dot{\mathbf{b}}^{(m)}, \dot{\mathbf{c}}^{(m)}), \\ &\ddot{b}_{k/m} + 2\frac{k\Omega}{m}\dot{c}_{k/m} - \left(\frac{k\Omega}{m}\right)^2 b_{k/m} = F_{1k}^{(m)}(a_0^{(m)}, \mathbf{b}^{(m)}, \mathbf{c}^{(m)}, \dot{a}_0^{(m)}, \dot{\mathbf{b}}^{(m)}, \dot{\mathbf{c}}^{(m)}), \\ &\ddot{c}_{k/m} - 2\frac{k\Omega}{m}\dot{b}_{k/m} - \left(\frac{k\Omega}{m}\right)^2 c_{k/m} = F_{2k}^{(m)}(a_0^{(m)}, \mathbf{b}^{(m)}, \mathbf{c}^{(m)}, \dot{a}_0^{(m)}, \dot{\mathbf{b}}^{(m)}, \dot{\mathbf{c}}^{(m)}) \\ &\text{for } k = 1, 2, \ldots, N \end{aligned} \tag{3.58}$$

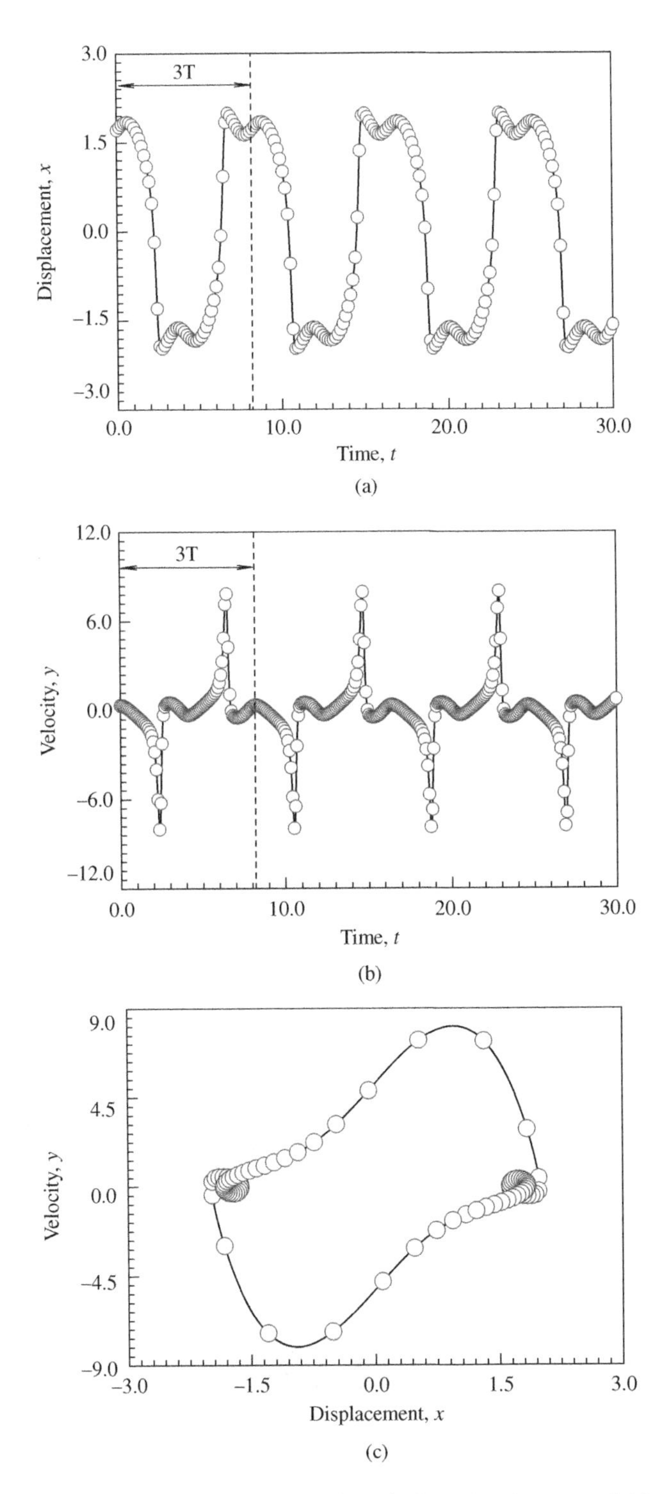

Figure 3.11 Period-3 motion: (a) displacement, (b) velocity, (c) trajectory, and (d) amplitude. Initial condition $(x_0, \dot{x}_0) = (1.707910, 0.330133)$, $(\alpha_1 = 5.0, \alpha_2 = 5.0, \alpha_3 = 1.0, Q_0 = 5.0)$

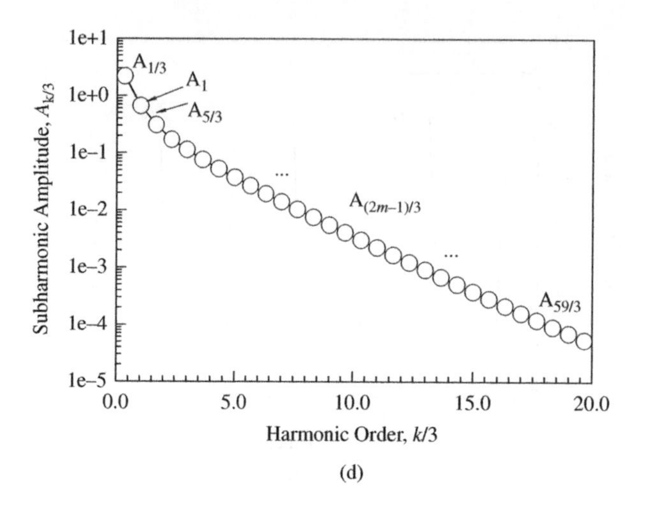

(d)

Figure 3.11 *(continued)*

where

$$
\begin{aligned}
&F_0^{(m)}(a_0^{(m)}, \mathbf{b}^{(m)}, \mathbf{c}^{(m)}, \dot{a}_0^{(m)}, \dot{\mathbf{b}}^{(m)}, \dot{\mathbf{c}}^{(m)}) \\
&\quad = \frac{1}{mT}\int_0^{mT} F(x^{(m)*}, \dot{x}^{(m)*}, t)dt \\
&\quad = \alpha_1 \dot{a}_0^{(m)} - \alpha_3 a_0^{(m)} - \alpha_2 f_1^{(m)} - \alpha_4 f_2^{(m)}, \\
&F_{1k}^{(m)}(a_0^{(m)}, \mathbf{b}^{(m)}, \mathbf{c}^{(m)}, \dot{a}_0^{(m)}, \dot{\mathbf{b}}^{(m)}, \dot{\mathbf{c}}^{(m)}) \\
&\quad = \frac{2}{mT}\int_0^{mT} F(x^{(m)*}, \dot{x}^{(m)*}, t)\cos\left(\frac{k}{m}\Omega t\right) dt \\
&\quad = \alpha_1\left(\dot{b}_{k/m} + \frac{k}{m}\Omega c_{k/m}\right) - \alpha_3 b_{k/m} + Q_0\delta_k^m - \alpha_2 f_{1k/m}^{(c)} - \alpha_4 f_{2k/m}^{(c)},
\end{aligned}
$$

$$
\begin{aligned}
&F_{2k}^{(m)}(a_0^{(m)}, \mathbf{b}^{(m)}, \mathbf{c}^{(m)}, \dot{a}_0^{(m)}, \dot{\mathbf{b}}^{(m)}, \dot{\mathbf{c}}^{(m)}) \\
&\quad = \frac{2}{mT}\int_0^{mT} F(x^{(m)*}, \dot{x}^{(m)*}, t)\sin\left(\frac{k}{m}\Omega t\right) dt \\
&\quad = \alpha_1\left(\dot{c}_{k/m} - \frac{k}{m}\Omega b_{k/m}\right) - \alpha_3 c_{k/m} - \alpha_2 f_{1k/m}^{(s)} - \alpha_4 f_{2k/m}^{(s)}
\end{aligned}
\tag{3.59}
$$

and

$$
\begin{aligned}
f_1^{(m)} &= \dot{a}_0^{(m)}(a_0^{(m)})^2 + \sum_{n=1}^{6}\sum_{l=1}^{N}\sum_{j=1}^{N}\sum_{i=1}^{N} f_{1(n)}^{(m,0)}(i,j,l), \\
f_2^{(m)} &= (a_0^{(m)})^3 + \sum_{n=1}^{3}\sum_{l=1}^{N}\sum_{j=1}^{N}\sum_{i=1}^{N} f_{2(n)}^{(m,0)}(i,j,l)
\end{aligned}
\tag{3.60}
$$

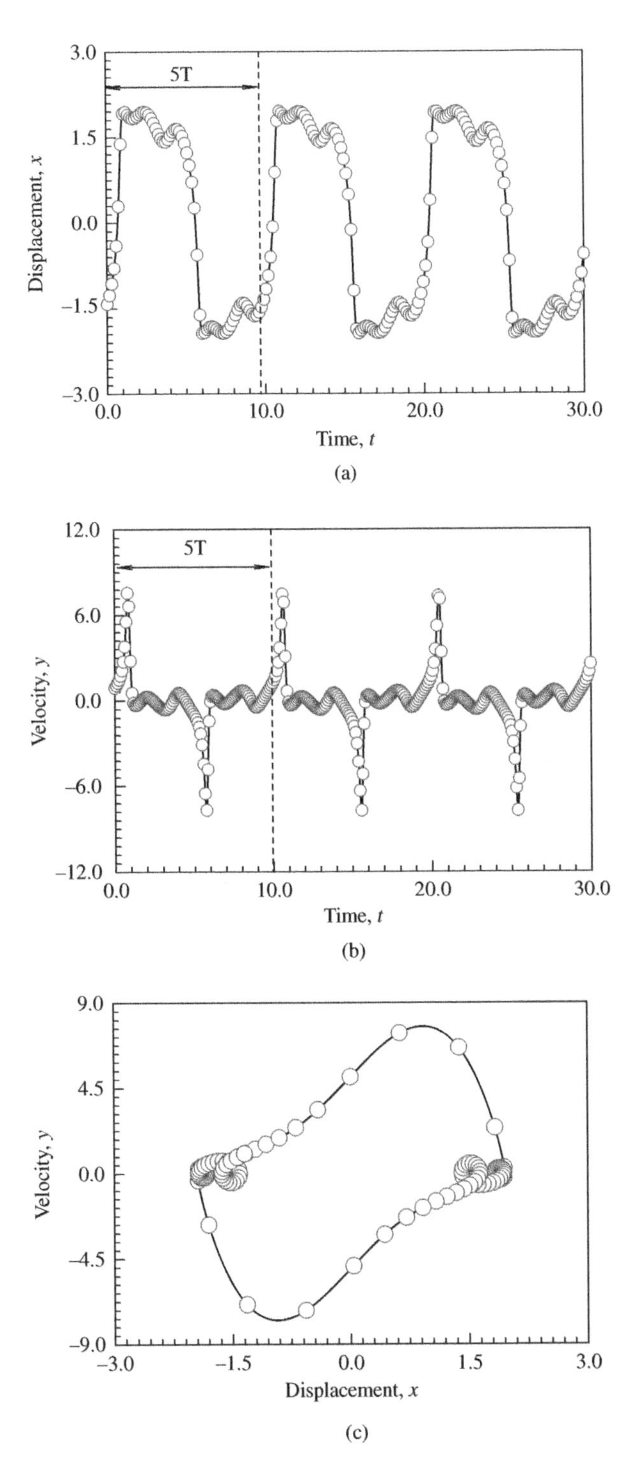

Figure 3.12 Period-5 motion: (a) displacement, (b) velocity, (c) trajectory, and (d) amplitude. Initial condition $(x_0, \dot{x}_0) = (-0.245784, 0.098412)$, $(\alpha_1 = 5.0, \alpha_2 = 5.0, \alpha_3 = 1.0, Q_0 = 5.0)$

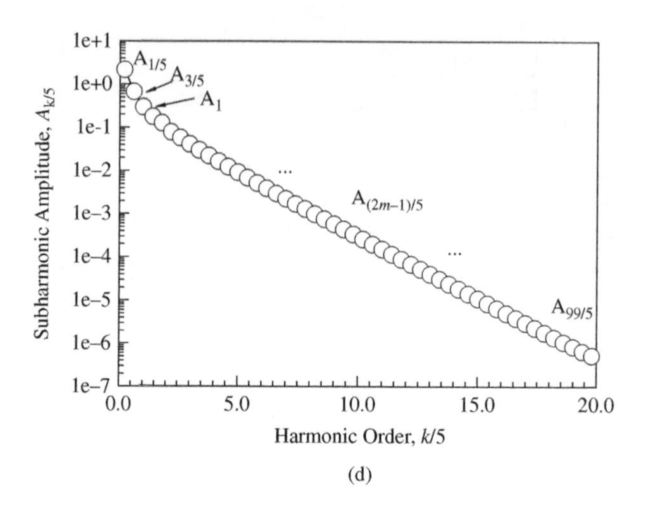

(d)

Figure 3.12 *(continued)*

with

$$\begin{aligned}
f_{1(1)}^{(m,0)}(i,j,l) &= \frac{1}{2N}\dot{a}_0^{(m)}(b_{i/m}b_{j/m} + c_{i/m}c_{j/m})\delta_{i-j}^0,\\
f_{1(2)}^{(m,0)}(i,j,l) &= \frac{1}{N}a_0^{(m)}\left[b_{i/m}\left(\dot{b}_{l/m} + \frac{l}{m}\Omega c_{l/m}\right) + c_{i/m}(\dot{c}_{l/m} - \frac{l}{m}\Omega b_{l/m})\right]\delta_{l-i}^0,\\
f_{1(3)}^{(m,0)}(i,j,l) &= \frac{1}{4}(b_{i/m}b_{j/m} - c_{i/m}c_{j/m})\left(\dot{b}_{l/m} + \frac{l}{m}\Omega c_{k/m}\right)\delta_{l-i-j}^0,\\
f_{1(4)}^{(m,0)}(i,j,l) &= \frac{1}{4}(b_{j/m}c_{i/m} + b_{i/m}c_{j/m})\left(\dot{c}_{l/m} - \frac{l}{m}\Omega b_{l/m}\right)\delta_{l-i-j}^0,\\
f_{1(5)}^{(m,0)}(i,j,l) &= \frac{1}{4}(b_{i/m}b_{j/m} + c_{i/m}c_{j/m})\left(\dot{b}_{l/m} + \frac{l}{m}\Omega c_{l/m}\right)(\delta_{l-i+j}^0 + \delta_{l+i-j}^0),\\
f_{1(6)}^{(m,0)}(i,j,l) &= \frac{1}{4}(b_{j/m}c_{i/m} - b_{i/m}c_{j/m})\left(\dot{c}_{l/m} - \frac{l}{m}\Omega b_{l/m}\right)(\delta_{l-i+j}^0 + \delta_{l+i-j}^0);
\end{aligned} \tag{3.61}$$

$$\begin{aligned}
f_{2(1)}^{(m,0)}(i,j,l) &= \frac{3}{2N}a_0^{(m)}(b_{i/m}b_{j/m}\delta_{i-j}^0 + c_{i/m}c_{j/m}\delta_{i-j}^0),\\
f_{2(2)}^{(m,0)}(i,j,l) &= \frac{1}{4}b_{i/m}b_{j/m}b_{l/m}(\delta_{i-j-l}^0 + \delta_{i-j+l}^0 + \delta_{i+j-l}^0),\\
f_{2(3)}^{(m,0)}(i,j,l) &= \frac{3}{4}b_{i/m}c_{j/m}c_{l/m}(\delta_{i+j-l}^0 + \delta_{i-j+l}^0 - \delta_{i-j-l}^0)
\end{aligned} \tag{3.62}$$

and

$$\begin{aligned}
f_{1k/m}^{(c)} &= \sum_{n=1}^{8}\sum_{l=1}^{N}\sum_{j=1}^{N}\sum_{i=1}^{N} f_{1(n)}^{(m,c)}(i,j,l,k),\\
f_{2k/m}^{(c)} &= \sum_{n=1}^{4}\sum_{l=1}^{N}\sum_{j=1}^{N}\sum_{i=1}^{N} f_{2(n)}^{(m,c)}(i,j,l,k)
\end{aligned} \tag{3.63}$$

with

$$
\begin{aligned}
f_{1(1)}^{(m,c)}(i,j,l,k) &= \frac{2}{N^2}\dot{a}_0^{(m)}a_0^{(m)}b_{i/m}\delta_i^k + \frac{1}{N^2}(a_0^{(m)})^2\left(\dot{b}_{l/m} + \frac{k}{m}\Omega c_{l/m}\right)\delta_l^k,\\
f_{1(2)}^{(m,c)}(i,j,l,k) &= \frac{1}{2N}\dot{a}_0^{(m)}\left[\left(b_{i/m}b_{j/m} + c_{i/m}c_{j/m}\right)\delta_{i-j}^k\right.\\
&\quad \left. + \left(b_{i/m}b_{j/m} - c_{i/m}c_{j/m}\right)\delta_{i+j}^k\right],\\
f_{1(3)}^{(m,c)}(i,j,l,k) &= \frac{1}{N}a_0^{(m)}b_{i/m}\left(\dot{b}_{l/m} + \frac{l}{m}\Omega c_{l/m}\right)(\delta_{|l-i|}^k + \delta_{l+i}^k),\\
f_{1(4)}^{(m,c)}(i,j,l,k) &= \frac{1}{N}a_0^{(m)}c_{i/m}\left(\dot{c}_{l/m} - \frac{l}{m}\Omega b_{l/m}\right)(\delta_{|l-i|}^k - \delta_{l+i}^k),\\
f_{1(5)}^{(m,c)}(i,j,l,k) &= \frac{1}{4}(b_{i/m}b_{j/m} - c_{i/m}c_{j/m})\left(\dot{b}_{l/m} + \frac{l}{m}\Omega c_{l/m}\right)(\delta_{|l-i-j|}^k + \delta_{l+i+j}^k),\\
f_{1(6)}^{(m,c)}(i,j,l,k) &= \frac{1}{4}(b_{j/m}c_{i/m} + b_{i/m}c_{j/m})\left(\dot{c}_{l/m} - \frac{l}{m}\Omega b_{l/m}\right)(\delta_{|l-i-j|}^k - \delta_{l+i+j}^k),\\
f_{1(7)}^{(m,c)}(i,j,l,k) &= \frac{1}{4}(b_{i/m}b_{j/m} + c_{i/m}c_{j/m})\left(\dot{b}_{l/m} + \frac{l}{m}\Omega c_{l/m}\right)(\delta_{|l-i-j|}^k + \delta_{l+i+j}^k),\\
f_{1(8)}^{(m,c)}(i,j,l,k) &= \frac{1}{4}(b_{j/m}c_{i/m} - b_{i/m}c_{j/m})\left(\dot{c}_{l/m} - \frac{l}{m}\Omega b_{l/m}\right)(\delta_{|l-i-j|}^k - \delta_{l+i+j}^k);
\end{aligned}
\tag{3.64}
$$

$$
\begin{aligned}
f_{2(1)}^{(m,c)}(i,j,l,k) &= \frac{3}{N^2}(a_0^{(m)})^2 b_{l/m}\delta_l^k + \frac{3}{2N}a_0^{(m)}b_{l/m}b_{j/m}(\delta_{|l-j|}^k + \delta_{l+j}^k),\\
f_{2(2)}^{(m,c)}(i,j,l,k) &= \frac{3}{2N}a_0^{(m)}c_{l/m}c_{j/m}(\delta_{|l-j|}^k - \delta_{l+j}^k),\\
f_{2(3)}^{(m,c)}(i,j,l,k) &= \frac{1}{4}b_{l/m}b_{j/m}b_{i/m}(\delta_{|l-j-i|}^k + \delta_{l+j+i}^k + \delta_{|l-j+i|}^k + \delta_{|l+j-i|}^k),\\
f_{2(4)}^{(m,c)}(i,j,l,k) &= \frac{3}{4}b_{l/m}c_{j/m}c_{i/m}(\delta_{|l+j-i|}^k - \delta_{l+j+i}^k + \delta_{|l-j+i|}^k - \delta_{|l-j-i|}^k)
\end{aligned}
\tag{3.65}
$$

and

$$
\begin{aligned}
f_{1k/m}^{(s)} &= \sum_{n=1}^{9}\sum_{l=1}^{N}\sum_{j=1}^{N}\sum_{i=1}^{N} f_{1(n)}^{(m,s)}(i,j,l,k),\\
f_{2k/m}^{(s)} &= \sum_{n=1}^{3}\sum_{l=1}^{N}\sum_{j=1}^{N}\sum_{i=1}^{N} f_{2(n)}^{(m,s)}(i,j,l,k)
\end{aligned}
\tag{3.66}
$$

with

$$
\begin{aligned}
f_{1(1)}^{(m,s)}(i,j,l,k) &= \frac{1}{N^2}\left(2\dot{a}_0^{(m)}a_0^{(m)}c_{i/m}\delta_i^k + (a_0^{(m)}\dot{c}_{l/m} - \frac{l}{m}\Omega b_{l/m})\delta_l^k\right],\\
f_2^{(m,2)}(i,j,l,k) &= \frac{1}{2N}\dot{a}_0^{(m)}(b_{j/m}c_{i/m} - b_{i/m}c_{j/m})\mathrm{sgn}(i-j)\delta_{|i-j|}^k,\\
f_3^{(m,2)}(i,j,l,k) &= \frac{1}{2N}\dot{a}_0^{(m)}(b_{j/m}c_{i/m} + b_{i/m}c_{j/m})\delta_{i+j}^k,\\
f_{1(4)}^{(m,s)}(i,j,l,k) &= \frac{1}{N}a_0^{(m)}b_{i/m}\left(\dot{c}_{l/m} - \frac{l}{m}\Omega b_{l/m}\right)[\delta_{l+i}^k + \mathrm{sgn}(l-i)\delta_{|l-i|}^k],
\end{aligned}
$$

$$f_{1(5)}^{(m,s)}(i,j,l,k) = \frac{1}{N}a_0^{(m)}c_{i/m}\left(\dot{b}_{l/m} + \frac{l}{m}\Omega c_{l/m}\right)[\delta_{l+i}^k - \text{sgn}(l-i)\delta_{|l-i|}^k],$$

$$f_{1(6)}^{(m,s)}(i,j,l,k) = \frac{1}{4}(b_{i/m}c_{j/m} + b_{j/m}c_{i/m})\left(\dot{b}_{l/m} + \frac{l}{m}\Omega c_{l/m}\right)$$
$$\times [\delta_{l+i+j}^k - \text{sgn}(l-i-j)\delta_{|l-i-j|}^k],$$

$$f_{1(7)}^{(m,s)}(i,j,l,k) = \frac{1}{4}(b_{i/m}b_{j/m} - c_{i/m}c_{j/m})\left(\dot{c}_{k/m} - \frac{l}{m}\Omega b_{l/m}\right)$$
$$\times [\delta_{l+i+j}^k + \text{sgn}(l-i-j)\delta_{|l-i-j|}^k],$$

$$f_{1(8)}^{(m,s)}(i,j,l,k) = \frac{1}{4}(b_{i/m}b_{j/m} + c_{i/m}c_{j/m})\left(\dot{c}_{l/m} - \frac{l}{m}\Omega b_{l/m}\right)$$
$$\times [\text{sgn}(l-i+j)\delta_{|l-i+j|}^k + \text{sgn}(l+i-j)\delta_{|l+i-j|}^k],$$

$$f_{1(9)}^{(m,s)}(i,j,l,k) = \frac{1}{4}(b_{i/m}c_{j/m} - b_{j/m}c_{i/m})\left(\dot{b}_{l/m} + \frac{l}{m}\Omega c_{l/m}\right)$$
$$\times [\text{sgn}(l-i+j)\delta_{|l-i+j|}^k - \text{sgn}(l+i-j)\delta_{|l+i-j|}^k]; \tag{3.67}$$

$$f_{2(1)}^{(m,s)}(i,j,l,k) = \frac{3}{N^2}(a_0^{(m)})^2 c_{l/m}\delta_l^k + \frac{3}{N}a_0^{(m)}b_{l/m}c_{j/m}[\delta_{l+j}^k - \text{sgn}(l-j)\delta_{|l-j|}^k],$$

$$f_{2(1)}^{(m,s)}(i,j,l,k) = \frac{1}{4}c_{l/m}c_{j/m}c_{i/m}\left[\text{sgn}\,(l-j+i)\,\delta_{|l-j+i|}^k - \delta_{l+j+i}^k\right.$$
$$\left. + \text{sgn}\,(l+j-i)\,\delta_{|l+j-i|}^k - \text{sgn}(l-j-i)\delta_{|l-j-i|}^k\right],$$

$$f_{2(3)}^{(m,s)}(i,j,l,k) = \frac{3}{4}b_{l/m}b_{j/m}c_{i/m}[\text{sgn}(l-j+i)\delta_{|l-j+i|}^k + \delta_{l+j+i}^k$$
$$- \text{sgn}(l+j-i)\delta_{|l+j-i|}^k - \text{sgn}(l-j-i)\delta_{|l-j-i|}^k)]. \tag{3.68}$$

Define

$$\begin{aligned}
\mathbf{z}^{(m)} &\triangleq (a_0^{(m)}, \mathbf{b}^{(m)}, \mathbf{c}^{(m)})^{\mathrm{T}} \\
&= (a_0^{(m)}, b_{1/m}, \ldots, b_{N/m}, c_{1/m}, \ldots, c_{N/m})^{\mathrm{T}} \\
&\equiv (z_0^{(m)}, z_1^{(m)}, \ldots, z_{2N}^{(m)})^{\mathrm{T}} \\
\mathbf{z}_1 &\triangleq \dot{\mathbf{z}} = (\dot{a}_0^{(m)}, \dot{\mathbf{b}}^{(m)}, \dot{\mathbf{c}}^{(m)})^{\mathrm{T}} \\
&= (\dot{a}_0^{(m)}, \dot{b}_{1/m}, \ldots, \dot{b}_{N/m}, \dot{c}_{1/m}, \ldots, \dot{c}_{N/m})^{\mathrm{T}} \\
&\equiv (\dot{z}_0^{(m)}, \dot{z}_1^{(m)}, \ldots, \dot{z}_{2N}^{(m)})^{\mathrm{T}}
\end{aligned} \tag{3.69}$$

where

$$\begin{aligned}
\mathbf{b}^{(m)} &= (b_{1/m}, \ldots, b_{N/m})^{\mathrm{T}}, \\
\mathbf{c}^{(m)} &= (c_{1/m}, \ldots, c_{N/m})^{\mathrm{T}}.
\end{aligned} \tag{3.70}$$

Equation (3.58) becomes

$$\dot{\mathbf{z}}^{(m)} = \mathbf{z}_1^{(m)} \text{ and } \dot{\mathbf{z}}_1^{(m)} = \mathbf{g}^{(m)}(\mathbf{z}^{(m)}, \mathbf{z}_1^{(m)}) \tag{3.71}$$

where

$$\mathbf{g}^{(m)}(\mathbf{z}^{(m)},\mathbf{z}_1^{(m)}) = \begin{pmatrix} F_0^{(m)}(\mathbf{z}^{(m)},\mathbf{z}_1^{(m)}) \\ \mathbf{F}_1^{(m)}(\mathbf{z}^{(m)},\mathbf{z}_1^{(m)}) - 2\mathbf{k}_1\frac{\Omega}{m}\dot{\mathbf{c}}^{(m)} + \mathbf{k}_2\left(\frac{\Omega}{m}\right)^2\mathbf{b}^{(m)} \\ \mathbf{F}_2^{(m)}(\mathbf{z}^{(m)},\mathbf{z}_1^{(m)}) + 2\mathbf{k}_1\frac{\Omega}{m}\dot{\mathbf{b}}^{(m)} + \mathbf{k}_2\left(\frac{\Omega}{m}\right)^2\mathbf{c}^{(m)} \end{pmatrix} \tag{3.72}$$

and

$$\begin{aligned} &\mathbf{k}_1 = diag(1,2,\ldots,N), \\ &\mathbf{k}_2 = diag(1,2^2,\ldots,N^2), \\ &\mathbf{F}_1^{(m)} = (F_{11}^{(m)},F_{12}^{(m)},\ldots,F_{1N}^{(m)})^{\mathrm{T}}, \\ &\mathbf{F}_2^{(m)} = (F_{21}^{(m)},F_{22}^{(m)},\ldots,F_{2N}^{(m)})^{\mathrm{T}} \\ &\text{for } N = 1,2,\ldots,\infty. \end{aligned} \tag{3.73}$$

Introducing

$$\mathbf{y}^{(m)} \equiv (\mathbf{z}^{(m)},\mathbf{z}_1^{(m)}) \text{ and } \mathbf{f}^{(m)} = (\mathbf{z}_1^{(m)},\mathbf{g}^{(m)})^{\mathrm{T}}, \tag{3.74}$$

Equation (3.71) becomes

$$\dot{\mathbf{y}}^{(m)} = \mathbf{f}^{(m)}(\mathbf{y}^{(m)}). \tag{3.75}$$

The steady-state solutions for periodic motions of the van del Pol-Duffing oscillator in Equation (3.52) can be obtained by setting $\dot{\mathbf{y}}^{(m)} = \mathbf{0}$, that is,

$$\begin{aligned} &F_0^{(m)}(\mathbf{z}^{(m)},\mathbf{0}) = 0, \\ &\mathbf{F}_1^{(m)}(\mathbf{z}^{(m)},\mathbf{0}) - \mathbf{k}_2\left(\frac{\Omega}{m}\right)^2\mathbf{b}^{(m)} = \mathbf{0}, \\ &\mathbf{F}_2^{(m)}(\mathbf{z}^{(m)},\mathbf{0}) - \mathbf{k}_2\left(\frac{\Omega}{m}\right)^2\mathbf{c}^{(m)} = \mathbf{0}. \end{aligned} \tag{3.76}$$

The $(2N+1)$ nonlinear equations in Equation (3.76) are solved by the Newton–Raphson method. In Luo (2012a), the linearized equation at equilibrium point $\mathbf{y}^* = (\mathbf{z}^*,\mathbf{0})^{\mathrm{T}}$ is given by

$$\Delta\dot{\mathbf{y}}^{(m)} = D\mathbf{f}(\mathbf{y}^{(m)^*})\Delta\mathbf{y}^{(m)} \tag{3.77}$$

where

$$D\mathbf{f}(\mathbf{y}^{(m)^*}) = \partial\mathbf{f}(\mathbf{y}^{(m)})/\partial\mathbf{y}^{(m)}|_{\mathbf{y}^{(m)*}}. \tag{3.78}$$

The corresponding eigenvalues are determined by

$$|D\mathbf{f}(\mathbf{y}^{(m)^*}) - \lambda\mathbf{I}_{2(2N+1)\times 2(2N+1)}| = 0. \tag{3.79}$$

where

$$D\mathbf{f}(\mathbf{y}^{(m)^*}) = \begin{bmatrix} \mathbf{0}_{(2N+1)\times(2N+1)} & \mathbf{I}_{(2N+1)\times(2N+1)} \\ \mathbf{G}_{(2N+1)\times(2N+1)} & \mathbf{H}_{(2N+1)\times(2N+1)} \end{bmatrix} \tag{3.80}$$

and

$$\mathbf{G} = \frac{\partial\mathbf{g}^{(m)}}{\partial\mathbf{z}^{(m)}} = (\mathbf{G}^{(0)},\mathbf{G}^{(c)},\mathbf{G}^{(s)})^{\mathrm{T}} \tag{3.81}$$

$$
\begin{aligned}
\mathbf{G}^{(0)} &= (G_0^{(0)}, G_1^{(0)}, \ldots, G_{2N}^{(0)}), \\
\mathbf{G}^{(c)} &= (\mathbf{G}_1^{(c)}, \mathbf{G}_2^{(c)}, \ldots, \mathbf{G}_N^{(c)})^{\mathrm{T}}, \\
\mathbf{G}^{(s)} &= (\mathbf{G}_1^{(s)}, \mathbf{G}_2^{(s)}, \ldots, \mathbf{G}_N^{(s)})^{\mathrm{T}}
\end{aligned} \tag{3.82}
$$

for $N = 1, 2, \ldots, \infty$ with

$$
\begin{aligned}
\mathbf{G}_k^{(c)} &= (G_{k0}^{(c)}, G_{k1}^{(c)}, \ldots, G_{k(2N)}^{(c)}), \\
\mathbf{G}_k^{(s)} &= (G_{k0}^{(s)}, G_{k1}^{(s)}, \ldots, G_{k(2N)}^{(s)})
\end{aligned} \tag{3.83}
$$

for $k = 1, 2, \ldots, N$. The corresponding components are

$$
\begin{aligned}
G_r^{(0)} &= -\alpha_3 \delta_0^r - \alpha_2 g_{1r}^{(m,0)} - \alpha_4 g_{2r}^{(m,0)}, \\
G_{kr}^{(c)} &= \left(\frac{k\Omega}{m}\right)^2 \delta_k^r + \alpha_1 \left(\frac{k}{m}\Omega\right) \delta_{k+N}^r - \alpha_3 \delta_k^r - \alpha_2 g_{kr}^{(m,c)} - \alpha_4 g_{2kr}^{(m,c)}, \\
G_{kr}^{(s)} &= \left(\frac{k\Omega}{m}\right)^2 \delta_{k+N}^r - \alpha_1 \left(\frac{k}{m}\Omega\right) \delta_k^r - \alpha_3 \delta_{k+N}^r - \alpha_2 g_{1kr}^{(m,s)} - \alpha_4 g_{2kr}^{(m,s)}
\end{aligned} \tag{3.84}
$$

where

$$
\begin{aligned}
g_{1r}^{(m,0)} &= g_{1(0)}^{(m,0)}(r) + \sum_{n=1}^{16} \sum_{l=1}^{N} \sum_{j=1}^{N} \sum_{i=1}^{N} g_{1(n)}^{(m,0)}(i,j,l,r), \\
g_{2r}^{(m,0)} &= g_{2(0)}^{(m,0)}(r) + \sum_{n=1}^{4} \sum_{l=1}^{N} \sum_{j=1}^{N} \sum_{i=1}^{N} g_{2(n)}^{(m,0)}(i,j,l,r).
\end{aligned} \tag{3.85}
$$

The corresponding nonlinear function terms for constant term are

$$
\begin{aligned}
g_{1(0)}^{(m,0)}(r) &= 2\dot{a}_0^{(m)} a_0^{(m)} \delta_r^0, \\
g_{1(1)}^{(m,0)}(i,j,l,r) &= \frac{1}{N} b_{i/m} \left(\dot{b}_{k/m} + \frac{l}{m}\Omega c_{l/m}\right) \delta_{l-i}^0 \delta_0^r, \\
g_{1(2)}^{(m,0)}(i,j,l,r) &= \frac{1}{N} c_{i/m} \left(\dot{c}_{l/m} - \frac{l}{m}\Omega b_{l/m}\right) \delta_{l-i}^0 \delta_0^r, \\
g_{1(3)}^{(m,0)}(i,j,l,r) &= \frac{1}{N} \left[\dot{a}_0^{(m)} b_{i/m} \delta_{i-j}^0 + \frac{1}{N} a_0^{(m)} (\dot{b}_{k/m} + \frac{l}{m}\Omega c_{l/m}) \delta_{l-i}^0\right] \delta_i^r, \\
g_{1(4)}^{(m,0)}(i,j,l,r) &= -\frac{1}{N}\frac{k}{m}\Omega a_0^{(m)} c_{i/m} \delta_{l-i}^0 \delta_l^r, \\
g_{1(5)}^{(m,0)}(i,j,l,r) &= \frac{1}{2} b_{j/m} \left(\dot{b}_{l/m} + \frac{l}{m}\Omega c_{l/m}\right) \delta_{l-i-j}^0 \delta_i^r, \\
g_{1(6)}^{(m,0)}(i,j,l,r) &= \frac{1}{2} c_{j/m} \left(\dot{c}_{l/m} - \frac{l}{m}\Omega b_{l/m}\right) \delta_{l-i-j}^0 \delta_i^r, \\
g_{1(7)}^{(m,0)}(i,j,l,r) &= -\frac{1}{4}\frac{l}{m}\Omega (b_{j/m} c_{i/m} + b_{i/m} c_{j/m}) \delta_{l-i-j}^0 \delta_l^r, \\
g_{1(8)}^{(m,0)}(i,j,l,r) &= \frac{1}{2} b_{j/m} \left(\dot{b}_{l/m} + \frac{l}{m}\Omega c_{l/m}\right) (\delta_{l-i+j}^0 + \delta_{l+i-j}^0) \delta_i^r, \\
g_{1(9)}^{(m,0)}(i,j,l,r) &= -\frac{1}{4}\frac{l}{m}\Omega (b_{j/m} c_{i/m} - b_{i/m} c_{j/m}) (\delta_{l-i+j}^0 + \delta_{l+i-j}^0) \delta_l^r,
\end{aligned}
$$

$$
\begin{aligned}
g_{1(10)}^{(m,0)}(i,j,l,r) &= \frac{1}{N}\left[\dot{a}_0^{(m)} c_{j/m}\delta_{i-j}^0\delta_{j+N}^r + \frac{l}{m}\Omega a_0^{(m)} b_{i/m}\delta_{l-i}^0\delta_{l+N}^r\right],\\
g_{1(11)}^{(m,0)}(i,j,l,r) &= \frac{1}{N}a_0^{(m)}\left(\dot{c}_{l/m} - \frac{l}{m}\Omega b_{l/m}\right)\delta_{l-i}^0\delta_{i+N}^r,\\
g_{1(12)}^{(m,0)}(i,j,l,r) &= -\frac{1}{2}c_{j/m}\left(\dot{b}_{l/m} + \frac{l}{m}\Omega c_{l/m}\right)\delta_{l-i-j}^0\delta_{i+N}^r,\\
g_{1(13)}^{(m,0)}(i,j,l,r) &= \frac{1}{4}\frac{l}{m}\Omega(b_{i/m}b_{j/m} - c_{i/m}c_{j/m})\delta_{l-i-j}^0\delta_{l+N}^r,\\
g_{1(14)}^{(m,0)}(i,j,l,r) &= \frac{1}{2}b_{j/m}\left(\dot{c}_{l/m} - \frac{l}{m}\Omega b_{l/m}\right)\delta_{l-i-j}^0\delta_{i+N}^r,\\
g_{1(15)}^{(m,0)}(i,j,l,r) &= \frac{1}{2}c_{j/m}\left(\dot{b}_{l/m} + \frac{l}{m}\Omega c_{l/m}\right)(\delta_{l-i+j}^0 + \delta_{l+i-j}^0)\delta_{i+N}^r,\\
g_{1(16)}^{(m,0)}(i,j,l,r) &= \frac{1}{4}\frac{l}{m}\Omega(b_{i/m}b_{j/m} + c_{i/m}c_{j/m})(\delta_{l-i+j}^0 + \delta_{l+i-j}^0)\delta_{l+N}^r;
\end{aligned}
\tag{3.86}
$$

$$
\begin{aligned}
g_{2(1)}^{(m,0)}(r) &= 3(a_0^{(m)})^2\delta_r^0,\\
g_{2(1)}^{(m,0)}(i,j,l,r) &= \frac{3}{2N}(b_{i/m}b_{j/m}\delta_r^0 + 2a_0^{(m)}b_{i/m}\delta_j^r)\delta_{i-j}^0,\\
g_{2(2)}^{(m,0)}(i,j,l,r) &= \frac{3}{2N}(c_{i/m}c_{j/m}\delta_r^0 + 2a_0^{(m)}c_{i/m}\delta_{j+N}^r)\delta_{i-j}^0,\\
g_{2(3)}^{(m,0)}(i,j,l,r) &= \frac{3}{4}b_{i/m}b_{j/m}\delta_l^r(\delta_{i-j-l}^0 + \delta_{i-j+l}^0 + \delta_{i+j-l}^0),\\
g_{2(4)}^{(m,0)}(i,j,l,r) &= \frac{3}{4}(c_{j/m}c_{l/m}\delta_i^r + 2b_{i/m}c_{j/m}\delta_{l+N}^r)(\delta_{i+j-l}^0 + \delta_{i-j+l}^0 - \delta_{i-j-l}^0).
\end{aligned}
\tag{3.87}
$$

The corresponding nonlinear terms for cosine are:

$$
\begin{aligned}
g_{1kr}^{(m,c)} &= \sum_{n=1}^{18}\sum_{l=1}^{N}\sum_{j=1}^{N}\sum_{i=1}^{N} g_{1(n)}^{(m,c)}(i,j,l,k,r),\\
g_{2kr}^{(m,c)} &= \sum_{n=1}^{5}\sum_{l=1}^{N}\sum_{j=1}^{N}\sum_{i=1}^{N} g_{2(n)}^{(m,c)}(i,j,l,k,r)
\end{aligned}
\tag{3.88}
$$

with

$$
\begin{aligned}
g_{1(1)}^{(m,c)}(i,j,l,k,r) &= \frac{2}{N^2}\left[\dot{a}_0^{(m)} b_{i/m}\delta_i^k\delta_0^r + a_0^{(m)}(\dot{b}_{l/m} + \frac{l}{m}\Omega c_{l/m})\delta_l^k\delta_0^r\right],\\
g_{1(2)}^{(m,c)}(i,j,l,k,r) &= \frac{1}{N}b_{i/m}\left(\dot{b}_{l/m} + \frac{l}{m}\Omega c_{l/m}\right)(\delta_{|l-i|}^k + \delta_{l+i}^k)\delta_0^r,\\
g_{1(3)}^{(m,c)}(i,j,l,k,r) &= \frac{1}{N}c_{i/m}\left(\dot{c}_{k/m} - \frac{l}{m}\Omega b_{l/m}\right)(\delta_{|l-i|}^k - \delta_{l+i}^k)\delta_0^r,\\
g_{1(4)}^{(m,c)}(i,j,l,k,r) &= \frac{2}{N^2}\dot{a}_0^{(m)}a_0^{(m)}\delta_i^k\delta_i^r + \frac{1}{N}\dot{a}_0^{(m)}b_{j/m}\delta_i^r(\delta_{|i-j|}^k + \delta_{i+j}^k)\delta_i^r,\\
g_{1(5)}^{(m,c)}(i,j,l,k,r) &= \frac{1}{N}a_0^{(m)}\left(\dot{b}_{l/m} + \frac{l}{m}\Omega c_{l/m}\right)(\delta_{|l-i|}^k + \delta_{l+i}^k)\delta_i^r,\\
g_{1(6)}^{(m,c)}(i,j,l,k,r) &= -\frac{1}{N}\frac{l}{m}\Omega a_0^{(m)}c_{i/m}(\delta_{|l-i|}^k - \delta_{l+i}^k)\delta_k^r,
\end{aligned}
$$

$$
\begin{aligned}
g_{1(7)}^{(m,c)}(i,j,l,k,r) &= \frac{1}{2}b_{j/m}\left(\dot{b}_{l/m} + \frac{l}{m}\Omega c_{l/m}\right)(\delta^k_{|l-i-j|} + \delta^k_{l+i+j})\delta^r_i,\\
g_{1(8)}^{(m,c)}(i,j,l,k,r) &= \frac{1}{2}c_{j/m}\left(\dot{c}_{l/m} - \frac{l}{m}\Omega b_{l/m}\right)(\delta^k_{|l-i-j|} - \delta^k_{l+i+j})\delta^r_i,\\
g_{1(9)}^{(m,c)}(i,j,l,k,r) &= -\frac{1}{4}\frac{l}{m}\Omega(b_{j/m}c_{i/m} + b_{i/m}c_{j/m})(\delta^k_{|l-i-j|} - \delta^k_{l+i+j})\delta^r_l,\\
g_{1(10)}^{(m,c)}(i,j,l,k,r) &= \frac{1}{2}b_{j/m}\left(\dot{b}_{l/m} + \frac{l}{m}\Omega c_{l/m}\right)(\delta^k_{|l-i+j|} + \delta^k_{|l+i-j|})\delta^r_i,\\
g_{1(11)}^{(m,c)}(i,j,l,k,r) &= -\frac{1}{4}\frac{l}{m}\Omega(b_{j/m}c_{i/m} - b_{i/m}c_{j/m})(\delta^k_{|l-i+j|} - \delta^k_{|l+i-j|})\delta^r_l,\\
g_{1(12)}^{(m,c)}(i,j,l,k,r) &= \frac{1}{N^2}\frac{l}{m}\Omega(a_0^{(m)})^2\delta^k_l\delta^r_{l+N} + \frac{1}{N}\dot{a}_0^{(m)}c_{j/m}(\delta^k_{|i-j|} - \delta^k_{i+j})\delta^r_{i+N},\\
g_{1(13)}^{(m,c)}(i,j,l,k,r) &= \frac{1}{N}\frac{k}{m}\Omega a_0^{(m)}b_{i/m}(\delta^k_{|l-i|} + \delta^k_{l+i})\delta^r_{l+N},\\
g_{1(14)}^{(m,c)}(i,j,l,k,r) &= -\frac{1}{2}c_j\left(\dot{b}_{l/m} + \frac{l}{m}\Omega c_{l/m}\right)(\delta^k_{|l-i-j|} + \delta^k_{l+i+j})\delta^r_{i+N},\\
g_{1(15)}^{(m,c)}(i,j,l,k,r) &= \frac{1}{4}\frac{l}{m}\Omega(b_{i/m}b_{j/m} - c_{i/m}c_{j/m})(\delta^k_{|l-i-j|} + \delta^k_{l+i+j})\delta^r_{l+N},\\
g_{1(16)}^{(m,c)}(i,j,l,k,r) &= \frac{1}{2}b_{j/m}\left(\dot{c}_{l/m} - \frac{l}{m}\Omega b_{l/m}\right)(\delta^k_{|l-i-j|} - \delta^k_{l+i+j})\delta^r_{i+N},\\
g_{1(17)}^{(m,c)}(i,j,l,k,r) &= \frac{1}{2}c_{j/m}\left(\dot{b}_{l/m} + \frac{l}{m}\Omega c_{l/m}\right)(\delta^k_{|l-i+j|} + \delta^k_{|l+i-j|})\delta^r_{i+N},\\
g_{1(19)}^{(m,c)}(i,j,l,k,r) &= \frac{1}{4}\frac{l}{m}\Omega(b_{i/m}b_{j/m} + c_{i/m}c_{j/m})(\delta^k_{|l-i+j|} + \delta^k_{|l+i-j|})\delta^r_{l+N};
\end{aligned}
\tag{3.89}
$$

$$
\begin{aligned}
g_{2(1)}^{(m,c)}(i,j,l,k,r) &= \frac{3}{N^2}\left[\left(a_0^{(m)}\right)^2\delta^r_l + \frac{2}{N^2}a_0^{(m)}b_{l/m}\delta^r_0\right]\delta^k_l,\\
g_{2(2)}^{(m,c)}(i,j,l,k,r) &= \frac{3}{2N}(b_{l/m}b_{j/m}\delta^r_0 + 2a_0^{(m)}b_{j/m}\delta^r_l)(\delta^k_{|l-j|} + \delta^k_{l+j}),\\
g_{2(3)}^{(m,c)}(i,j,l,k,r) &= \frac{3}{2N}(c_{l/m}c_{j/m}\delta^r_0 + a_0^{(m)}c_{j/m}\delta^r_{l+N})(\delta^k_{|l-j|} - \delta^k_{l+j}),\\
g_{2(4)}^{(m,c)}(i,j,l,k,r) &= \frac{3}{4}b_{j/m}b_{i/m}\delta^r_l(\delta^k_{|l-j-i|} + \delta^k_{l+j+i} + \delta^k_{|l-j+i|} + \delta^k_{|l+j-i|}),\\
g_{2(5)}^{(m,c)}(i,j,l,k,r) &= \frac{3}{4}(c_{j/m}c_{i/m}\delta^r_l + 2b_{l/m}c_{i/m}\delta^r_{j+N})\\
&\quad\times(\delta^k_{|l+j-i|} - \delta^k_{l+j+i} + \delta^k_{|l-j+i|} - \delta^k_{|l-j-i|}).
\end{aligned}
\tag{3.90}
$$

The corresponding nonlinear function terms for sine are

$$
\begin{aligned}
g_{1kr}^{(m,s)} &= \sum_{n=1}^{21}\sum_{l=1}^{N}\sum_{j=1}^{N}\sum_{i=1}^{N} g_{1(n)}^{(m,s)}(i,j,l,k,r),\\
g_{2kr}^{(m,s)} &= \sum_{n=1}^{4}\sum_{l=1}^{N}\sum_{j=1}^{N}\sum_{i=1}^{N} g_{2(n)}^{(m,s)}(i,j,l,k,r),
\end{aligned}
\tag{3.91}
$$

with

$$g_{1(1)}^{(m,s)}(i,j,l,k,r) = \frac{2}{N^2}\left[\dot{a}_0^{(m)} c_{i/m}\delta_i^k\delta_0^r + a_0^{(m)}(\dot{c}_{l/m} - \frac{l}{m}\Omega b_{l/m})\delta_l^k\delta_0^r\right],$$

$$g_{1(2)}^{(m,s)}(i,j,l,k,r) = \frac{1}{N}b_{i/m}\left(\dot{c}_{l/m} - \frac{l}{m}\Omega b_{l/m}\right)(\delta_{l+i}^k + \text{sgn}(l-i)\delta_{|l-i|}^k)\delta_0^r,$$

$$g_{1(3)}^{(m,s)}(i,j,l,k,r) = \frac{1}{N}c_{i/m}\left(\dot{b}_{l/m} + \frac{l}{m}\Omega c_{l/m}\right)(\delta_{l+i}^k - \text{sgn}(l-i)\delta_{|l-i|}^k)\delta_0^r,$$

$$g_{1(4)}^{(m,s)}(i,j,l,k,r) = -\frac{1}{N^2}\frac{l}{m}\Omega(a_0^{(m)})^2\delta_l^k\delta_l^r + \frac{1}{2N}\dot{a}_0^{(m)} c_{i/m}(\text{sgn}(i-j)\delta_{|i-j|}^k + \delta_{i+j}^k)\delta_j^r,$$

$$g_{1(5)}^{(m,s)}(i,j,l,k,r) = \frac{1}{2N}\dot{a}_0^{(m)} c_{j/m}(\delta_{i+j}^k - \text{sgn}(i-j)\delta_{|i-j|}^k)\delta_i^r,$$

$$g_{1(6)}^{(m,s)}(i,j,l,k,r) = \frac{1}{N}a_0^{(m)}\left(\dot{c}_{l/m} - \frac{l}{m}\Omega b_{l/m}\right)(\delta_{l+i}^k + \text{sgn}(l-i)\delta_{|l-i|}^k)\delta_i^r,$$

$$g_{1(7)}^{(m,s)}(i,j,l,k,r) = -\frac{1}{N}\frac{l}{m}\Omega a_0^{(m)} b_{i/m}[\delta_{l+i}^k + \text{sgn}(l-i)\delta_{|l-i|}^k]\delta_l^r,$$

$$g_{1(8)}^{(m,s)}(i,j,l,k,r) = \frac{1}{2}c_{j/m}\left(\dot{b}_{l/m} + \frac{l}{m}\Omega c_{l/m}\right)(\delta_{l+i+j}^k - \text{sgn}(l-i-j)\delta_{|l-i-j|}^k]\delta_i^r,$$

$$g_{1(9)}^{(m,s)}(i,j,l,k,r) = \frac{1}{2}b_{j/m}\left(\dot{c}_{l/m} - \frac{l}{m}\Omega b_{l/m}\right)[\delta_{l+i+j}^k + \text{sgn}(l-i-j)\delta_{|l-i-j|}^k]\delta_i^r,$$

$$g_{1(10)}^{(m,s)}(i,j,l,k,r) = -\frac{1}{4}\frac{l}{m}\Omega(b_{i/m}b_{j/m} - c_{i/m}c_{j/m}) \times [\delta_{l+i+j}^k + \text{sgn}(l-i-j)\delta_{|l-i-j|}^k]\delta_l^r,$$

$$g_{1(11)}^{(m,s)}(i,j,l,k,r) = \frac{1}{4}b_{j/m}\left(\dot{c}_{l/m} - \frac{l}{m}\Omega b_{l/m}\right)[\text{sgn}(l-i+j)\delta_{|l-i+j|}^k + \text{sgn}(l+i-j)\delta_{|l+i-j|}^k]\delta_i^r,$$

$$g_{1(12)}^{(m,s)}(i,j,l,k,r) = -\frac{1}{4}\frac{l}{m}\Omega(b_{i/m}b_{j/m} + c_{i/m}c_{j/m})[\text{sgn}(l-i+j)\delta_{|l-i+j|}^k + \text{sgn}(l+i-j)\delta_{|l+i-j|}^k]\delta_l^r,$$

$$g_{1(13)}^{(m,s)}(i,j,l,k,r) = \frac{2}{N^2}\dot{a}_0^{(m)} a_0^{(m)}\delta_i^k\delta_{i+N}^r + \frac{1}{2N}\dot{a}_0^{(m)} b_{j/m}[\text{sgn}(i-j)\delta_{|i-j|}^k + \delta_{i+j}^k]\delta_{i+N}^q,$$

$$g_{1(14)}^{(m,s)}(i,j,l,k,r) = \frac{1}{2N}\dot{a}_0^{(m)} b_{i/m}[\delta_{i+j}^k - \text{sgn}(i-j)\delta_{|i-j|}^k]\delta_{j+N}^r,$$

$$g_{1(15)}^{(m,s)}(i,j,l,k,r) = \frac{1}{N}a_0^{(m)}\left(\dot{b}_{l/m} + \frac{l}{m}\Omega c_{l/m}\right)[\delta_{l+i}^k - \text{sgn}(l-i)\delta_{|l-i|}^k]\delta_{i+N}^r,$$

$$g_{1(16)}^{(m,s)}(i,j,l,k,r) = \frac{1}{N}\frac{l}{m}\Omega a_0^{(m)} c_{i/m}[\delta_{l+i}^k - \text{sgn}(l-i)\delta_{|l-i|}^k]\delta_{l+N}^r,$$

$$
\begin{aligned}
g_{1(17)}^{(m,s)}(i,j,l,k,r) &= \frac{1}{2}b_{j/m}\left(\dot{b}_{l/m} + \frac{l}{m}\Omega c_{l/m}\right) \\
&\quad \times [\delta_{l+i+j}^{k} - \operatorname{sgn}(l-i-j)\delta_{|l-i-j|}^{k}]\delta_{i+N}^{r}, \\
g_{1(18)}^{(m,s)}(i,j,l,k,r) &= \frac{1}{4}\frac{l}{m}\Omega(b_{i/m}c_{j/m} + b_{j/m}c_{i/m}) \\
&\quad \times [\delta_{l+i+j}^{k} - \operatorname{sgn}(l-i-j)\delta_{|l-i-j|}^{k}]\delta_{l+N}^{r}, \\
g_{1(19)}^{(m,s)}(i,j,l,k,r) &= -\frac{1}{2}c_{j/m}\left(\dot{c}_{l/m} - \frac{l}{m}\Omega b_{l/m}\right) \\
&\quad \times [\delta_{l+i+j}^{k} + \operatorname{sgn}(l-i-j)\delta_{|l-i-j|}^{k}]\delta_{i+N}^{r}, \\
g_{1(20)}^{(m,s)}(i,j,l,k,r) &= \frac{1}{2}c_{j/m}\left(\dot{c}_{l/m} - \frac{l}{m}\Omega b_{l/m}\right)[\operatorname{sgn}(l-i+j)\delta_{|l-i+j|}^{k} \\
&\quad + \operatorname{sgn}(l+i-j)\delta_{|l+i-j|}^{k}]\delta_{i+N}^{r}, \\
g_{1(21)}^{(m,s)}(i,j,l,k,r) &= \frac{1}{4}\frac{l}{m}\Omega(b_{i/m}c_{j/m} - b_{j/m}c_{i/m})[\operatorname{sgn}(l-i+j)\delta_{|l-i+j|}^{k}, \\
&\quad - \operatorname{sgn}(l+i-j)\delta_{|l+i-j|}^{k}]\delta_{l+N}^{r};
\end{aligned}
\tag{3.92}
$$

and

$$
\begin{aligned}
g_{2(1)}^{(m,s)}(i,j,l,k,r) &= 3\frac{a_0^{(m)}}{N^2}\left[a_0^{(m)}\delta_{l+N}^{r} + 2c_{l/m}\delta_0^{r}\right]\delta_l^{k}, \\
g_{2(2)}^{(m,s)}(i,j,l,k,r) &= \frac{3}{N}(a_0^{(m)}c_{j/m}\delta_l^{r} + a_0^{(m)}b_{l/m}\delta_{j+N}^{r} + b_{l/m}c_{j/m}\delta_0^{r}) \\
&\quad \times [\delta_{l+j}^{k} - \operatorname{sgn}(l-j)\delta_{|l-j|}^{k}], \\
g_{2(3)}^{(m,s)}(i,j,l,k,r) &= \frac{3}{4}c_{j/m}c_{i/m}\delta_{l+N}^{r}[\operatorname{sgn}(l-j+i)\delta_{|l-j+i|}^{k} - \delta_{l+j+i}^{k} \\
&\quad + \operatorname{sgn}(l+j-i)\delta_{|l+j-i|}^{k} - \operatorname{sgn}(l-j-i)\delta_{|l-j-i|}^{k}], \\
g_{2(4)}^{(m,s)}(i,j,l,k,r) &= \frac{3}{4}(b_{l/m}b_{j/m}\delta_{i+N}^{r} + 2b_{j/m}c_{i/m}\delta_l^{r})[\operatorname{sgn}(l-j+i)\delta_{|l-j+i|}^{k} + \delta_{l+j+i}^{k} \\
&\quad - \operatorname{sgn}(l+j-i)\delta_{|l+j-i|}^{k} - \operatorname{sgn}(l-j-i)\delta_{|l-j-i|}^{k})]
\end{aligned}
\tag{3.93}
$$

for $r = 0, 1, \ldots, 2N$.

$$
\mathbf{H} = \frac{\partial \mathbf{g}^{(m)}}{\partial \mathbf{z}_1^{(m)}} = (\mathbf{H}^{(0)}, \mathbf{H}^{(c)}, \mathbf{H}^{(s)})^{\mathrm{T}}. \tag{3.94}
$$

where

$$
\begin{aligned}
\mathbf{H}^{(0)} &= (H_0^{(0)}, H_1^{(0)}, \ldots, H_{2N}^{(0)}), \\
\mathbf{H}^{(c)} &= (\mathbf{H}_1^{(c)}, \mathbf{H}_2^{(c)}, \ldots, \mathbf{H}_N^{(c)})^{\mathrm{T}}, \\
\mathbf{H}^{(s)} &= (\mathbf{H}_1^{(s)}, \mathbf{H}_2^{(s)}, \ldots, \mathbf{H}_N^{(s)})^{\mathrm{T}}
\end{aligned}
\tag{3.95}
$$

for $N = 1, 2, \ldots \infty$, with

$$\mathbf{H}_k^{(c)} = (H_{k0}^{(c)}, H_{k1}^{(c)}, \ldots, H_{k(2N)}^{(c)}),$$
$$\mathbf{H}_k^{(s)} = (H_{k0}^{(s)}, H_{k1}^{(s)}, \ldots, H_{k(2N)}^{(s)}) \tag{3.96}$$

for $k = 1, 2, \ldots N$. The corresponding components are

$$H_r^{(0)} = \alpha_1 \delta_0^r - \alpha_2 h_{1r}^{(m,0)}$$
$$H_{kr}^{(c)} = -2\left(\frac{k\Omega}{m}\right)\delta_{k+N}^r + \alpha_1 \delta_k^r - \alpha_2 h_{1kr}^{(m,c)},$$
$$H_{kr}^{(s)} = 2\left(\frac{k\Omega}{m}\right)\delta_k^r + \alpha_1 \delta_{k+N}^r - \alpha_2 h_{1kr}^{(m,s)} \tag{3.97}$$

for $r = 0, 1, \ldots, 2N$.

$$h_{1r}^{(m,0)} = (a_0^{(m)})^2 \delta_0^r + \sum_{n=1}^{6}\sum_{l=1}^{N}\sum_{j=1}^{N}\sum_{i=1}^{N} h_{1(n)}^{(m,0)}(i,j,l,r) \tag{3.98}$$

with

$$h_{1(1)}^{(m,0)}(i,j,l,r) = \frac{1}{2N} b_{i/m} b_{j/m} \delta_{i-j}^0 \delta_0^r,$$
$$h_{1(2)}^{(m,0)}(i,j,l,r) = \frac{1}{2N} c_{i/m} c_{j/m} \delta_{i-j}^0 \delta_0^r + \frac{1}{N} a_0^{(m)} b_{i/m} \delta_{l-i}^0 \delta_l^r,$$
$$h_{1(3)}^{(m,0)}(i,j,l,r) = \frac{1}{4}(b_{i/m} b_{j/m} - c_{i/m} c_{j/m}) \delta_{l-i-j}^0 \delta_l^r,$$
$$h_{1(4)}^{(m,0)}(i,j,l,r) = \frac{1}{4}(b_{i/m} b_{j/m} + c_{i/m} c_{j/m})(\delta_{l-i+j}^0 + \delta_{l+i-j}^0)\delta_l^r,$$
$$h_{1(5)}^{(m,0)}(i,j,l,r) = \frac{a_0}{N} c_{i/m} \delta_{l-i}^0 \delta_{l+N}^r + \frac{1}{4}(b_{j/m} c_{i/m} + b_{i/m} c_{j/m}) \delta_{l-i-j}^0 \delta_{l+N}^r,$$
$$h_{1(6)}^{(m,0)}(i,j,l,r) = \frac{1}{4}(b_{j/m} c_{i/m} - b_{i/m} c_{j/m})(\delta_{l-i+j}^0 + \delta_{l+i-j}^0)\delta_{l+N}^r; \tag{3.99}$$

and

$$h_{1kr}^{(m,c)} = \sum_{n=1}^{9}\sum_{l=1}^{N}\sum_{j=1}^{N}\sum_{i=1}^{N} h_{1(n)}^{(m,c)}(i,j,l,k,r) \tag{3.100}$$

with

$$h_{1(1)}^{(m,c)}(i,j,l,k,r) = \frac{2}{N^2} a_0^{(m)} b_{i/m} \delta_l^k \delta_0^r,$$
$$h_{1(2)}^{(m,c)}(i,j,l,k,r) = \frac{1}{2N} b_{i/m} b_{j/m} (\delta_{|l-i|}^k + \delta_{l+i}^k)\delta_0^r,$$
$$h_{1(3)}^{(m,c)}(i,j,l,k,r) = \frac{1}{2N} c_{i/m} c_{j/m} (\delta_{|l-i|}^k - \delta_{l+i}^k)\delta_0^r,$$
$$h_{1(4)}^{(m,c)}(i,j,l,k,r) = \frac{1}{N^2}(a_0^{(m)})^2 \delta_l^k \delta_l^r + \frac{1}{N} a_0^{(m)} b_{i/m} (\delta_{|l-i|}^k + \delta_{l+i}^k)\delta_l^r,$$

$$
\begin{aligned}
h_{1(5)}^{(m,c)}(i,j,l,k,r) &= \frac{1}{4}(b_{i/m}b_{j/m} - c_{i/m}c_{j/m})(\delta_{|l-i-j|}^{k} + \delta_{l+i+j}^{k})\delta_{l}^{r},\\
h_{1(6)}^{(m,c)}(i,j,l,k,r) &= \frac{1}{4}(b_{i/m}b_{j/m} + c_{i/m}c_{j/m})(\delta_{|l-i+j|}^{k} + \delta_{|l+i-j|}^{k})\delta_{l}^{r},\\
h_{1(7)}^{(m,c)}(i,j,l,k,r) &= \frac{1}{N}a_{0}^{(m)}c_{i/m}(\delta_{|l-i|}^{k} - \delta_{l+i}^{k})\delta_{l+N}^{r},\\
h_{1(8)}^{(m,c)}(i,j,l,k,r) &= \frac{1}{4}(b_{j/m}c_{i/m} + b_{i/m}c_{j/m})(\delta_{|l-i-j|}^{k} - \delta_{l+i+j}^{k})\delta_{l+N}^{r},\\
h_{1(9)}^{(m,c)}(i,j,l,k,r) &= \frac{1}{4}(b_{j/m}c_{i/m} - b_{i/m}c_{j/m})(\delta_{|l-i+j|}^{k} - \delta_{|l+i-j|}^{k})\delta_{l+N}^{r};
\end{aligned} \tag{3.101}
$$

and

$$
h_{1kr}^{(m,s)} = \sum_{n=1}^{6}\sum_{l=1}^{N}\sum_{j=1}^{N}\sum_{i=1}^{N} h_{1(n)}^{(m,s)}(i,j,l,k,r) \tag{3.102}
$$

with

$$
\begin{aligned}
h_{1(1)}^{(m,s)}(i,j,l,k,r) &= \frac{2}{N^2}a_{0}^{(m)}c_{i/m}\delta_{i}^{k}\delta_{0}^{r},\\
h_{1(2)}^{(m,s)}(i,j,l,k,r) &= \frac{1}{2N}b_{j/m}c_{i/m}[\text{sgn}(i-j)\delta_{|i-j|}^{k} + \delta_{i+j}^{k}]\delta_{0}^{r},\\
h_{1(3)}^{(m,s)}(i,j,l,k,r) &= \frac{1}{2N}b_{i/m}c_{j/m}[\delta_{i+j}^{k} - \text{sgn}(i-j)\delta_{|i-j|}^{k}]\delta_{0}^{r},\\
h_{1(4)}^{(m,s)}(i,j,l,k,r) &= \frac{1}{N}a_{0}^{(m)}c_{i/m}[\delta_{l+i}^{k} - \text{sgn}(l-i)\delta_{|l-i|}^{k}]\delta_{l}^{r},\\
h_{1(5)}^{(m,s)}(i,j,l,k,r) &= \frac{1}{4}(b_{i/m}c_{j/m} + b_{j/m}c_{i/m})[\delta_{l+i+j}^{k} - \text{sgn}(l-i-j)\delta_{|l-i-j|}^{k}]\delta_{l}^{r},\\
h_{1(6)}^{(m,s)}(i,j,l,k,r) &= \frac{1}{4}(b_{i/m}c_{j/m} - b_{j/m}c_{i/m})[\text{sgn}(l-i+j)\delta_{|l-i+j|}^{k},\\
&\quad - \text{sgn}(l+i-j)\delta_{|l+i-j|}^{k}]\delta_{l}^{r},
\end{aligned} \tag{3.103}
$$

From Luo (2012a); the eigenvalues of $D\mathbf{f}(\mathbf{y}^{(m)*})$ are classified as

$$
(n_1, n_2, n_3 | n_4, n_5, n_6). \tag{3.104}
$$

The corresponding boundary between the stable and unstable solution is given by the saddle-node bifurcation and Hopf bifurcation.

3.2.2 Analytical Predictions

As in Luo and Lakeh (2013b), the harmonic amplitude varying with excitation frequency Ω are illustrated. The harmonic amplitude and phase are defined by

$$
A_{k/m} \equiv \sqrt{b_{k/m}^2 + c_{k/m}^2} \text{ and } \varphi_{k/m} = \arctan\frac{c_{k/m}}{b_{k/m}}. \tag{3.105}
$$

The corresponding solution in Equation (3.55) becomes

$$x^*(t) = a_0^{(m)} + \sum_{k=1}^{N} A_{k/m} \cos\left(\frac{k}{m}\Omega t + \varphi_{k/m}\right). \tag{3.106}$$

Consider system parameters as

$$\alpha_1 = 1.0, \alpha_2 = 1.0, \alpha_3 = 1.0, \alpha_4 = 2.5. \tag{3.107}$$

Through the aforesaid parameters, the independent bifurcation trees of symmetric and asymmetric periodic motions to chaos can be observed and there are quasi-periodic or chaotic motions in a gap between two bifurcation trees. In addition to bifurcation trees, independent periodic motions also exist as in the periodically forced, van der Pol-Duffing oscillator, and such periodic motions are bounded between two saddle-node bifurcations, which are embedded in quasi-periodic motion or chaotic motion. Solid and dashed curves represent stable and unstable periodic solutions, respectively.

The frequency-amplitude curves for an independent period-4 motion within the range of $\Omega \in (11.2263, 11.4966)$ with $Q_0 = 50.0$ are presented in Figure 3.13(i)–(vi) for $a_0^{(m)}$ and $A_{k/4}$ $(k = 1, 2, 3, 4, 28)$. In Figure 3.13(i), the constant term $a_0^{(m)}$ is presented for $m = 4$. There are two sets of period-4 motions in the left and right sides of the y-axis. For the two sets of periodic motions, the other harmonic amplitudes are the same, as shown in Figure 3.13(ii)–(vi). However, the harmonic phase are different (i.e., $\varphi^L_{k/m} = \varphi^R_{k/m} + (k+1)\pi$). The main harmonic amplitudes are in the ranges of $A_{1/4} \in (1.74, 1.82)$, $A_{1/2} \in (0.12, 0.18)$, $A_{3/4} \in (0.087, 0.99)$, $A_1 \in (0.41, 0.44)$, and $A_7 \in (10^{-9}, 3 \times 10^{-9})$. The harmonic amplitude $A_{1/4}$ plays an important role in period-4 motions. For $A_{28/4} = A_7$, the harmonic amplitude is in quantity level of 10^{-9}. Thus we can keep 28 harmonic terms to get approximate period-4 motion with accuracy tolerance of $\varepsilon = 10^{-8}$. The period-4 motion has two coexisting solutions. One solution is unstable and the other is stable. The stable and unstable solutions meet and vanish at two saddle-node bifurcation points of $\Omega^{(4)}_{cr1} \approx 11.2263$ and $\Omega^{(4)}_{cr2} \approx 11.4966$, which are located at the right and left ends of frequency-amplitude curves. Beyond the two saddle-node bifurcations, quasi-periodic and chaotic motion can be observed.

An independent bifurcation tree from the period-2 motion to chaos for $\Omega \in (4.8587, 6.2359)$ with $Q_0 = 20.0$ is presented in Figure 3.14(i)–(xii) for $a_0^{(m)}$ and $A_{k/8}$ $(k = 1, 2, \ldots, 8, 16, 24, 56)$, respectively. The motions of the bifurcation tree lies between two saddle-nodes of $\Omega^{(2)}_{\mathrm{cr1}} \approx 4.8587$ and $\Omega^{(2)}_{\mathrm{cr2}} \approx 6.2359$ for period-2 motion. Therefore the period-2 motion exists in the frequency range of $\Omega \in (4.8587, 6.2359)$. Beyond the left and right ends of the period-2 motion, there exist two zones of quasi-periodic or chaos. The period-2 motion possesses two coexisting solutions. One solution between the two saddle-node points is completely unstable, and the other one has two segments of stable solutions plus one segment of unstable solution. On the stable branch of period-2 motion, the Hopf bifurcation occur at $\Omega_{\mathrm{cr}} \approx 4.975, 5.731$. Once the Hopf bifurcation of period-2 motion occurs, the stable period-2 motion become unstable, and the onset of stable period-4 motion exists. Thus the Hopf bifurcations of period-2 motion are the saddle-node bifurcations of the period-4 motion. The period-4 motion exists in $\Omega \in (4.8587, 6.2359)$. When the Hopf bifurcation of period-4 motion occurs, the stable period-4 motion becomes unstable and the onset of stable period-8 motion will appear. The Hopf bifurcations of the period-4 motion occur at $\Omega \approx 5.03, 5.47$, which are the saddle-node bifurcation of period-8 motions. Once again, the Hopf bifurcations of the stable period-8 motion occur, the stable period-8 motion becomes unstable, and the stable period-16 motion

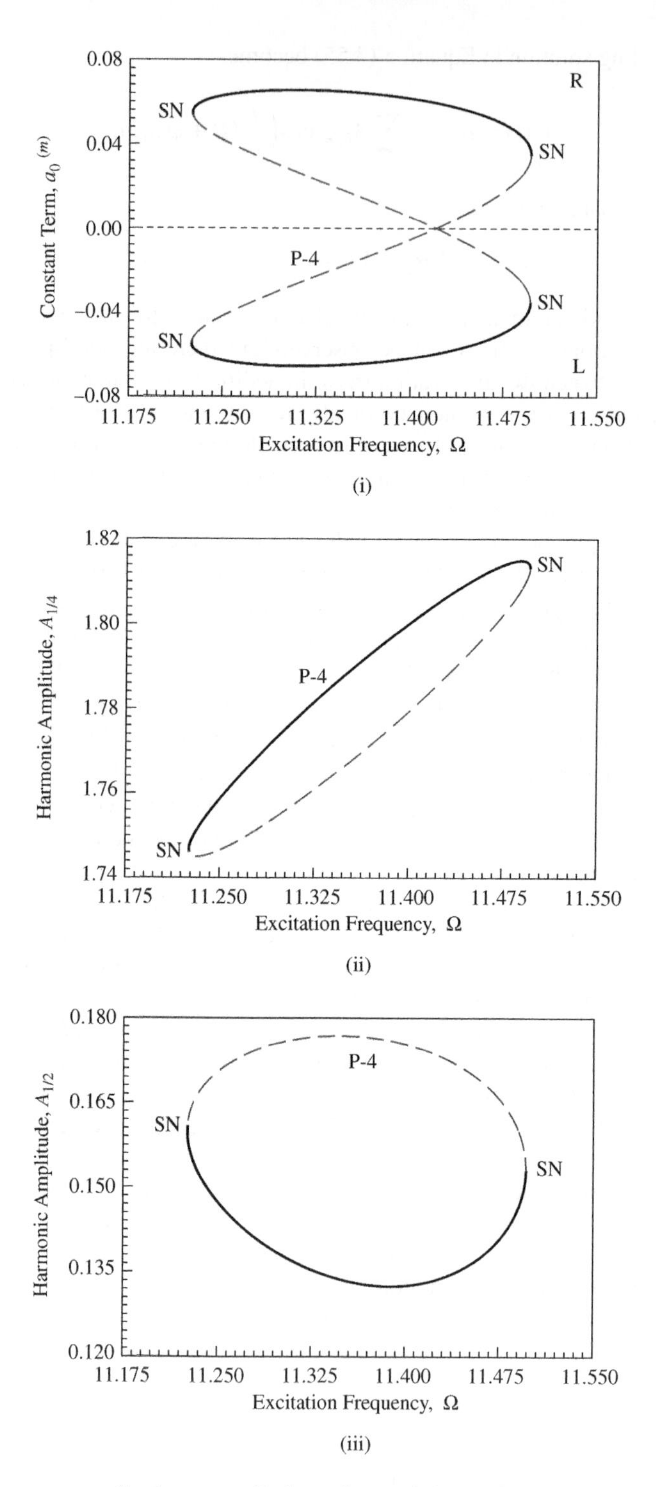

Figure 3.13 Frequency-amplitude curves of independent period-4 motions based on 28 harmonic terms (HB28) in the van der Pol-Duffing oscillator: (i) $a_0^{(m)}$ and (ii)–(vi) $A_{k/m}$ ($k = 1, 2, 3, 4, 28$, $m = 4$) ($\alpha_1 = 1.0$, $\alpha_2 = 1.0$, $\alpha_3 = 1.0$, $\alpha_4 = 2.5$, $Q_0 = 50.0$)

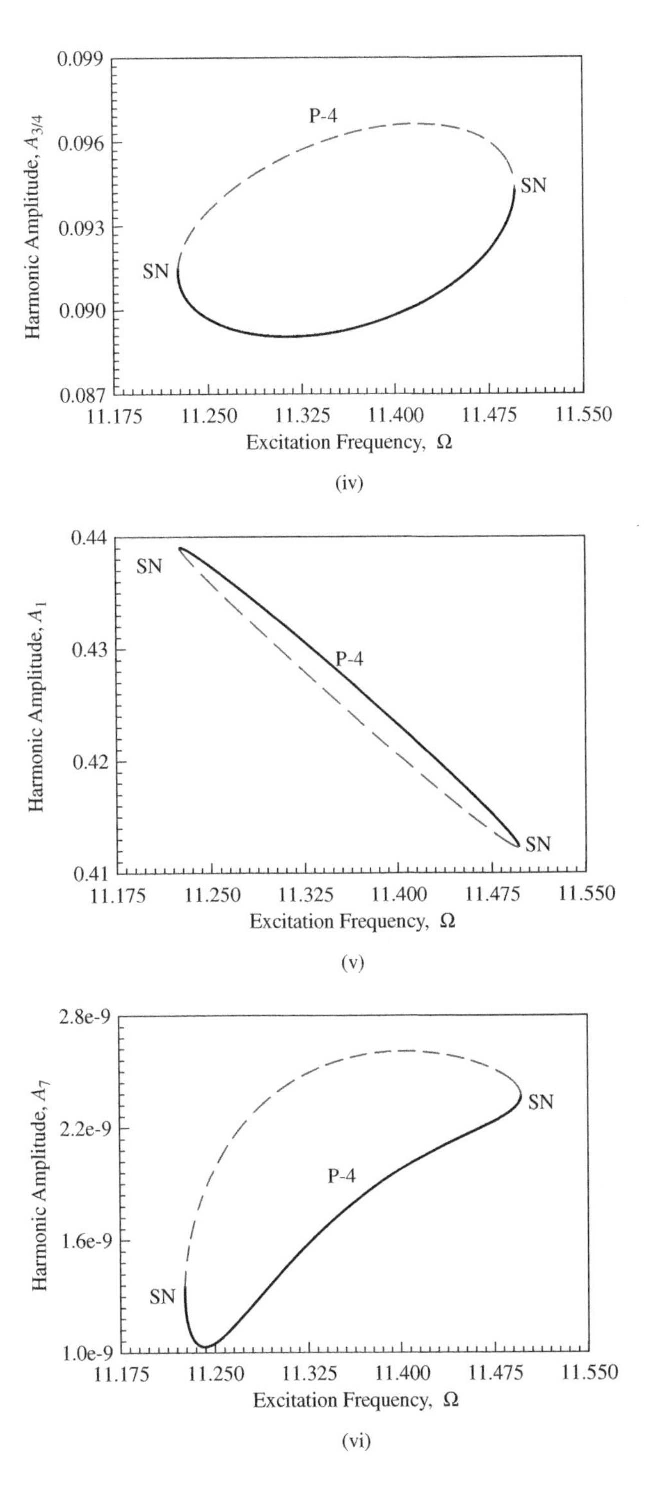

Figure 3.13 *(continued)*

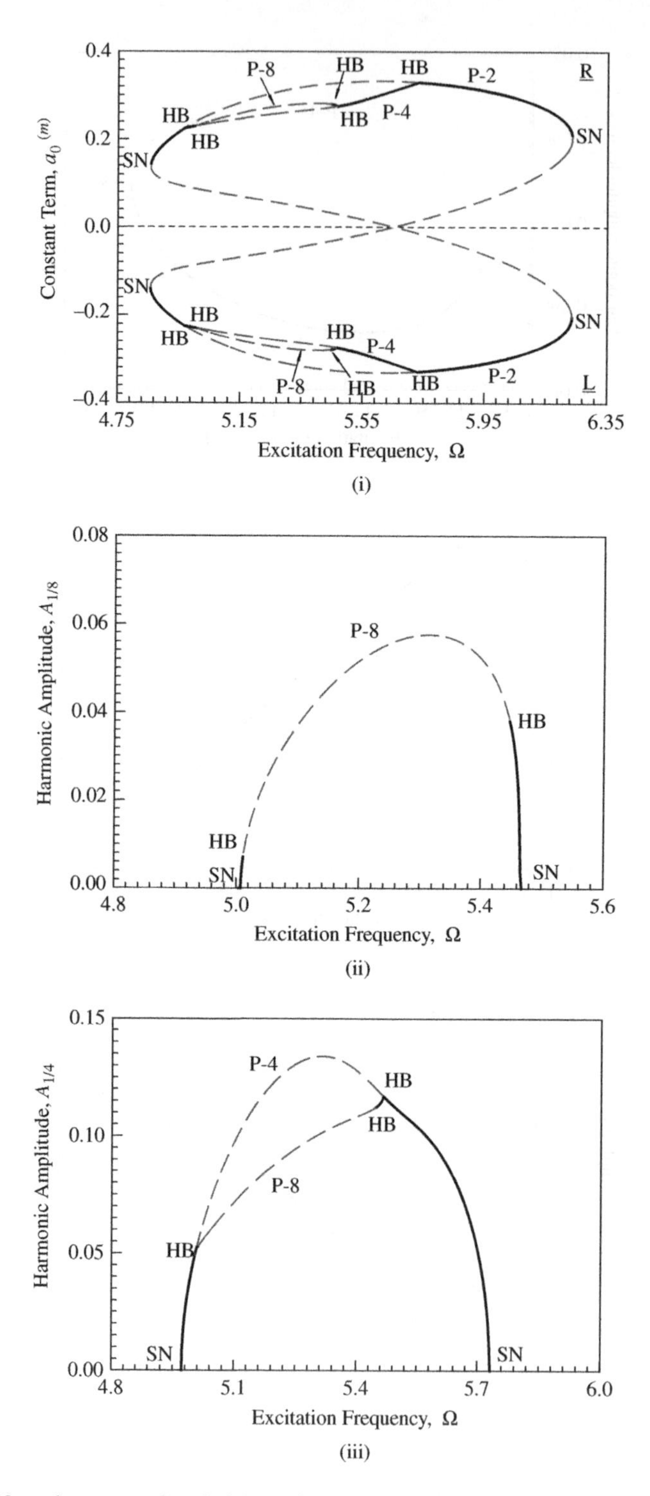

Figure 3.14 Bifurcation trees of period-2 motion to period-8 motion based on 56 harmonic terms (HB56) in the van der Pol-Duffing oscillator: (i) $a_0^{(m)}$ and (ii)–(xii) $A_{k/m}$ $(k = 1, 2, \ldots, 8, 16, 24, 56, m = 8)$ $(\alpha_1 = 1.0, \alpha_2 = 1.0, \alpha_3 = 1.0, \alpha_4 = 2.5, Q_0 = 20.0)$

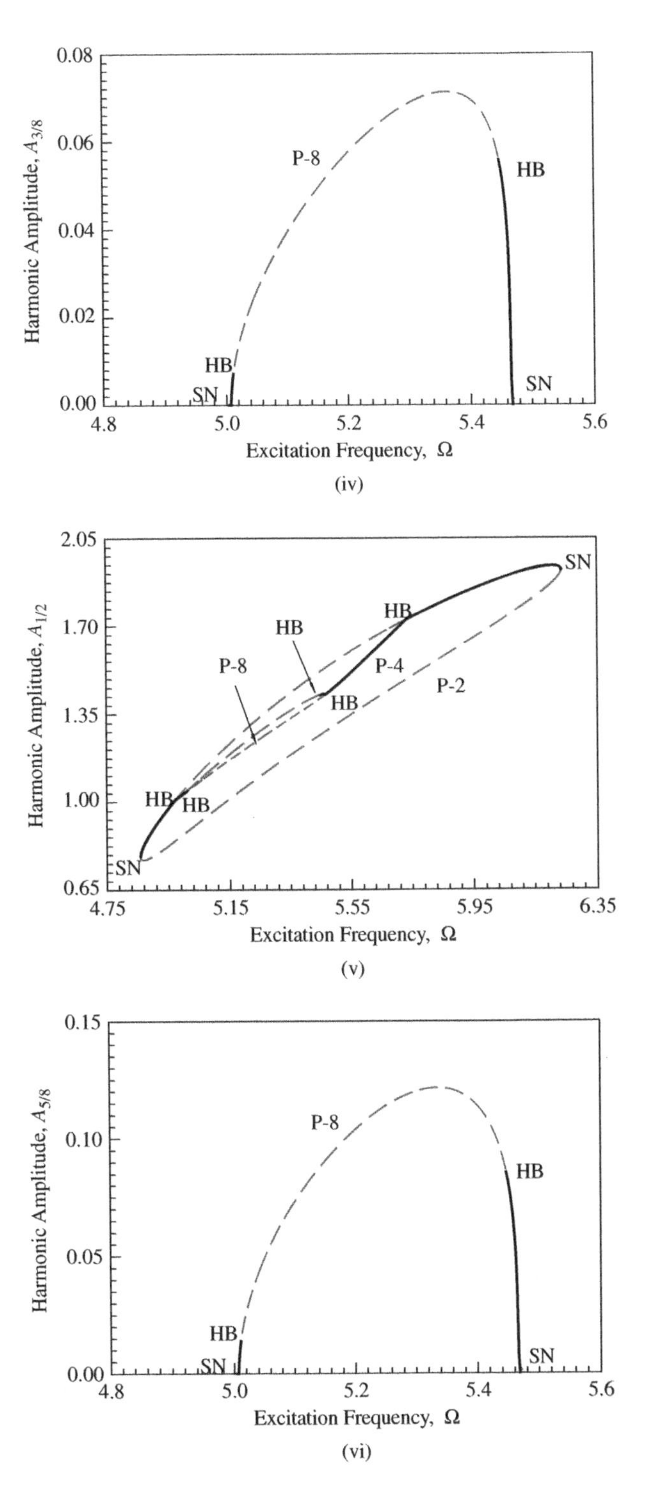

Figure 3.14 *(continued)*

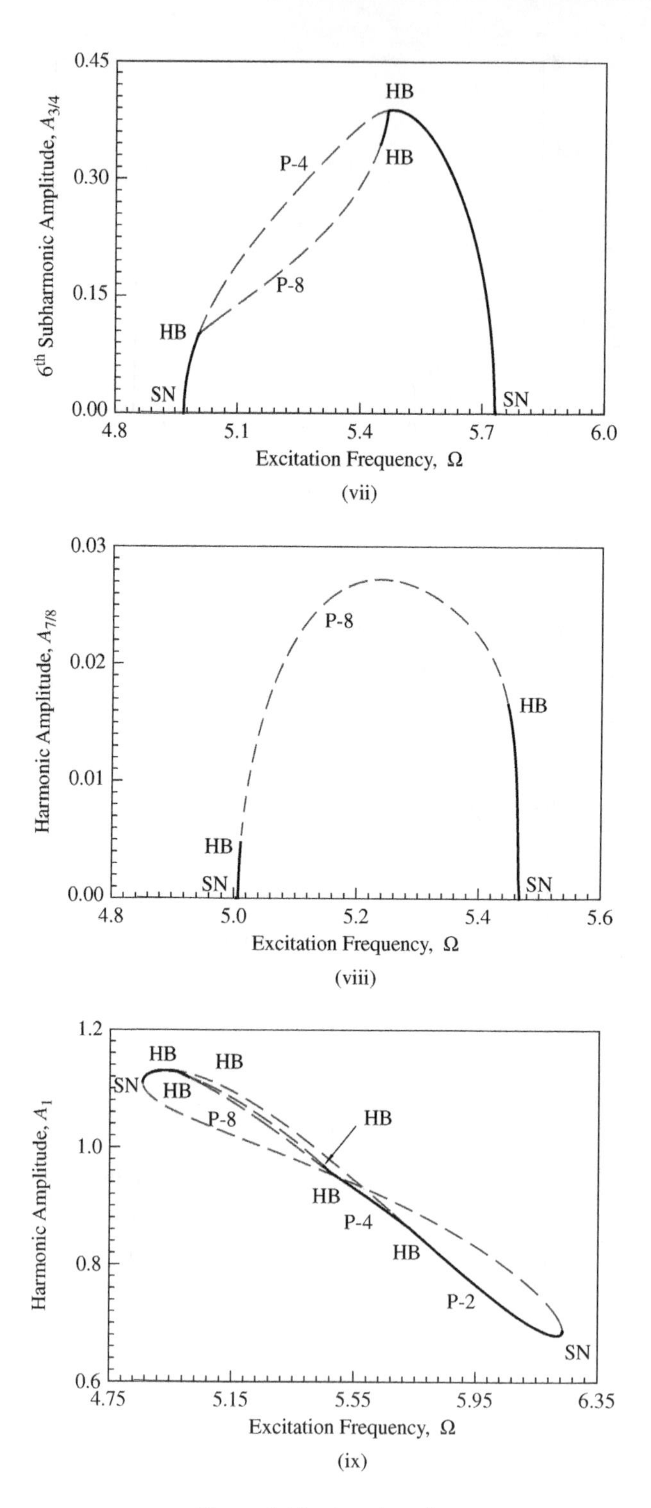

Figure 3.14 *(continued)*

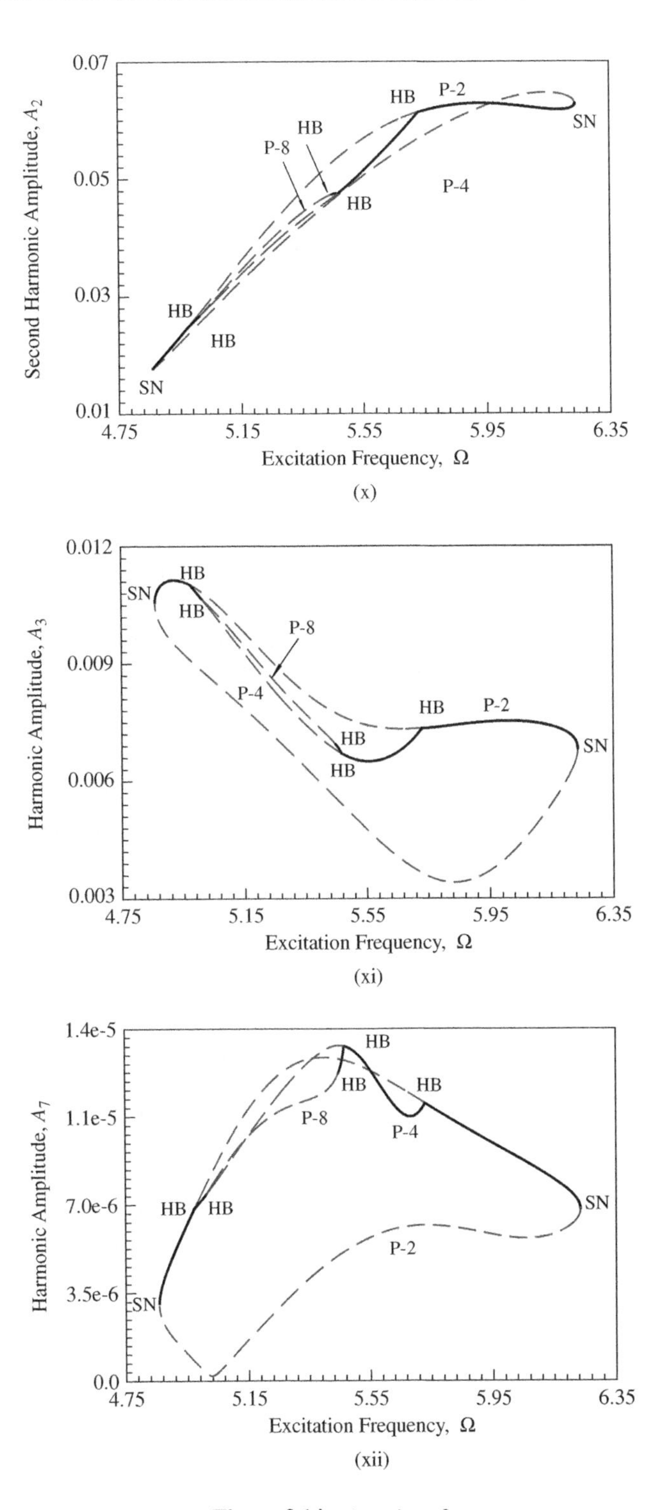

Figure 3.14 *(continued)*

will appear. The Hopf bifurcations of stable period-8 motion occur at $\Omega \approx 5.032, 5.445$. Continuously, the bifurcation tree of period-2 motion to chaos can be achieved. In Figure 3.14(i), the constant term $a_0^{(m)}$ is presented for $m = 2, 4, 8$. There are two sets of period-m motions in the left and right sides of the y-axis. For the two sets of period-m motions, the other harmonic amplitudes are the same, as shown in Figure 3.14(ii)–(xii). However, the harmonic phase are different (i.e., $\varphi_{k/m}^{L} = \varphi_{k/m}^{R} + (k+1)\pi$, $m = 2, 4, 8$). In Figure 3.14(ii), the harmonic amplitude of $A_{1/8} < 0.06$ for period-8 motion is presented and $A_{1/8} = 0$ for period-2 and period-4 motions. In Figure 3.14(iii), the harmonic amplitude $A_{1/4} < 0.15$ is presented for period-4 and period-8 motions. The harmonic amplitude of $A_{3/8} < 0.08$ similar to $A_{1/8}$ for period-8 motion only is presented in Figure 3.14(iv). In Figure 3.14(v), the harmonic amplitude of $A_{1/2} \in (0.65, 2.05)$ for period-m motions ($m = 2, 4, 8$) is presented. The harmonic amplitudes of $A_{5/8} < 0.15$ and $A_{7/8} < 0.03$ for period-8 motion are presented in Figure 3.14(vi) and (viii), respectively. Both of the two harmonics have similar frequency-amplitude curves. In Figure 3.14(vii), the harmonic amplitude of $A_{3/4} < 0.45$ is presented. The primary harmonic amplitude of $A_1 \in (0.6, 1.2)$ is presented in Figure 3.14(ix) for period-m motions ($m = 2, 4, 8$). To avoid abundant illustrations, the harmonic amplitudes of $A_2 \in (0.01, 0.07)$ and $A_3 \in (0.003, 0.012)$ are presented in Figure 3.14(x) and (xi) for period-m motions ($m = 2, 4, 8$). Finally, the harmonic amplitude of $A_7 < 1.5 \times 10^{-5}$ is presented. Similarly, the other higher-order harmonic amplitudes can be computed and illustrated.

An independent bifurcation tree of period-5 to chaos for $\Omega \in (4.58729, 5.3648)$ with $Q_0 = 20.0$ is illustrated through period-5 to period-10 motion. The constant term $a_0^{(m)}$ and harmonic amplitudes $A_{k/10}$ ($k = 1, 2, \ldots, 10, 70$) is presented in Figure 3.15(i)–(xii), respectively. The solutions of the bifurcation tree lie in the range of $\Omega \in (4.58729, 5.3648)$. The bifurcation tree of period-5 motion begins from two saddle-node bifurcations of stable symmetric period-5 motion. The saddle-node bifurcations of symmetric period-5 motion occur at $\Omega_{cr}^{(5)} \approx 4.58729, 5.3648$. Beyond the left and right ends of the period-5 motion, there exist two zones of quasi-periodic or chaos. The symmetric period-5 motion has two coexisting solutions. One solution branch is fully unstable, and another solution branch possesses stable symmetric period-5 motions. Once the saddle-node bifurcation of stable symmetric period-5 motion occur at $\Omega_{cr} \approx 4.616, 5.278$, the stable symmetric period-5 motion becomes unstable and asymmetric period-5 motion will appear. The period-5 motion possesses two coexisting asymmetric solutions on the left and right sides of y-axis. The saddle-node bifurcation of stable asymmetric period-5 motion also occur at $\Omega_{cr} \approx 4.616, 5.278$. The saddle-node bifurcation of the stable asymmetric period-5 motion in its middle segments occur at $\Omega_{cr} \approx 4.84578, 5.01559$. The Hopf bifurcation of stable asymmetric period-5 motion occurs at $\Omega_{cr} \approx 4.622, 5.23$, and the Hopf bifurcation of stable period-5 motion occurs at $\Omega_{cr} \approx 4.84595, 5.01515$. Once the Hopf bifurcation of stable period-5 motion occurs, the stable asymmetric period-5 motion become unstable and the onset of stable period-10 motion exists. Thus the Hopf bifurcations of period-5 motion are the saddle-node bifurcations of the period-10 motion. The Hopf bifurcation of stable asymmetric period-5 motion is the saddle-node bifurcation of period-10 motion. The Hopf bifurcations of period-10 motion are at $\Omega_{cr} \approx 4.623, 4.84597, 5.01501, 5.214$. The period-10 motion becomes unstable and period-20 motion will appear. Continuously, if this bifurcation process repeats over and over, and the bifurcation tree of period-5 motion to chaos can be achieved. In Figure 3.15(i), the constant term $a_0^{(m)}$ for asymmetric period-5 motions and period-10 motion are presented. For symmetric period-5 motion, we have $a_0^{(5)} = 0$. The corresponding asymmetric period-5 motion possesses two coexisting solutions on the left and right sides of y-axis. For the two sets of periodic motions, the other harmonic amplitudes are same, as shown in Figure 3.15(ii)–(xii). However, the harmonic phase are

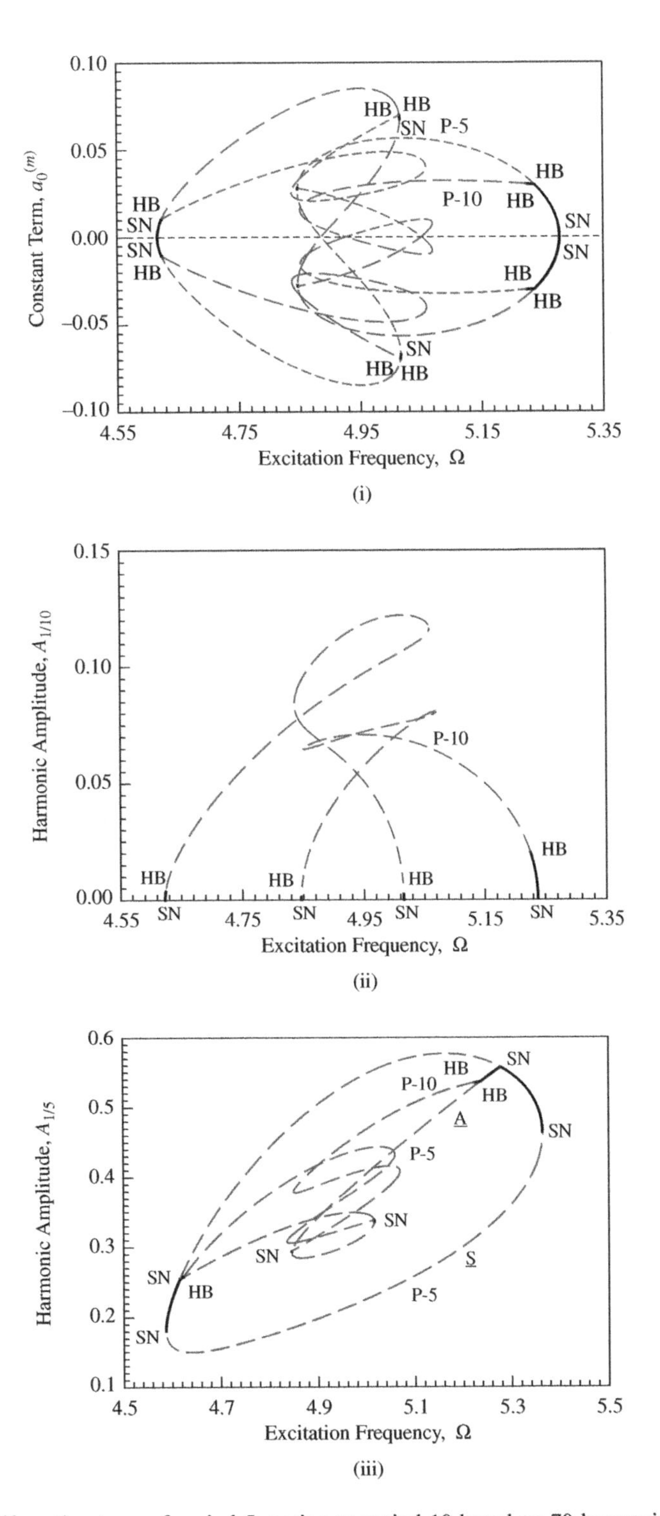

Figure 3.15 Bifurcation trees of period-5 motion to period-10 based on 70 harmonic terms (HB70) in the van der Pol-Duffing oscillator: (i) $a_0^{(m)}$ and (ii)–(xii) $A_{k/m}$ ($k = 1, 2, \ldots, 10, 70, m = 10$) ($\alpha_1 = 1.0$, $\alpha_2 = 1.0$, $\alpha_3 = 1.0$, $\alpha_4 = 2.5$, $Q_0 = 20.0$)

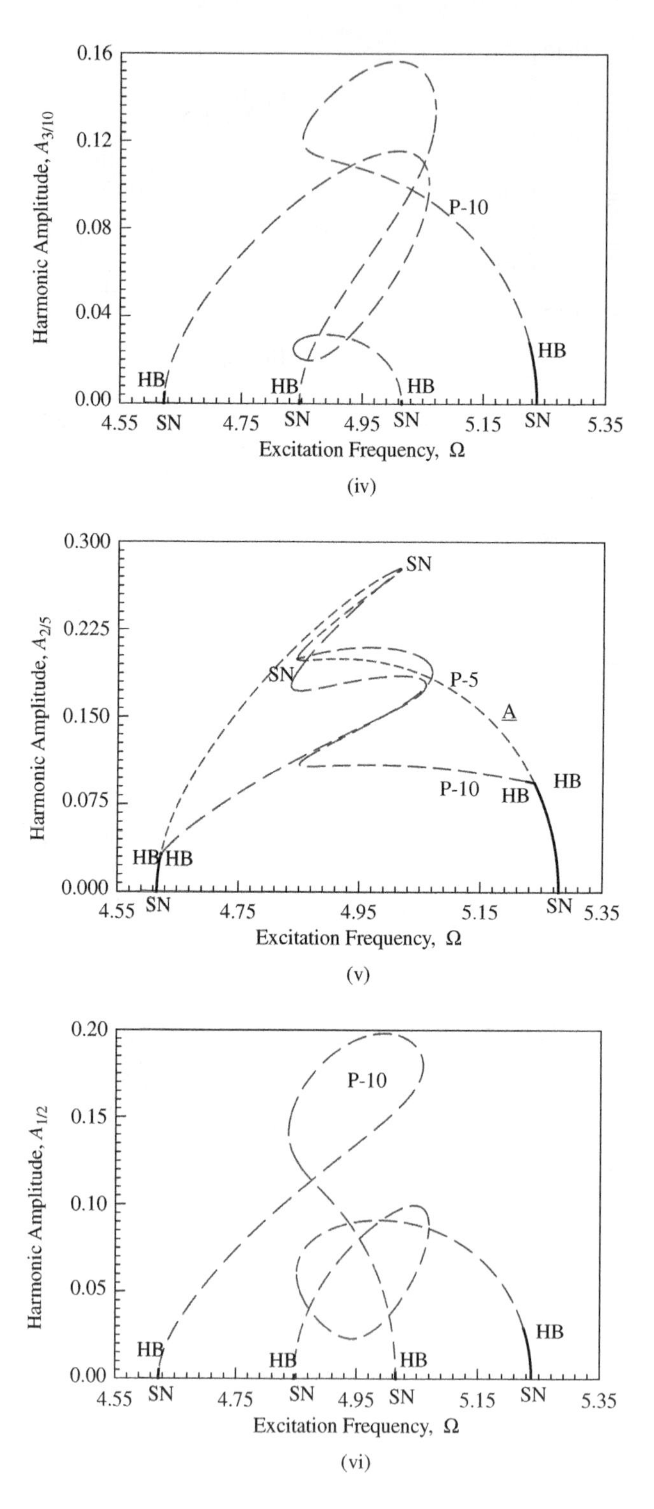

Figure 3.15 (*continued*)

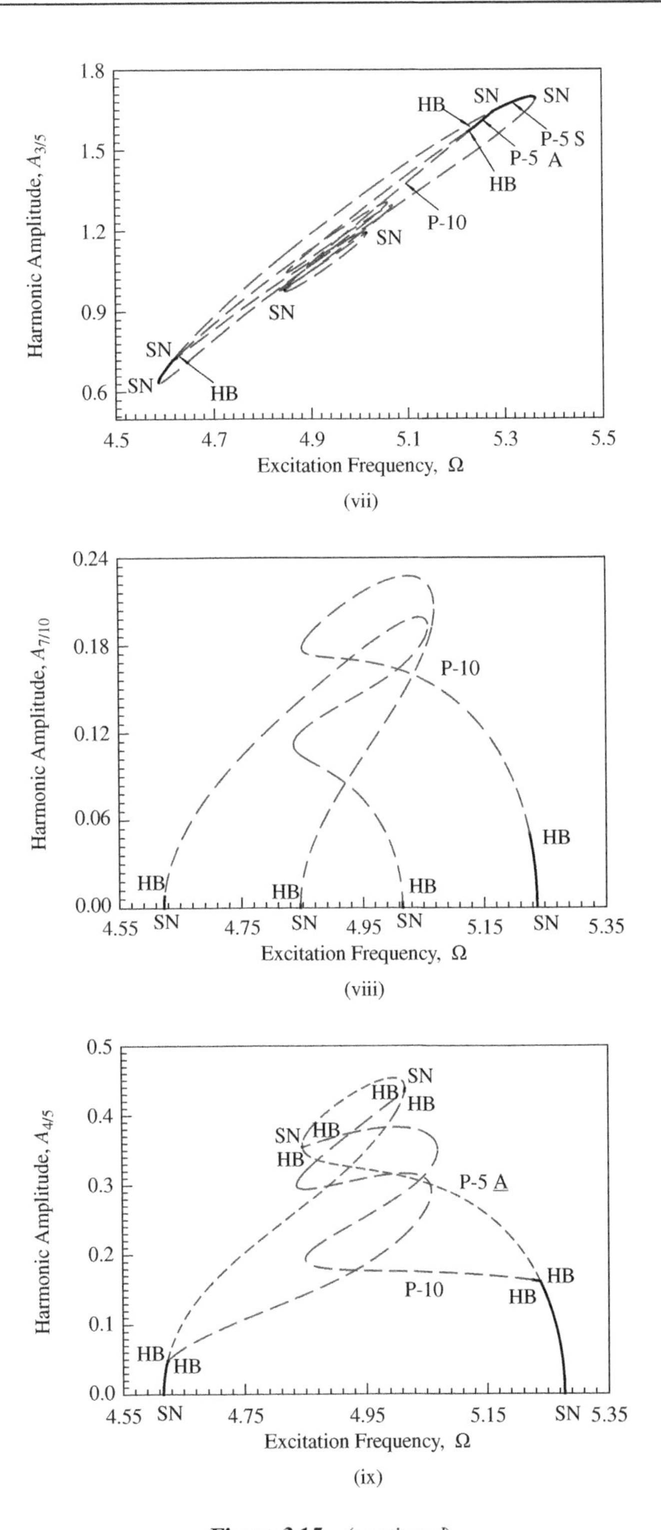

Figure 3.15 (*continued*)

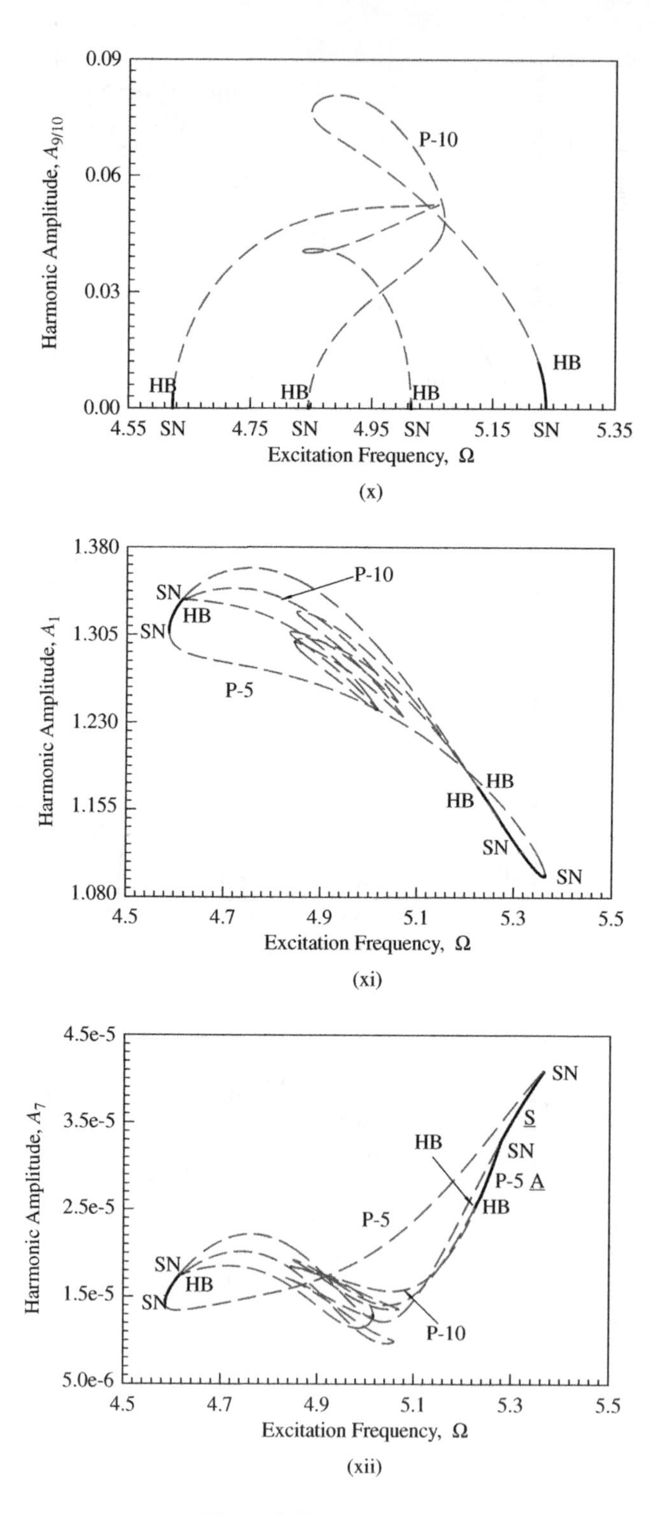

Figure 3.15 (*continued*)

different (i.e., $\varphi_{k/m}^{L} = \varphi_{k/m}^{R} + (k + 1)\pi, m = 5, 10$). In Figure 3.15(ii), the harmonic amplitude $A_{1/10} < 0.15$ is for period-10 motion only. The saddle-node bifurcation and Hopf bifurcation are clearly observed for period-10 motion. In Figure 3.15(iii), the harmonic amplitude $A_{1/5} \in (0.1, 0.6)$ is for symmetric and asymmetric period-5 motions and period-10 motions. In Figure 3.15(iv), the harmonic amplitude $A_{3/10} < 0.16$ also is for period-10 motions only. In Figure 3.15(v), the harmonic amplitude $A_{2/5} < 0.3$ is for asymmetric period-5 motions and period-10 motions. For symmetric period-5 motion, we have $A_{2/5} = 0$. Similarly, the harmonic amplitudes $A_{5/10} = A_{1/2} < 0.2$, $A_{7/10} < 0.24$, and $A_{9/10} < 0.09$ are presented for period-10 motions only in Figure 3.15(vi), (vii), and (x), respectively. The harmonic amplitude $A_{3/5} \in (0.6, 1.6)$ is presented in Figure 3.15(vii) for symmetric and asymmetric period-5 motions and period-10 motion. In Figure 3.15(ix), the harmonic amplitude $A_{4/5} < 0.5$ is for asymmetric period-5 motions and period-10 motion. In Figure 3.15(xi), the primary harmonic amplitude $A_1 \in (1.0, 1.38)$ is presented for symmetric and asymmetric period-5 motions and period-10 motion. To avoid abundant illustrations, other harmonic amplitudes will not be plotted herein. Finally, to show the accuracy tolerance of analytical solution, the harmonic amplitude $A_{70/10} = A_7 \in (5.0 \times 10^{-6}, 4.5 \times 10^{-5})$ is presented in Figure 3.15(xii) for symmetric and asymmetric period-5 motions and period-10 motion.

3.2.3 *Numerical Illustrations*

To illustrate period-*m* motions in the van der Pol-Duffing oscillator, numerical simulations, and analytical solutions will be presented. The initial conditions for numerical simulations are computed from approximate analytical solutions of periodic solutions. In all plots, circular symbols gives approximate solutions, and solid curves give numerical simulation results. The numerical solutions of periodic motions are generated via the symplectic scheme.

An independent period-4 motion is presented in Figure 3.16 for $\Omega = 11.25$ and $Q_0 = 30$. The analytical solution of the period-4 motion has 28 harmonic terms. With Equation (3.103), the analytical solution gives the initial condition of $(x_0, y_0) \approx (-1.418330, 3.343110)$. The time-histories of displacement and velocity for the period-4 motion are presented in Figure 3.16(a) and (b), respectively. Four periods in the two plots are labeled. The trajectory in phase plane is presented for 40 periods in Figure 3.16(c). The analytical and numerical results of the stable period-4 motion match very well. The analytical amplitude spectrum is presented in Figure 3.16(d). The main amplitudes are $a_0^{(4)} \approx 0.062826$, $A_{1/4} \approx 1.758588, A_{2/4} \approx 0.147344, A_{3/4} \approx 0.089692, A_{4/4} \approx 0.437385, A_{5/4} \approx 5.606028\text{e-}3$, $A_{6/4} \approx 0.024583$, $A_{7/4} \approx 7.269280\text{e-}3$, $A_{8/4} \approx 2.802221\text{e-}3$, $A_{9/4} \approx 3.709644\text{e-}3$, $A_{k/4} \sim 10^{-4}$ $(k = 10, 11, 12)$, 10^{-5} $(k = 13, 14, 15, 16)$, 10^{-6} $(k = 17, 18, 19)$, 10^{-7} $(k = 20, 21, 22)$, 10^{-8} $(k = 23, 24, \ldots, 28)$.

From the bifurcation tree of period-2 motion to chaos, asymmetric period-2 motion to period-8 motion are presented. To avoid too many illustrations, only trajectories and harmonic amplitude spectrums are presented in Figure 3.17 for $Q_0 = 20.0$ with $\Omega = 6, 5.5, 5.45$. In Figure 3.17(i) and (ii), the trajectory and harmonic amplitude spectrum of a stable period-2 motion are presented with $\Omega = 6.0$. for over 40 periods. The analytical solution of the stable period-2 motion possesses 14 harmonic terms (HB14), and the initial condition computed by such an analytical solution is $(x_0, y_0) \approx (-1.171200, 5.441010)$. The analytical and numerical results of the stable period-2 motion match very well in phase plane. The main harmonic amplitudes of the analytical solution are $a_0^{(2)} \approx 0.303081$,

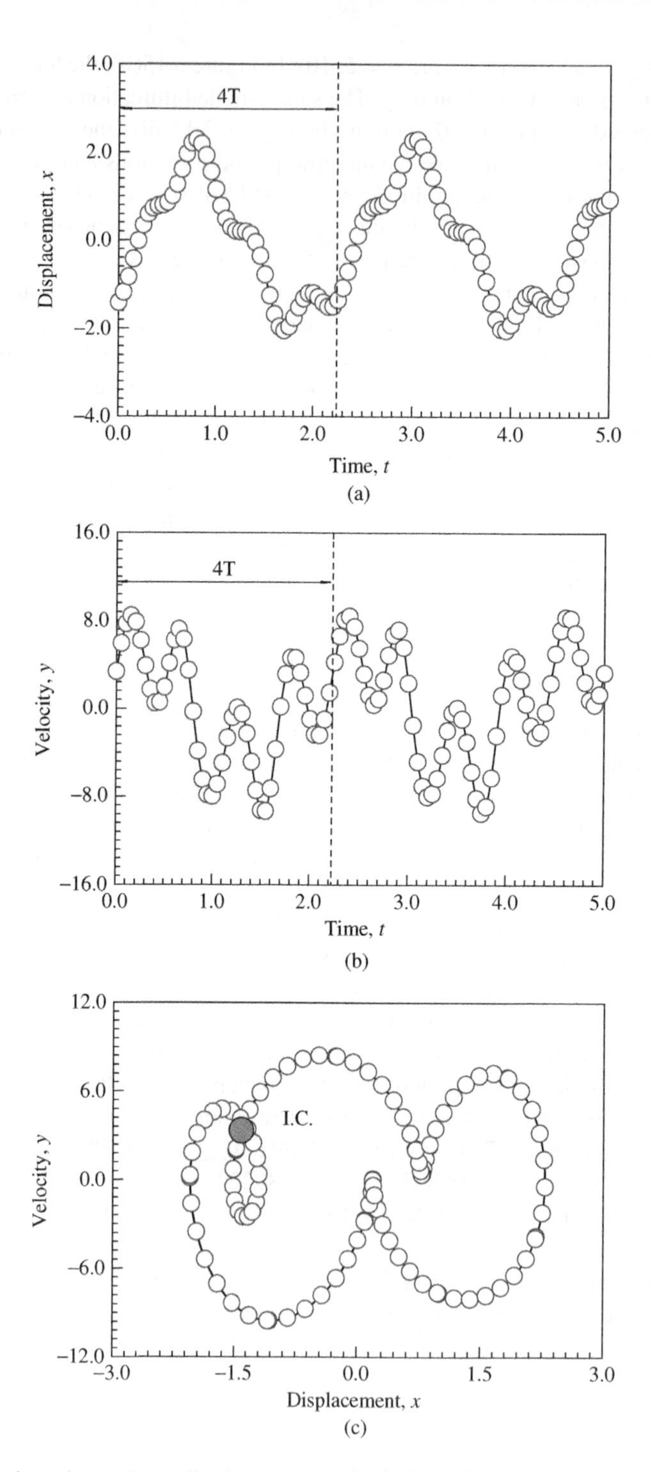

Figure 3.16 Trajectories and amplitude spectrum for independent period-4 motion. ($\Omega = 11.25$): (a) displacement, (b) velocity, (c) trajectory, and (d) amplitude. Initial condition $(x_0, \dot{x}_0) \approx (-1.418330, 3.343110)$, $(\alpha_1 = 1, \alpha_2 = 1, \alpha_3 = 1, \alpha_4 = 2.5, Q_0 = 30)$

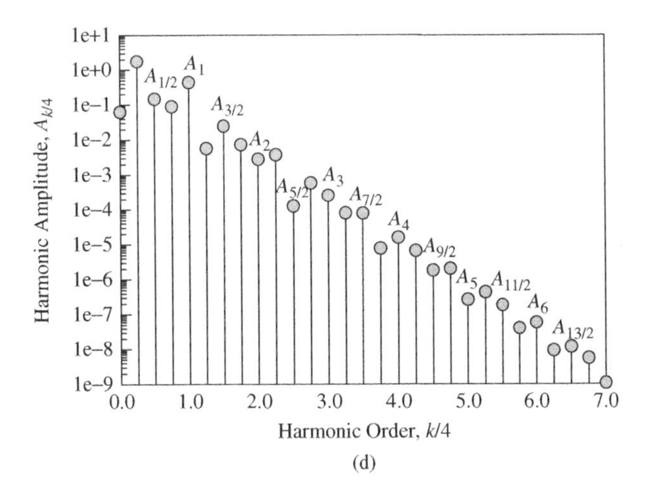

(d)

Figure 3.16 (*continued*)

$A_{1/2} \approx 1.87232$, $A_{2/2} \approx 0.745388$, $A_{3/2} \approx 0.066281$, $A_{4/2} \approx 0.062689$, $A_{5/2} \approx 0.023664$, $A_{6/2} \approx 7.530870\text{e-}3$, $A_{7/2} \approx 3.682520\text{e-}3$, $A_{8/2} \approx 1.517430\text{e-}3$, $A_{k/2} \sim 10^{-4}$ $(k = 9 \sim 11)$, 10^{-5} $(k = 12, 13)$, 10^{-6} $(k = 14)$. In Figure 3.17(iii) and (iv), the trajectory and harmonic amplitude spectrum of a stable period-4 motion are presented for $\Omega = 5.5$. The analytical solution of the stable period-4 motion 28 harmonic terms (HB28) are employed, and the initial condition given by the analytical solution is $(x_0, y_0) \approx (-0.982057, 2.569100)$. The analytical and numerical results of the stable period-4 motion match very well in phase plane. The main harmonic amplitudes in the analytical period-4 motion are $a_0^{(4)} \approx 0.279543$, $A_{1/4} \approx 0.111339$, $A_{2/4} \approx 1.460673$, $A_{3/4} \approx 0.385979$, $A_{4/4} \approx 0.947514$, $A_{5/4} \approx 0.022090$, $A_{6/4} \approx 0.031976$, $A_{7/4} \approx 0.011718$, $A_{8/4} \approx 0.048921$, $A_{9/4} \approx 0.027270$, $A_{10/4} \approx 0.031543$, $A_{11/4} \approx 8.251962\text{e-}3$, $A_{12/4} \approx 6.585974\text{e-}3$, $A_{13/4} \approx 2.150779\text{e-}3$, $A_{14/4} \approx 3.022180\text{e-}3$, $A_{15/4} \approx 2.175560\text{e-}3$, $A_{16/4} \approx 1.965215\text{e-}3$, $A_{k/4} \sim 10^{-4}$ $(k = 17, 18, \ldots, 22)$, 10^{-5} $(k = 23, 24, \ldots, 28)$. In Figure 3.17(v) and (vi), the trajectory and harmonic amplitude spectrum of a stable period-8 motion are presented for $\Omega = 5.45$. For the period-8 motion, the analytical solutions possesses 56 harmonic terms (HB56), and the initial condition given by such an analytical solution is $(x_0, y_0) \approx (-1.078110, 2.806350)$. The analytical and numerical results of the stable period-8 motion match very well. The main harmonic amplitudes in the analytical solutions are $a_0^{(8)} \approx 0.278813$, $A_{1/8} \approx 0.036213$, $A_{2/8} \approx 0.112648$, $A_{3/8} \approx 0.053490$, $A_{4/8} \approx 1.426320$, $A_{5/8} \approx 0.081141$, $A_{6/8} \approx 0.348923$, $A_{7/8} \approx 0.015831$, $A_{8/8} \approx 0.968262$, $A_{9/8} \approx 9.895090\text{e-}3$, $A_{11/8} \approx 4.675900\text{e-}3$, $A_{12/8} \approx 0.031936$, $A_{13/8} \approx 7.470730\text{e-}4$, $A_{14/8} \approx 9.737430\text{e-}3$, $A_{15/8} \approx 3.559140\text{e-}3$, $A_{16/8} \approx 0.047411$, $A_{17/8} \approx 5.469130\text{e-}3$, $A_{18/8} \approx 0.024255$, $A_{19/8} \approx 2.077560\text{e-}3$, $A_{20/8} \approx 0.032349$, $A_{k/8} < 10^{-2}$ $(k = 21, 22, \ldots, 32)$, 10^{-3} $(k = 33, 34, \ldots, 44)$, 10^{-4} $(k = 45, 46, \ldots, 56)$.

From the bifurcation tree of period-5 motion to chaos, the trajectories and harmonic amplitude spectrums for symmetric and asymmetric period-5 motions to period-10 motion are presented in Figure 3.18 for $Q_0 = 20.0$ with $\Omega = 5.32, 5.25, 5.232$. In Figure 3.18(i) and (ii), the trajectory and harmonic amplitude spectrum of a stable symmetric period-5 motion are presented for $\Omega = 5.32$. The analytical solution of the stable symmetric period-5 motion has

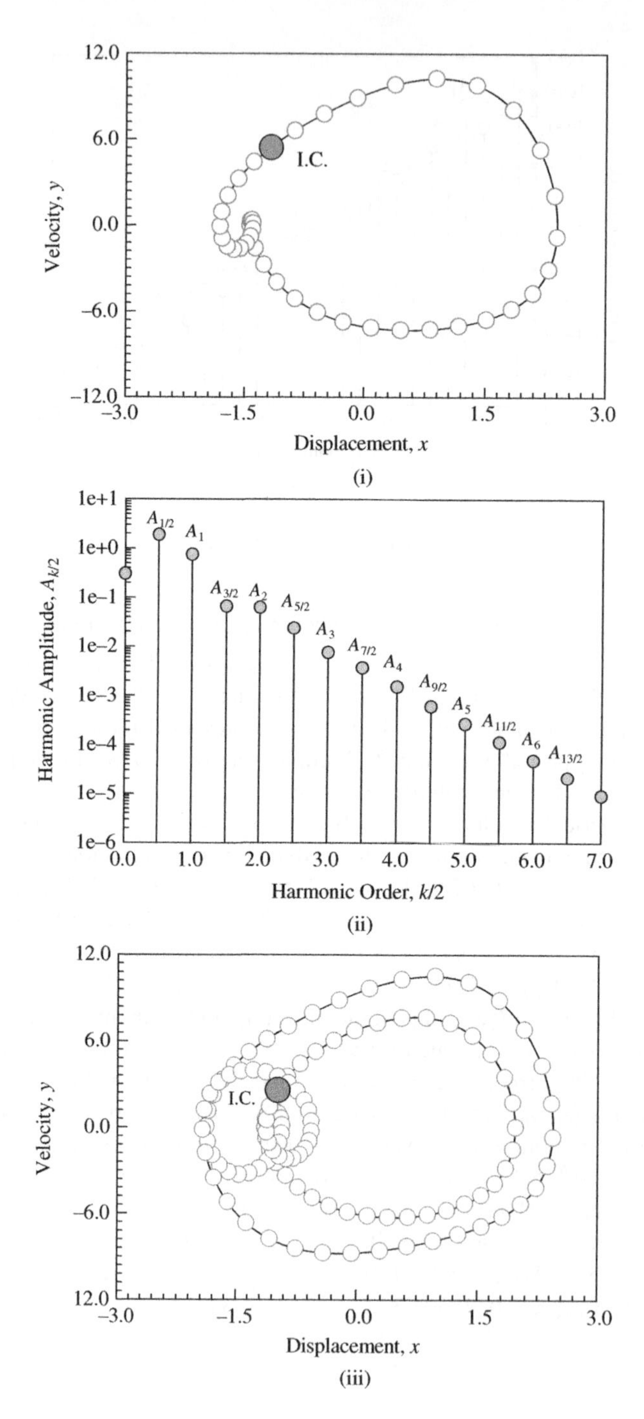

Figure 3.17 Trajectories and amplitude spectrum for period-2 motion to period-8 motion. Period-2 motion ($\Omega = 6.0$): (i) trajectory and (ii) amplitude $((x_0, \dot{x}_0) = (-1.171200, 5.441010))$; period-4 motion ($\Omega = 5.5$): (iii) trajectory and (iv) amplitude $((x_0, \dot{x}_0) \approx (-0.982057, 2.569100))$; period-8 motion ($\Omega = 5.45$): (v) trajectory and (vi) amplitude $((x_0, \dot{x}_0) \approx (-1.078110, 2.806350))$ $(\alpha_1 = 1, \alpha_2 = 1, \alpha_3 = 1, \alpha_4 = 2.5, Q_0 = 20)$

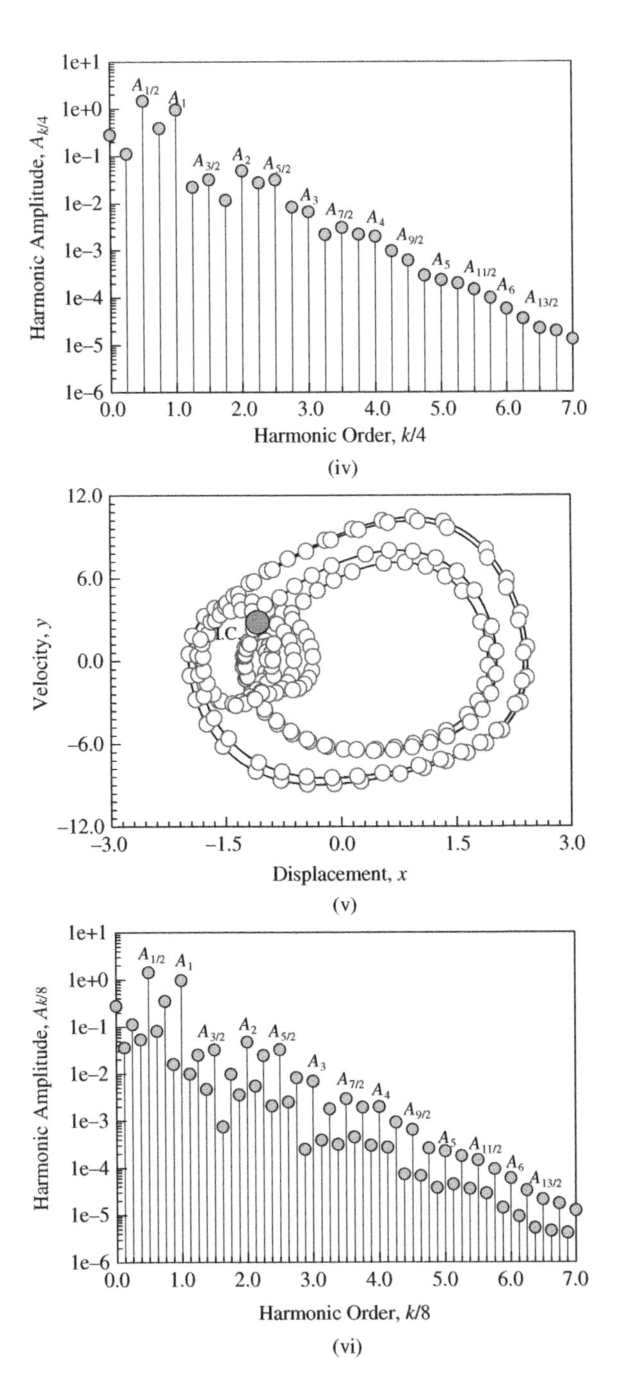

Figure 3.17 *(continued)*

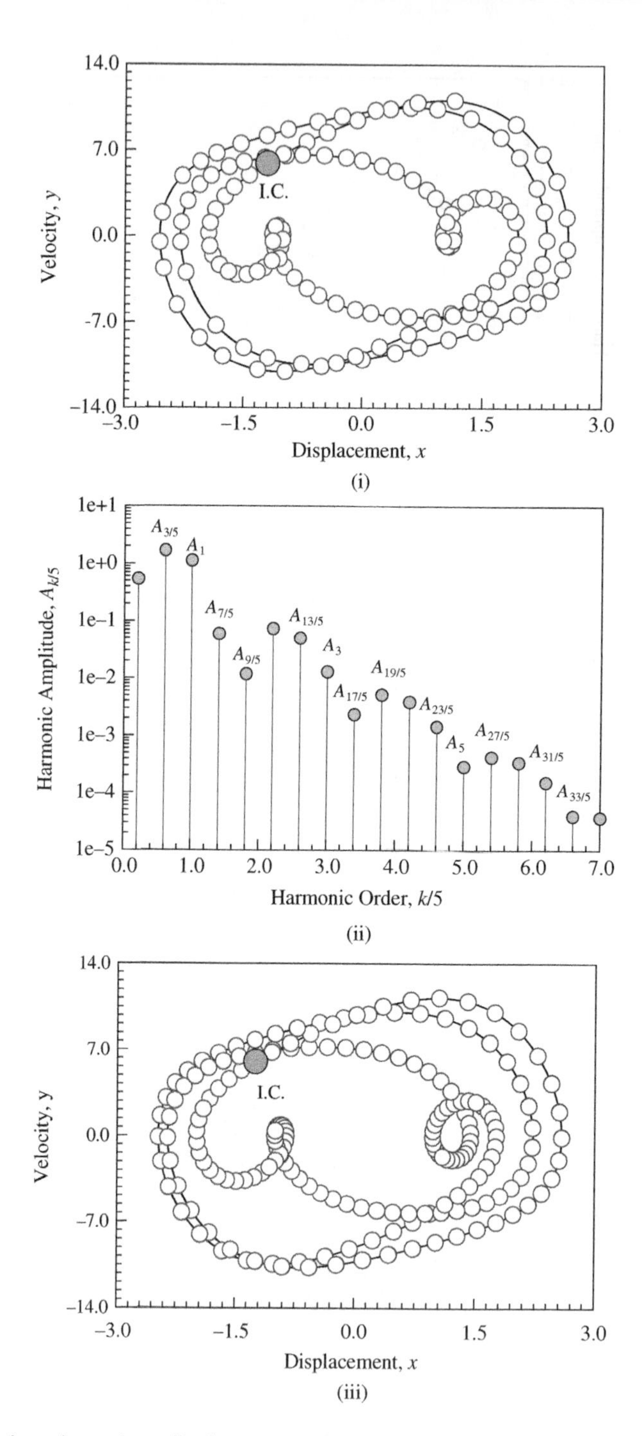

Figure 3.18 Trajectories and amplitude spectrum for period-5 motion to period-10 motion. Symmetric period-5 motion ($\Omega = 5.32$): (i) trajectory and (ii) amplitude $((x_0, \dot{x}_0) \approx (-1.205950, 5.925630))$; asymmetric period-5 motion ($\Omega = 5.25$): (iii) trajectory and (iv) amplitude $((x_0, \dot{x}_0) \approx (-1.248010, 6.021840))$; period-10 motion ($\Omega = 5.232$): (v) trajectory and (vi) amplitude $((x_0, \dot{x}_0) \approx (-1.167170, -4.407840))$ $(\alpha_1 = 1, \alpha_2 = 1, \alpha_3 = 1, \alpha_4 = 2.5, Q_0 = 20)$

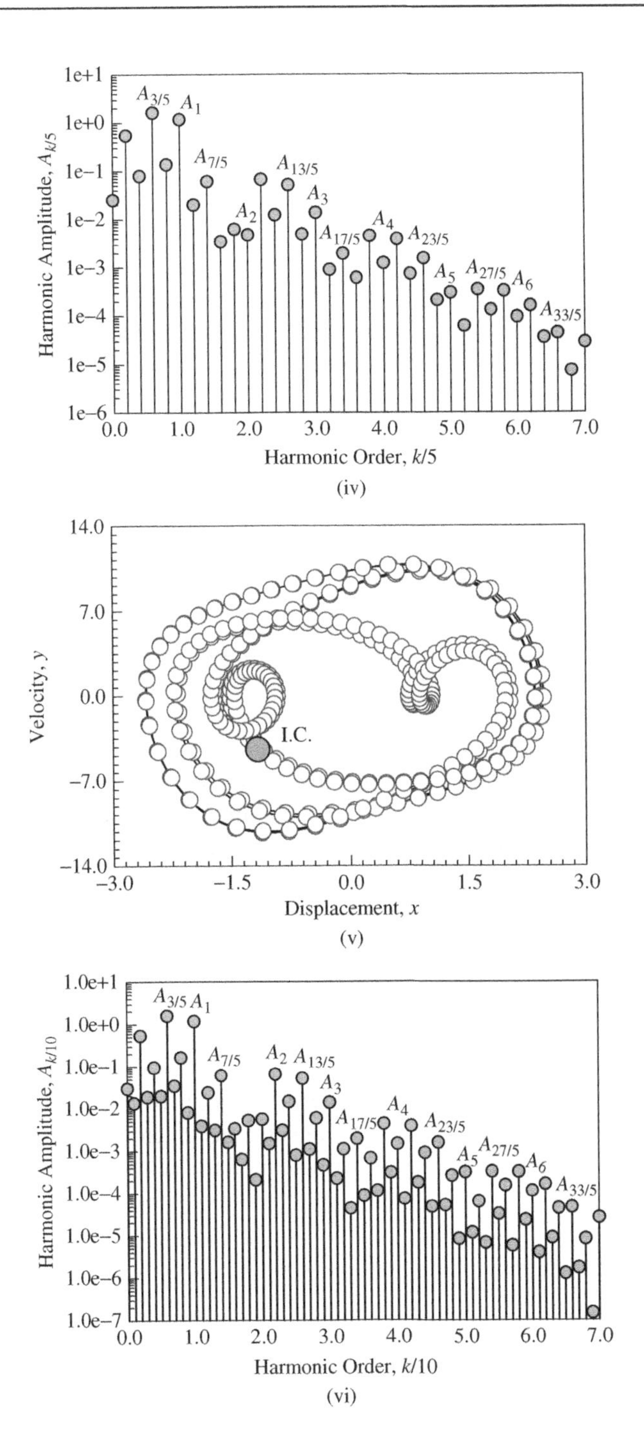

Figure 3.18 *(continued)*

35 harmonic terms (HB35). The initial condition of $(x_0, y_0) \approx (-1.205950, 5.925630)$ is computed from the analytical solution. The analytical and numerical results of symmetric period-5 motion match very well in phase plane. The main amplitudes of the analytical solutions are $A_{1/5} \approx 0.535450$, $A_{3/5} \approx 1.678394$, $A_{5/5} \approx 1.116459$, $A_{7/5} \approx 0.058676$, $A_{9/5} \approx 0.011727$, $A_{11/5} \approx 0.072620$, $A_{13/5} \approx 0.049625$, $A_{15/5} \approx 0.012822$, $A_{(2l-1)/5} \sim 10^{-3}$ $(l = 9, 10, 11, 12)$, 10^{-4} $(l = 13, 14, 15, 16)$, 10^{-5} $(l = 17, 18)$, and $a_0 = A_{(2l)/5} = 0$ $(l = 1, 2, \ldots, 17)$. On the bifurcation tree of period-5 motion to chaos, the periodic motion changes from the symmetric to asymmetric motion, and the saddle-node bifurcation of the symmetric motion occurs. In Figure 3.18(iii) and (iv), the trajectory and harmonic amplitude spectrum of a stable asymmetric period-5 motion are presented for $\Omega = 5.25$. The analytic solution of asymmetric period-5 motion also has 35 harmonic terms (HB35), and the corresponding initial condition of $(x_0, y_0) \approx (-1.248010, 6.021840)$ is computed from the analytical solution. The analytical and numerical results of the stable asymmetric period-5 motion match very well, as shown in Figure 3.18(iii). In Figure 3.18(iv), the analytical harmonic amplitude spectrum is presented, and the main amplitudes are $a_0^{(5)} \approx 0.025203$, $A_{1/5} \approx 0.542670$, $A_{2/5} \approx 0.078797$, $A_{3/5} \approx 1.601706$, $A_{4/5} \approx 0.136438$, $A_{5/5} \approx 1.159625$, $A_{6/5} \approx 0.020138$, $A_{7/5} \approx 0.061094$, $A_{8/5} \approx 3.462145\text{e-}3$, $A_{9/5} \approx 6.244835\text{e-}3$, $A_{10/5} \approx 4.778128\text{e-}3$, $A_{11/5} \approx 0.066821$, $A_{12/5} \approx 0.012324$, $A_{13/5} \approx 0.051326$, $A_{14/5} \approx 4.873882\text{e-}3$, $A_{k/5} \sim 10^{-2}$ $(k = 16, 17, \ldots, 23)$, 10^{-3} $(k = 24, 25, \ldots, 31)$, 10^{-4} $(k = 32, 33, \ldots, 35)$. In Figure 3.18(v) and (vi), the trajectory and harmonic amplitude spectrum of a stable period-10 motion are presented for $\Omega = 5.232$. The analytical solution of period-10 motion experiences 70 harmonic terms (HB70) and the initial condition generated by the analytical condition is $(x_0, y_0) \approx (-1.167170, -4.407840)$. The analytical and numerical results of the stable period-10 motion match very well in phase plane, as shown in Figure 3.18(v). The analytical harmonic amplitude spectrum is presented in Figure 3.18(vi), and the main amplitudes are $a_0^{(10)} \approx 0.030100$, $A_{1/10} \approx 0.013532$, $A_{2/10} \approx 0.535514$, $A_{3/10} \approx 0.018776$, $A_{4/10} \approx 0.094357$, $A_{5/10} \approx 0.019686$, $A_{6/10} \approx 1.576209$, $A_{7/10} \approx 0.034363$, $A_{8/10} \approx 0.162137$, $A_{9/10} \approx 7.965958\text{e-}3$, $A_{10/10} \approx 1.170793$, $A_{11/10} \approx 3.851542\text{e-}3$, $A_{12/10} \approx 0.023713$, $A_{13/10} \approx 3.069268\text{e-}3$, $A_{14/10} \approx 0.060352$, $A_{15/10} \approx 1.602452\text{e-}3$, $A_{16/10} \approx 3.348646\text{e-}3$, $A_{17/10} \approx 6.250436\text{e-}4$, $A_{18/10} \approx 5.200222\text{e-}3$, $A_{19/10} \approx 2.078449\text{e-}4$, $A_{20/10} \approx 5.563784\text{e-}3$, $A_{21/10} \approx 1.486147\text{e-}3$, $A_{22/10} \approx 0.064863$, $A_{23/10} \approx 3.053377\text{e-}3$, $A_{24/10} \approx 0.014578$, $A_{25/10} \approx 7.864651\text{e-}4$, $A_{26/10} \approx 0.051585$, $A_{27/10} \approx 1.096789\text{e-}3$, $A_{28/10} \approx 5.917919\text{e-}3$, $A_{29/10} \approx 4.613564\text{e-}4$, $A_{30/10} \approx 0.013886$, $A_{k/10} \sim 10^{-2}$ $(k = 31, 32, \ldots, 46)$, 10^{-3} $(k = 47, 48, \ldots, 62)$, 10^{-4} $(k = 63, 64, \ldots, 70)$. From afore-presented periodic motions, the analytical solutions given in this chapter are accurate, and the corresponding bifurcation trees of periodic motion to chaos are presented.

4

Parametric Nonlinear Oscillators

In this chapter, analytical solutions for period-m motions in parametric forced nonlinear oscillators are discussed. The bifurcation trees of periodic motions to chaos in a parametric oscillator with quadratic nonlinearity are discussed analytically. Nonlinear behaviors of such periodic motions are characterized through frequency-amplitude curves. This investigation shows that period-1 motions exist in parametric nonlinear systems and the corresponding bifurcation trees to chaos exist as well. In addition, analytical solutions for periodic motions in a Mathieu-Duffing oscillator are presented. The frequency-amplitude characteristics of asymmetric period-1 and symmetric period-2 motions are discussed. Period-1 asymmetric and period-2 symmetric motions are illustrated for a better understanding of periodic motions in the Mathieu-Duffing oscillator.

4.1 Parametric, Quadratic Nonlinear Oscillators

In this section, periodic motions in a parametric oscillator with quadratic nonlinearity will be discussed. Period-m motions in a parametrically forced, quadratic nonlinear oscillator will be discussed. The appropriate analytical solutions will also be presented with finite harmonic terms based on the prescribed accuracy of harmonic amplitudes. The analytical bifurcation tree for period-1 motion to chaos will be determined. Period-2 and period-4 motions will be illustrated.

4.1.1 Analytical Solutions

Consider a parametric, quadratic, nonlinear oscillator

$$\ddot{x} + \delta\dot{x} + (\alpha + Q_0 \cos \Omega t)x + \beta x^2 = 0 \tag{4.1}$$

where δ is the linear damping coefficient. α and β are linear and quadratic spring coefficients, respectively. Q_0 and Ω are parametric excitation amplitude and frequency, respectively. Equation (4.1) can be expressed in a standard form as

$$\ddot{x} = F(x, \dot{x}, t) \tag{4.2}$$

where

$$F(x, \dot{x}, t) = -\delta\dot{x} - (\alpha + Q_0 \cos \Omega t)x - \beta x^2. \tag{4.3}$$

Analytical Routes to Chaos in Nonlinear Engineering, First Edition. Albert C. J. Luo.

In Luo (2012a), the analytical solution of period-m motion with $\theta = \Omega t$ in Equation (4.1) is

$$x^{(m)*}(t) = a_0^{(m)}(t) + \sum_{k=1}^{N} b_{k/m}(t)\cos\left(\frac{k}{m}\theta\right) + c_{k/m}(t)\sin\left(\frac{k}{m}\theta\right). \tag{4.4}$$

where $a_0^{(m)}(t)$, $b_{k/m}(t)$, and $c_{k/m}(t)$ vary with time. The first and second order of derivatives of $x^*(t)$ are

$$\begin{aligned}\dot{x}^{(m)} &= \dot{a}_0^{(m)} + \sum_{k=1}^{N}\left(\dot{b}_{k/m} + \frac{k\Omega}{m}c_{k/m}\right)\cos\left(\frac{k}{m}\theta\right)\\ &\quad + \left(\dot{c}_{k/m} - \frac{k\Omega}{m}b_{k/m}\right)\sin\left(\frac{k}{m}\theta\right),\end{aligned} \tag{4.5}$$

$$\begin{aligned}\ddot{x}^{(m)} &= \ddot{a}_0^{(m)} + \sum_{k=1}^{N}\left(\ddot{b}_{k/m} + 2\frac{k\Omega}{m}\dot{c}_{k/m} - \left(\frac{k\Omega}{m}\right)^2 b_{k/m}\right)\cos\left(\frac{k}{m}\theta\right)\\ &\quad + \left(\ddot{c}_{k/m} - 2\frac{k\Omega}{m}\dot{b}_{k/m} - \left(\frac{k\Omega}{m}\right)^2 c_{k/m}\right)\sin\left(\frac{k}{m}\theta\right).\end{aligned} \tag{4.6}$$

Substitution of Equations (4.4)–(4.6) into Equation (4.2) and averaging for the harmonic terms of $\cos(k\theta/m)$ and $\sin(k\theta/m)$ $(k = 0, 1, 2, \ldots)$ gives

$$\begin{aligned}&\ddot{a}_0^{(m)} = F_0^{(m)}(a_0^{(m)}, \mathbf{b}^{(m)}, \mathbf{c}^{(m)}, \dot{a}_0^{(m)}, \dot{\mathbf{b}}^{(m)}, \dot{\mathbf{c}}^{(m)}),\\ &\ddot{b}_{k/m} + 2\frac{k\Omega}{m}\dot{c}_{k/m} - \left(\frac{k\Omega}{m}\right)^2 b_{k/m}\\ &\quad = F_{1k}^{(m)}(a_0^{(m)}, \mathbf{b}^{(m)}, \mathbf{c}^{(m)}, \dot{a}_0^{(m)}, \dot{\mathbf{b}}^{(m)}, \dot{\mathbf{c}}^{(m)}),\\ &\ddot{c}_{k/m} - 2\frac{k\Omega}{m}\dot{b}_{k/m} - \left(\frac{k\Omega}{m}\right)^2 c_{k/m}\\ &\quad = F_{2k}^{(m)}(a_0^{(m)}, \mathbf{b}^{(m)}, \mathbf{c}^{(m)}, \dot{a}_0^{(m)}, \dot{\mathbf{b}}^{(m)}, \dot{\mathbf{c}}^{(m)}).\\ &\text{for } k = 1, 2, \ldots, N\end{aligned} \tag{4.7}$$

where

$$\begin{aligned}&F_0^{(m)}(a_0^{(m)}, \mathbf{b}^{(m)}, \mathbf{c}^{(m)}, \dot{a}_0^{(m)}, \dot{\mathbf{b}}^{(m)}, \dot{\mathbf{c}}^{(m)})\\ &\quad = \frac{1}{mT}\int_0^{mT} F(x^{(m)*}, \dot{x}^{(m)*}, t)dt\\ &\quad = -\delta\dot{a}_0^{(m)} - \alpha a_0^{(m)} - \beta(a_0^{(m)})^2 - \frac{1}{2}Q_0 a_{k/m}\delta_m^k - \frac{\beta}{2}\sum_{i=1}^{N}(b_{i/m}^2 + c_{i/m}^2),\\ &F_{1k}^{(m)}(a_0^{(m)}, \mathbf{b}^{(m)}, \mathbf{c}^{(m)}, \dot{a}_0^{(m)}, \dot{\mathbf{b}}^{(m)}, \dot{\mathbf{c}}^{(m)})\\ &\quad = \frac{2}{mT}\int_0^{mT} F(x^{(m)*}, \dot{x}^{(m)*}, t)\cos\left(\frac{k}{m}\Omega t\right)dt\\ &\quad = -\delta\left(\dot{b}_{k/m} + c_{k/m}\frac{k\Omega}{m}\right) - \alpha b_{k/m} - 2\beta a_0^{(m)} b_{k/m} - f_{1k/m},\end{aligned}$$

$$\begin{aligned}
&F_{2k}^{(m)}(a_0^{(m)}, \mathbf{b}^{(m)}, \mathbf{c}^{(m)}, \dot{a}_0^{(m)}, \dot{\mathbf{b}}^{(m)}, \dot{\mathbf{c}}^{(m)}) \\
&= \frac{2}{mT}\int_0^{mT} F(x^{(m)*}, \dot{x}^{(m)*}, t)\sin\left(\frac{k}{m}\Omega t\right) dt \\
&= -\delta\left(\dot{c}_{k/m} - b_{k/m}\frac{k\Omega}{m}\right) - \alpha c_{k/m} - 2\beta a_0^{(m)} c_{k/m} - f_{2k/m},
\end{aligned} \tag{4.8}$$

and

$$\begin{aligned}
f_{1k/m} &= a_0^{(m)} Q_0 \delta_m^k + \frac{1}{2} Q_0 \sum_{i=1}^{N} a_{i/m}(\delta_{i+m}^k + \delta_{m-i}^k + \delta_{i-m}^k) \\
&+ \beta \sum_{i=1}^{N}\sum_{j=1}^{N}\left[\left(b_{i/m}b_{j/m} + c_{i/m}c_{j/m}\right)\delta_{j-i}^k \right. \\
&\left. + \frac{1}{2}\left(b_{i/m}b_{j/m} - c_{i/m}c_{j/m}\right)\delta_{i+j}^k\right], \\
f_{2k/m} &= \frac{1}{2} Q_0 \sum_{i=1}^{N} b_{i/m}(\delta_{i+m}^k + \delta_{i-m}^k - \delta_{m-i}^k) \\
&+ \beta \sum_{i=1}^{N}\sum_{j=1}^{N} b_{i/m}c_{j/m}(\delta_{i+j}^k + \delta_{j-i}^k - \delta_{i-j}^k).
\end{aligned} \tag{4.9}$$

Define

$$\begin{aligned}
\mathbf{z}^{(m)} &\triangleq (a_0^{(m)}, \mathbf{b}^{(m)}, \mathbf{c}^{(m)})^{\mathrm{T}} \\
&= (a_0^{(m)}, b_{1/m}, \ldots, b_{N/m}, c_{1/m}, \ldots, c_{N/m})^{\mathrm{T}} \\
&\equiv (z_0^{(m)}, z_1^{(m)}, \ldots, z_{2N}^{(m)})^{\mathrm{T}}, \\
\mathbf{z}_1 &\triangleq \dot{\mathbf{z}} = (\dot{a}_0^{(m)}, \dot{\mathbf{b}}^{(m)}, \dot{\mathbf{c}}^{(m)})^{\mathrm{T}} \\
&= (\dot{a}_0^{(m)}, \dot{b}_{1/m}, \ldots, \dot{b}_{N/m}, \dot{c}_{1/m}, \ldots, \dot{c}_{N/m})^{\mathrm{T}} \\
&\equiv (\dot{z}_0^{(m)}, \dot{z}_1^{(m)}, \ldots, \dot{z}_{2N}^{(m)})^{\mathrm{T}}
\end{aligned} \tag{4.10}$$

where

$$\begin{aligned}
\mathbf{b}^{(m)} &= (b_{1/m}, \ldots, b_{N/m})^{\mathrm{T}}, \\
\mathbf{c}^{(m)} &= (c_{1/m}, \ldots, c_{N/m})^{\mathrm{T}}.
\end{aligned} \tag{4.11}$$

Equation (4.7) is rewritten as

$$\dot{\mathbf{z}}^{(m)} = \mathbf{z}_1^{(m)} \text{ and } \dot{\mathbf{z}}_1^{(m)} = \mathbf{g}^{(m)}(\mathbf{z}^{(m)}, \mathbf{z}_1^{(m)}) \tag{4.12}$$

where

$$\mathbf{g}^{(m)}(\mathbf{z}^{(m)}, \mathbf{z}_1^{(m)}) = \begin{pmatrix} F_0^{(m)}(\mathbf{z}^{(m)}, \mathbf{z}_1^{(m)}) \\ \mathbf{F}_1^{(m)}(\mathbf{z}^{(m)}, \mathbf{z}_1^{(m)}) - 2\mathbf{k}_1\frac{\Omega}{m}\dot{\mathbf{c}}^{(m)} + \mathbf{k}_2\left(\frac{\Omega}{m}\right)^2\mathbf{b}^{(m)} \\ \mathbf{F}_2^{(m)}(\mathbf{z}^{(m)}, \mathbf{z}_1^{(m)}) + 2\mathbf{k}_1\frac{\Omega}{m}\dot{\mathbf{b}}^{(m)} + \mathbf{k}_2\left(\frac{\Omega}{m}\right)^2\mathbf{c}^{(m)} \end{pmatrix} \tag{4.13}$$

and

$$\begin{aligned}
&\mathbf{k}_1 = diag(1,2,\ldots,N),\\
&\mathbf{k}_2 = diag(1,2^2,\ldots,N^2),\\
&\mathbf{F}_1^{(m)} = (F_{11}^{(m)}, F_{12}^{(m)}, \ldots, F_{1N}^{(m)})^{\mathrm{T}},\\
&\mathbf{F}_2^{(m)} = (F_{21}^{(m)}, F_{22}^{(m)}, \ldots, F_{2N}^{(m)})^{\mathrm{T}}\\
&\text{for } N = 1,2,\ldots,\infty.
\end{aligned} \tag{4.14}$$

Introducing

$$\mathbf{y}^{(m)} \equiv (\mathbf{z}^{(m)}, \mathbf{z}_1^{(m)}) \text{ and } \mathbf{f}^{(m)} = (\mathbf{z}_1^{(m)}, \mathbf{g}^{(m)})^{\mathrm{T}}, \tag{4.15}$$

Equation (4.12) becomes

$$\dot{\mathbf{y}}^{(m)} = \mathbf{f}^{(m)}(\mathbf{y}^{(m)}). \tag{4.16}$$

The steady-state solutions for periodic motion in Equation (4.1) can be obtained by setting $\dot{\mathbf{y}}^{(m)} = \mathbf{0}$, that is,

$$\begin{aligned}
&F_0^{(m)}(\mathbf{z}^{(m)}, \mathbf{0}) = 0,\\
&\mathbf{F}_1^{(m)}(\mathbf{z}^{(m)}, \mathbf{0}) - \mathbf{k}_2\left(\frac{\Omega}{m}\right)^2 \mathbf{b}^{(m)} = \mathbf{0},\\
&\mathbf{F}_2^{(m)}(\mathbf{z}^{(m)}, \mathbf{0}) - \mathbf{k}_2\left(\frac{\Omega}{m}\right)^2 \mathbf{c}^{(m)} = \mathbf{0}.
\end{aligned} \tag{4.17}$$

The $(2N+1)$ nonlinear equations in Equation (4.17) are solved by the Newton–Raphson method. As in Luo (2012a), the linearized equation at equilibrium point $\mathbf{y}^* = (\mathbf{z}^*, \mathbf{0})^{\mathrm{T}}$ is given by

$$\Delta\dot{\mathbf{y}}^{(m)} = D\mathbf{f}(\mathbf{y}^{(m)*})\Delta\mathbf{y}^{(m)} \tag{4.18}$$

where

$$D\mathbf{f}(\mathbf{y}^{(m)*}) = \partial\mathbf{f}(\mathbf{y}^{(m)})/\partial\mathbf{y}^{(m)}|_{\mathbf{y}^{(m)*}}. \tag{4.19}$$

The corresponding eigenvalues are determined by

$$|D\mathbf{f}(\mathbf{y}^{(m)*}) - \lambda\mathbf{I}_{2(2N+1)\times 2(2N+1)}| = 0. \tag{4.20}$$

where

$$D\mathbf{f}(\mathbf{y}^{(m)*}) = \begin{bmatrix} \mathbf{0}_{(2N+1)\times(2N+1)} & \mathbf{I}_{(2N+1)\times(2N+1)} \\ \mathbf{G}_{(2N+1)\times(2N+1)} & \mathbf{H}_{(2N+1)\times(2N+1)} \end{bmatrix} \tag{4.21}$$

and

$$\mathbf{G} = \frac{\partial\mathbf{g}^{(m)}}{\partial\mathbf{z}^{(m)}} = (\mathbf{G}^{(0)}, \mathbf{G}^{(c)}, \mathbf{G}^{(s)})^{\mathrm{T}}, \tag{4.22}$$

$$\begin{aligned}
&\mathbf{G}^{(0)} = (G_0^{(0)}, G_1^{(0)}, \ldots, G_{2N}^{(0)}),\\
&\mathbf{G}^{(c)} = (\mathbf{G}_1^{(c)}, \mathbf{G}_2^{(c)}, \ldots, \mathbf{G}_N^{(c)})^{\mathrm{T}},\\
&\mathbf{G}^{(s)} = (\mathbf{G}_1^{(s)}, \mathbf{G}_2^{(s)}, \ldots, \mathbf{G}_N^{(s)})^{\mathrm{T}}
\end{aligned} \tag{4.23}$$

for $N = 1,2,\ldots\infty$ with

$$\begin{aligned}
&\mathbf{G}_k^{(c)} = (G_{k0}^{(c)}, G_{k1}^{(c)}, \ldots, G_{k(2N)}^{(c)}),\\
&\mathbf{G}_k^{(s)} = (G_{k0}^{(s)}, G_{k1}^{(s)}, \ldots, G_{k(2N)}^{(s)})
\end{aligned} \tag{4.24}$$

for $k = 1, 2, \ldots N$. The corresponding components are

$$
\begin{aligned}
G_r^{(0)} &= -\alpha\delta_0^r - \frac{1}{2}Q_0\delta_m^r - \beta g_{2r}^{(0)}, \\
G_{kr}^{(c)} &= \left(\frac{k\Omega}{m}\right)^2 \delta_k^r - \alpha\delta_k^r - \delta\frac{k\Omega}{m}\delta_{k+N}^r - \frac{1}{2}Q_0\delta_m^k\delta_0^r \\
&\quad - \frac{1}{4}Q_0\sum_{i=1}^{N}(\delta_{i+m}^k + \delta_{m-i}^k + \delta_{i-m}^k)\delta_i^r - \beta g_{2kr}^{(c)}, \\
G_{kr}^{(s)} &= \left(\frac{k\Omega}{m}\right)^2 \delta_{k+N}^r + \delta\frac{k\Omega}{m}\delta_k^r - \alpha\delta_{k+N}^r \\
&\quad - \frac{1}{4}Q_0\sum_{i=1}^{N}(\delta_{i+m}^k + \delta_{i-m}^k - \delta_{m-i}^k)\delta_{i+N}^r - \beta g_{2kr}^{(s)}
\end{aligned}
\tag{4.25}
$$

where

$$
g_{2r}^{(0)} = 2a_0^{(m)}\delta_0^r + b_{k/m}\delta_k^r + c_{k/m}\delta_{k+N}^r, \tag{4.26}
$$

$$
\begin{aligned}
g_{2kr}^{(c)} &= 2b_{k/m}\delta_r^0 + 2a_0^{(m)}\delta_k^r + \sum_{i=1}^{N}\sum_{j=1}^{N} b_{j/m}(\delta_{j-i}^k + \delta_{i-j}^k + \delta_{i+j}^k)\delta_i^r \\
&\quad + c_{j/m}(\delta_{j-i}^k + \delta_{i-j}^k - \delta_{i+j}^k)\delta_{i+N}^r,
\end{aligned}
\tag{4.27}
$$

$$
\begin{aligned}
g_{2kr}^{(s)} &= 2c_{k/m}\delta_0^r + 2a_0^{(m)}\delta_{k+N}^r + \sum_{i=1}^{N}\sum_{j=1}^{N} c_{j/m}(\delta_{i+j}^k + \delta_{j-i}^k - \delta_{i-j}^k)\delta_i^r \\
&\quad + b_{i/m}(\delta_{i+j}^k + \delta_{j-i}^k - \delta_{i-j}^k)\delta_{j+N}^r
\end{aligned}
\tag{4.28}
$$

for $r = 0, 1, \ldots, 2N$.

$$
\mathbf{H} = \frac{\partial \mathbf{g}^{(m)}}{\partial \mathbf{z}_1^{(m)}} = (\mathbf{H}^{(0)}, \mathbf{H}^{(c)}, \mathbf{H}^{(s)})^{\mathrm{T}} \tag{4.29}
$$

where

$$
\begin{aligned}
\mathbf{H}^{(0)} &= (H_0^{(0)}, H_1^{(0)}, \ldots, H_{2N}^{(0)}), \\
\mathbf{H}^{(c)} &= (\mathbf{H}_1^{(c)}, \mathbf{H}_2^{(c)}, \ldots, \mathbf{H}_N^{(c)})^{\mathrm{T}}, \\
\mathbf{H}^{(s)} &= (\mathbf{H}_1^{(s)}, \mathbf{H}_2^{(s)}, \ldots, \mathbf{H}_N^{(s)})^{\mathrm{T}}
\end{aligned}
\tag{4.30}
$$

for $N = 1, 2, \ldots \infty$, with

$$
\begin{aligned}
\mathbf{H}_k^{(c)} &= (H_{k0}^{(c)}, H_{k1}^{(c)}, \ldots, H_{k(2N)}^{(c)}), \\
\mathbf{H}_k^{(s)} &= (H_{k0}^{(s)}, H_{k1}^{(s)}, \ldots, H_{k(2N)}^{(s)})
\end{aligned}
\tag{4.31}
$$

for $k = 1, 2, \ldots N$. The corresponding components are

$$
\begin{aligned}
H_r^{(0)} &= -\delta\delta_0^r, \\
H_{kr}^{(c)} &= -2\frac{k\Omega}{m}\delta_{k+N}^r - \delta\delta_k^r, \\
H_{kr}^{(s)} &= 2\frac{k\Omega}{m}\delta_k^r - \delta\delta_{k+N}^r
\end{aligned}
\tag{4.32}
$$

for $r = 0, 1, \ldots, 2N$. The eigenvalues of $D\mathbf{f}(\mathbf{y}^{(m)*})$ are classified as

$$(n_1, n_2, n_3 | n_4, n_5, n_6). \tag{4.33}$$

The corresponding boundary between the stable and unstable solutions is given by the saddle-node bifurcation (SN) and Hopf bifurcation (HB).

4.1.2 Analytical Routes to Chaos

The curves of amplitude varying with excitation frequency Ω are illustrated. The harmonic amplitude and phase are defined by

$$A_{k/m} \equiv \sqrt{b_{k/m}^2 + c_{k/m}^2} \text{ and } \varphi_{k/m} = \arctan \frac{c_{k/m}}{b_{k/m}}. \tag{4.34}$$

The corresponding solution in Equation (4.1) becomes

$$x^*(t) = a_0^{(m)} + \sum_{k=1}^{N} A_{k/m} \cos\left(\frac{k}{m}\Omega t - \varphi_{k/m}\right). \tag{4.35}$$

Consider system parameters as

$$\delta = 0.5, \alpha = 5, \beta = 20. \tag{4.36}$$

In all frequency-amplitude curves, the acronyms "SN" and "HB" represent the saddle-node and Hopf bifurcations, respectively. Solid curves represent stable period-m motions. Long dashed, short dashed, and chain curves represent unstable period-1, period-2, and period-4 motions, respectively.

As in Luo and Yu (2013d), consider a bifurcation tree of period-2 motion to chaos through period-2 to period-4 motion in parametrically excited, quadratic nonlinear oscillator. Using the parameters in Equation (4.36), the frequency-amplitude curves based on 40 harmonic terms of period-2 to period-4 motion are presented in Figures 4.1–4.3 for $Q_0 = 10$. In Figure 4.1, a global view of frequency-amplitude curve for period-2 to period-4 motion is presented. In Figure 4.1(i), constant $a_0^{(m)}$ versus excitation frequency Ω is presented, and all the values of $a_0^{(m)}$ are less than zero. $a_0^{(m)} \in (-0.17, 0.0)$ is for period-2 and period-4 motion. The saddle-node bifurcations of period-2 motion are its onset points at $\Omega_{cr} \approx 2.6, 6.4$. The Hopf bifurcations of period-2 motions yield the onset of period-4 motions at $\Omega_{cr} \approx 2.59, 5.95$ for the large branch and $\Omega_{cr} \approx 2.92, 3.0$ for the small branch. In Figure 4.1(ii), the harmonic amplitude $A_{1/4}$ versus excitation frequency Ω is presented for period-4 motion. We have $A_{1/4} \in (0.0, 0.2)$. The Hopf bifurcation points of period-4 motions are the saddle-node bifurcation points for period-8 motions. Since the stable period-8 motion exists in the short ranges, period-8 motions will not presented herein. In Figure 4.1(iii), the harmonic amplitudes $A_{1/2}$ versus excitation frequency Ω are presented for period-2 and period-4 motions, as for constant $a_0^{(m)}$ in Figure 4.1(i). We have $A_{1/2} \in (0.0, 0.3)$. In Figure 4.1(iv), the harmonic amplitude $A_{3/4}$ versus excitation frequency Ω is presented for period-4 motion in the range of $A_{3/4} \in (0.0, 0.12)$. The frequency-amplitude curves in (Ω, A_1) are illustrated in Figure 4.1(v) for period-2 and period-4 motions in the range of $A_1 \in (0.0, 0.3)$. To avoid

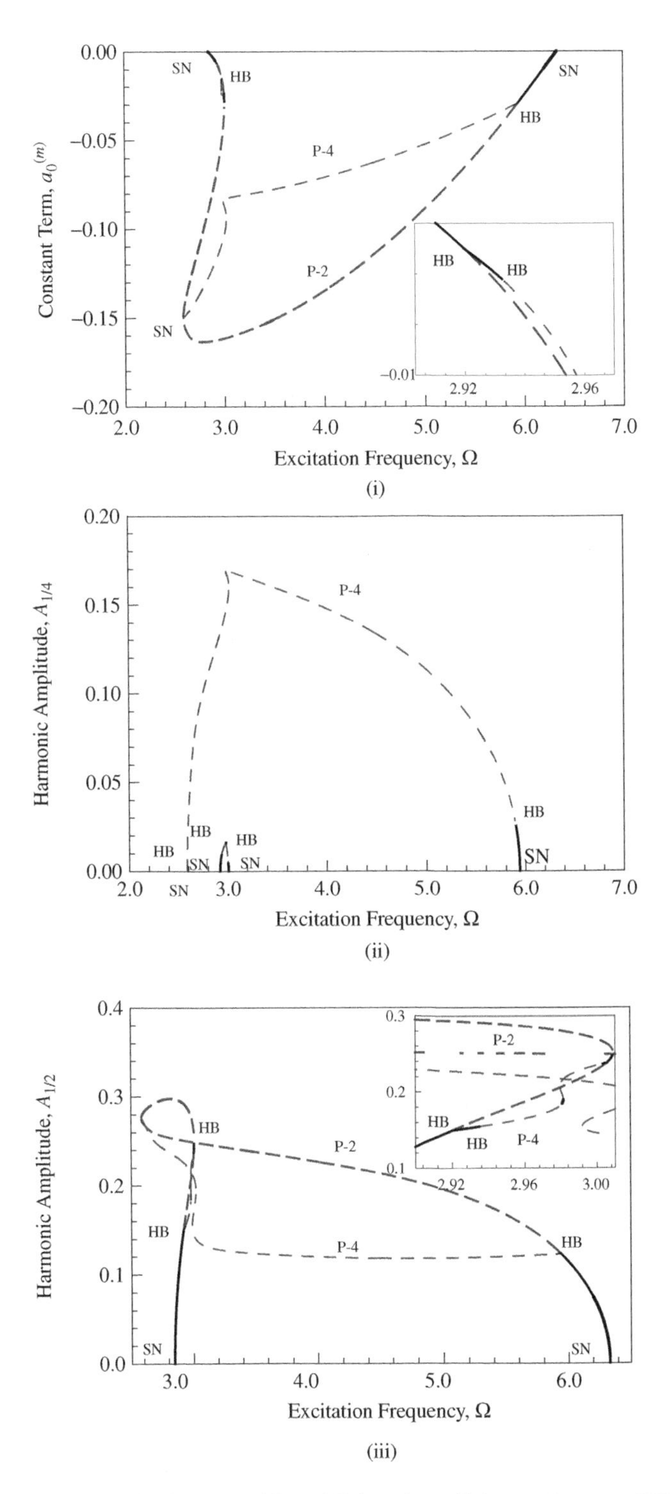

Figure 4.1 Frequency-amplitude curves ($Q_0 = 10$) based on 40 harmonic terms (HB40) of period-2 motion to period-4 motion in the parametric, nonlinear quadratic oscillator: (i) constant term $a_0^{(m)}$, (ii)–(viii) harmonic amplitude $A_{k/m}$ ($k = 1, 2, \ldots, 4, 8, 12, 40, m = 4$). ($\delta = 0.5, \alpha = 5, \beta = 20$)

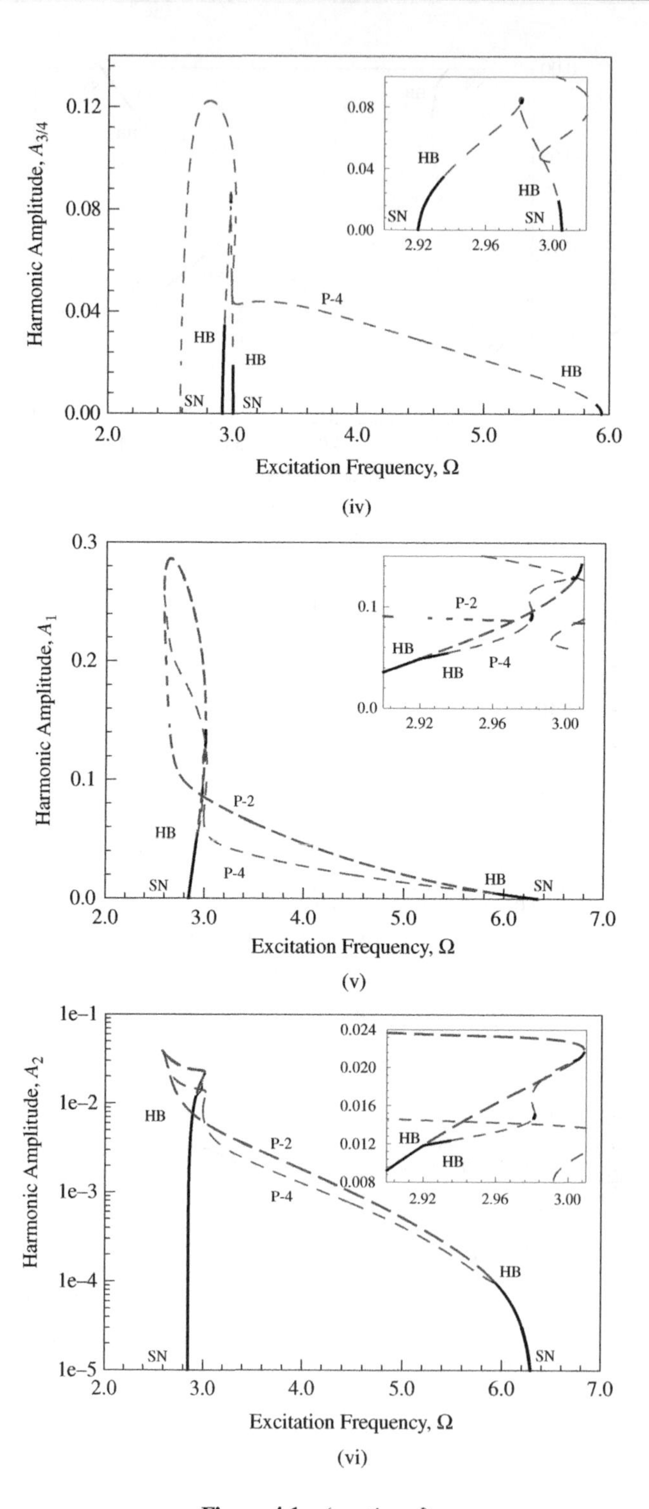

Figure 4.1 *(continued)*

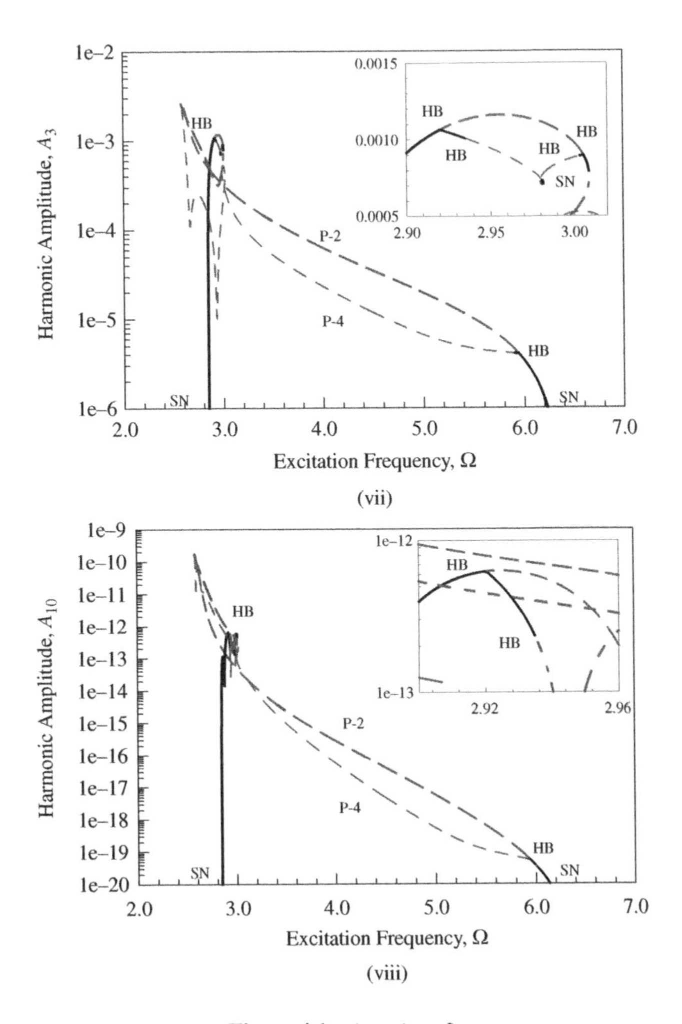

(vii)

(viii)

Figure 4.1 *(continued)*

abundant illustrations, the harmonic amplitude $A_{k/4}$ (mod$(k,4) \neq 0$) will not be presented. To further show harmonic term effects, the harmonic amplitudes A_2, A_3, A_{10} are presented in Figure 4.1(vi)–(viii). $A_2 < 10^{-1}$, $A_3 < 10^{-2}$, and $A_{10} < 10^{-9}$ are observed. Thus, effects of the higher order harmonic terms on period-2 and period-4 motions can be ignored.

To clearly illustrate the bifurcation trees of period-2 motions, the bifurcation tree-1 of period-2 motion to period-4 motion based on 40 harmonic terms is presented in Figure 4.2 for $Q_0 = 10$ within the range of $\Omega \in (5.80, 6.0)$. In Figure 4.2(i), the constant versus excitation frequency is presented. The bifurcation tree is very clearly presented. In Figure 4.2(ii), the frequency-amplitude for period-4 motion in the bifurcation tree-1 is presented. $A_{1/4} = 0$ for period-2 motion. The bifurcation tree-1 for harmonic amplitude $A_{1/2}$ is presented in Figure 4.2(iii). For the zoomed range, the quantity level of the harmonic amplitudes is $A_{1/2} \sim 10^{-1}$. As similar to $A_{1/4}$, the harmonic amplitude $A_{3/4}$ for period-4 motion in the

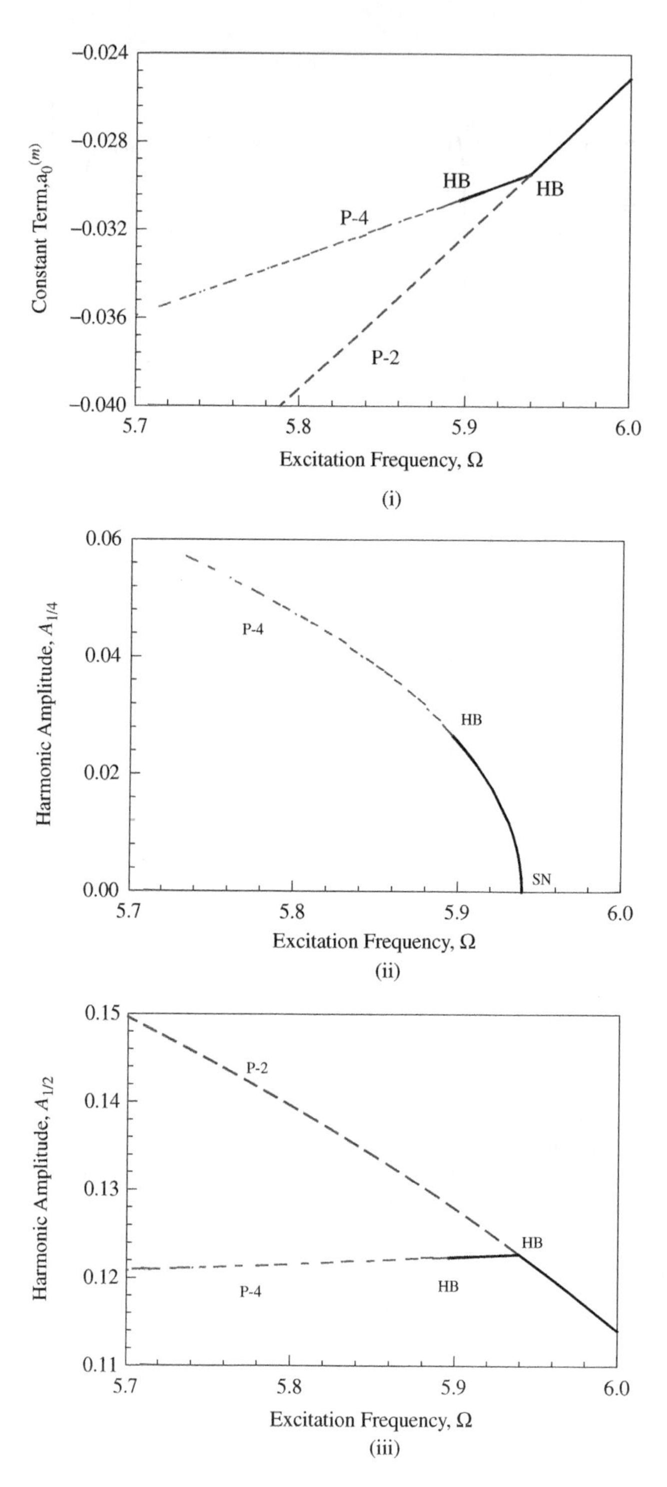

Figure 4.2 Bifurcation tree-1 of period-2 motion ($Q_0 = 10$) to period-4 motion based on 40 harmonic terms (HB40) in the parametric, nonlinear quadratic oscillator: (i) constant term $a_0^{(m)}$, (ii)–(viii) harmonic amplitude $A_{k/m}$ ($k = 1, 2, \ldots, 4, 8, 12, 40$, $m = 4$). ($\delta = 0.5$, $\alpha = 5$, $\beta = 20$)

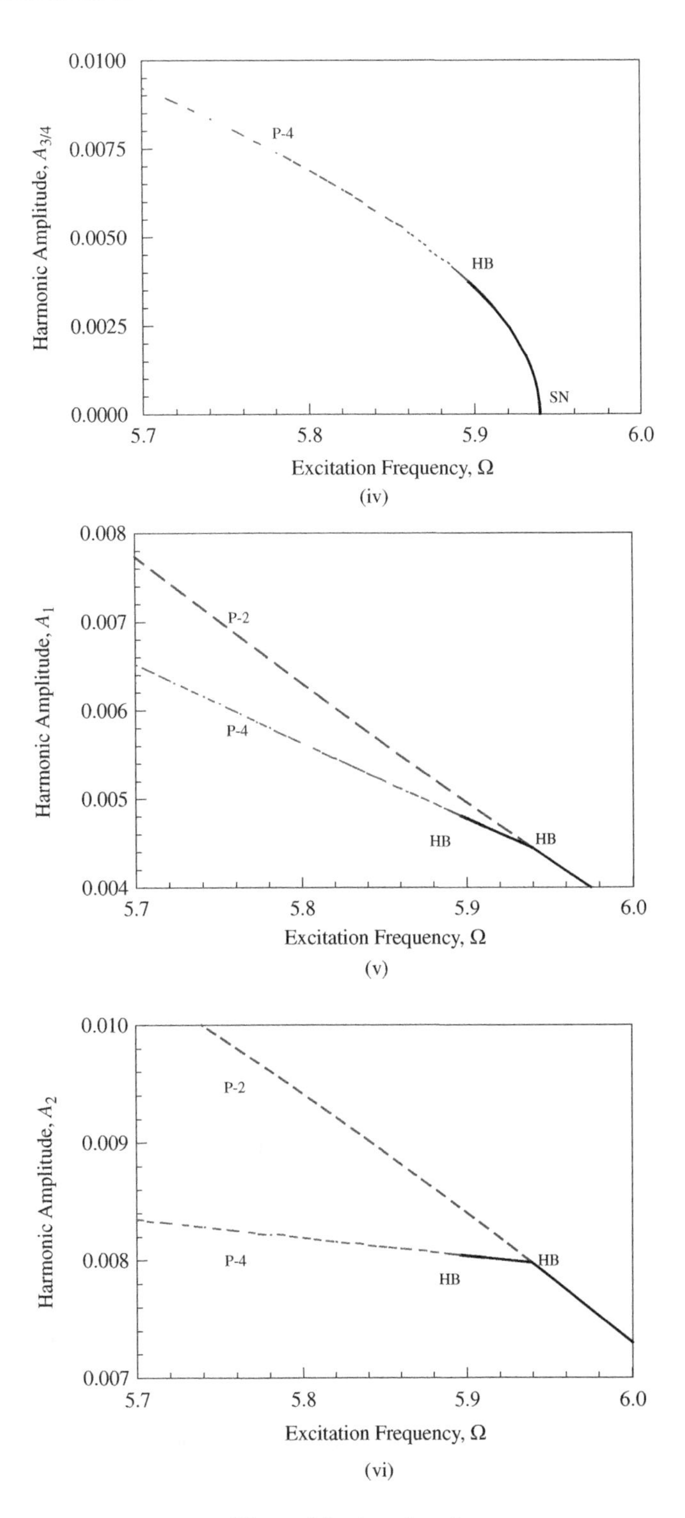

Figure 4.2 (*continued*)

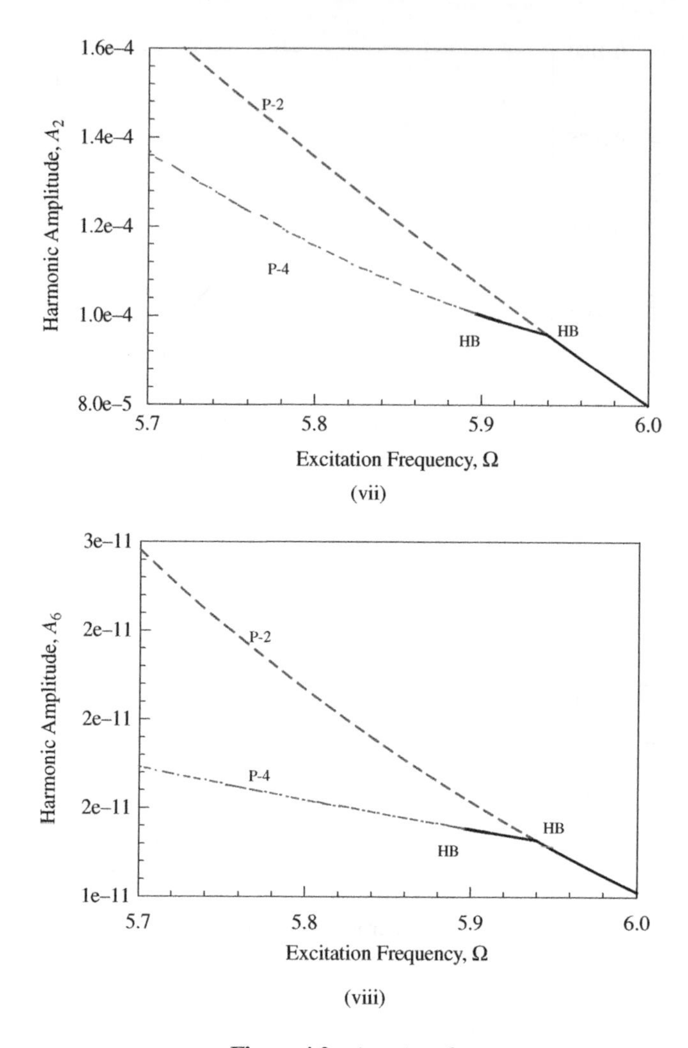

Figure 4.2 (*continued*)

bifurcation tree-1 is presented in Figure 4.2(iv). The harmonic amplitude A_1 for the bifurcation tree-1 of period-2 motion to period-4 is given in Figure 4.2(v). For the zoomed range, the quantity level of the harmonic amplitude is $A_1 \sim 7 \times 10^{-3}$. For the period-2 motion, the harmonic amplitude $A_{k/2}$ $(k = 2l - 1,\ l = 1, 2, \ldots)$ is very important. Thus, the harmonic amplitude $A_{3/2}$ for bifurcation tree-1 is presented in Figure 4.2(vi). For the zoomed range, the quantity level of the harmonic amplitudes is $A_{3/2} \sim 10^{-2}$. The harmonic amplitude with the range of $A_2 \sim 10^{-4}$ is presented in Figure 4.2(vii). To avoid abundant plots, the harmonic amplitude $A_6 \sim 3 \times 10^{-11}$ for the bifurcation is presented in Figure 4.2(viii). Other harmonic amplitude can be similarly presented.

The second bifurcation tree of period-2 motion to period-4 motion is in the narrow range of $\Omega \in (2.90, 3.02)$, which is shown in Figure 4.3. In this bifurcation tree, the period-4 motion has four parts of stable solutions and three parts of unstable solution on the same curve.

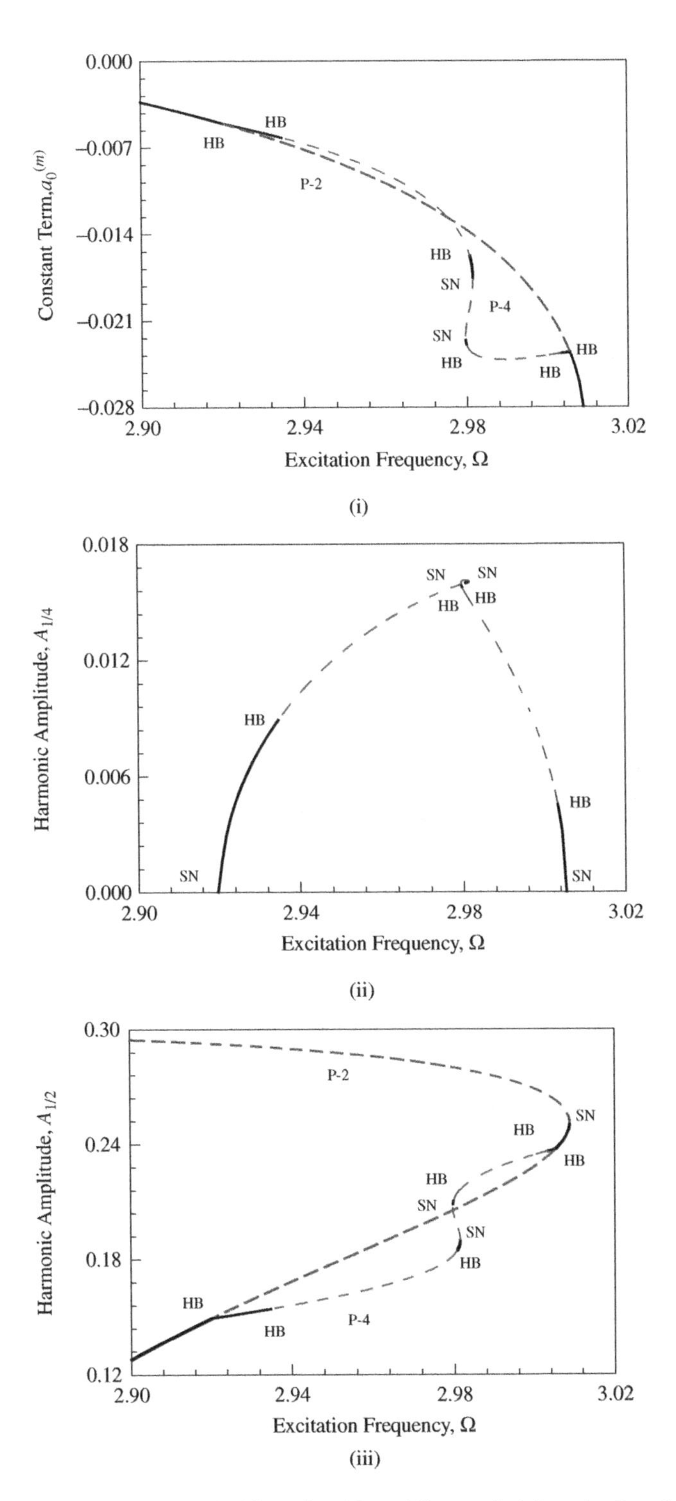

Figure 4.3 Bifurcation tree-2 of period-2 motion ($Q_0 = 10$) to period-4 motion based on 40 harmonic terms (HB40) in the parametric, nonlinear quadratic oscillator: (i) constant term $a_0^{(m)}$, (ii)–(viii) harmonic amplitude $A_{k/m}$ ($k = 1, 2, \dots, 4, 8, 12, 40$, $m = 4$). ($\delta = 0.5$, $\alpha = 5$, $\beta = 20$)

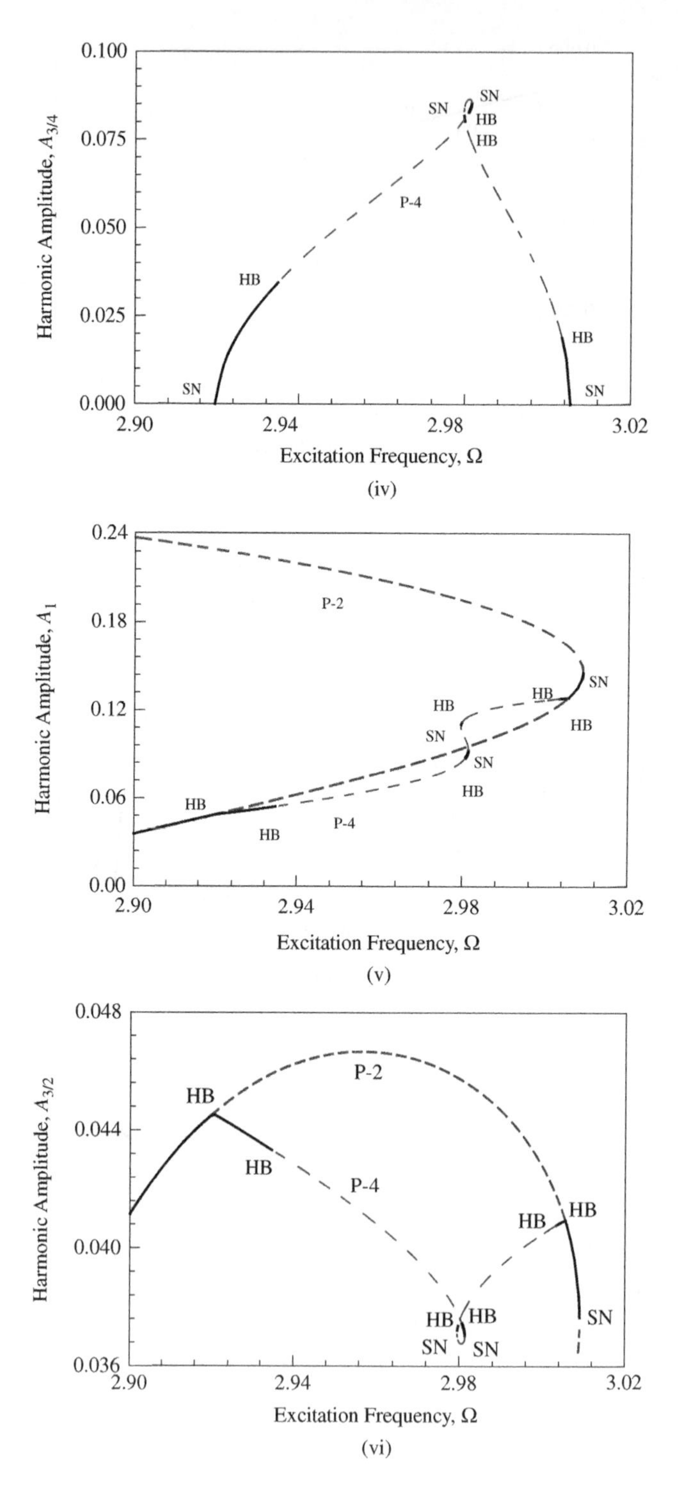

Figure 4.3 (*continued*)

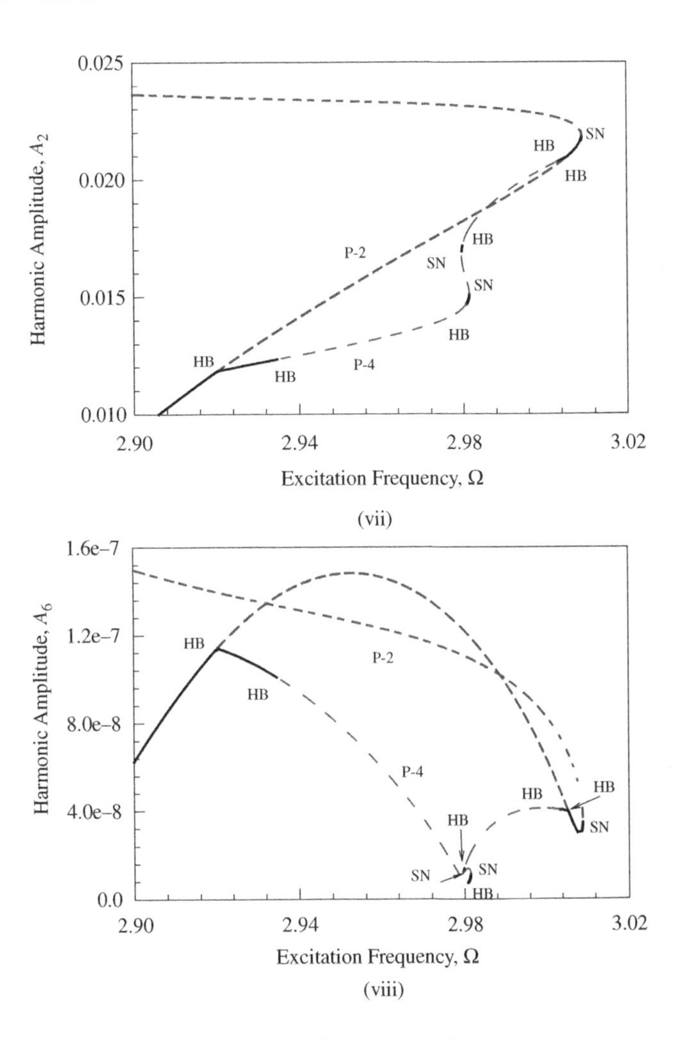

Figure 4.3 *(continued)*

In Figure 4.3(i), constant $a_0^{(m)} \sim -3 \times 10^{-2}$ is presented. The four segments of stable solutions and three segments of unstable solutions of period-4 motions are clearly observed. The harmonic amplitude $A_{1/4} \sim 2 \times 10^{-2}$ is presented in Figure 4.3(ii) and the two segments of stable solutions in the middle of frequency ranges are very tiny, which is difficult to observe. The harmonic amplitude $A_{1/2} \sim 0.3$ is presented in Figure 4.3(iii) for the bifurcation tree of period-2 to period-4 motion. The harmonic amplitude $A_{3/4} \sim 10^{-1}$ in Figure 4.3(iv) is similar to the harmonic amplitude $A_{1/4}$. The harmonic amplitude $A_1 \sim 0.24$ is similar to $A_{1/2}$, as shown in Figure 4.3(v). To avoid abundant illustrations, the harmonic amplitudes $A_{k/4}$ ($k = 4l + 1, l = 1, 2, \ldots$) will not be presented, which are similar to $A_{1/4}$. The harmonic amplitude $A_{3/2} \sim 5 \times 10^{-2}$ is presented in Figure 4.3(vi), which is different from $A_{1/2}$. The harmonic amplitude $A_2 \sim 2.5 \times 10^{-2}$ is similar to A_1, as shown in Figure 4.3(vii). For this bifurcation tree, the harmonic amplitude $A_6 \sim 2 \times 10^{-7}$ is presented in Figure 4.3(viii).

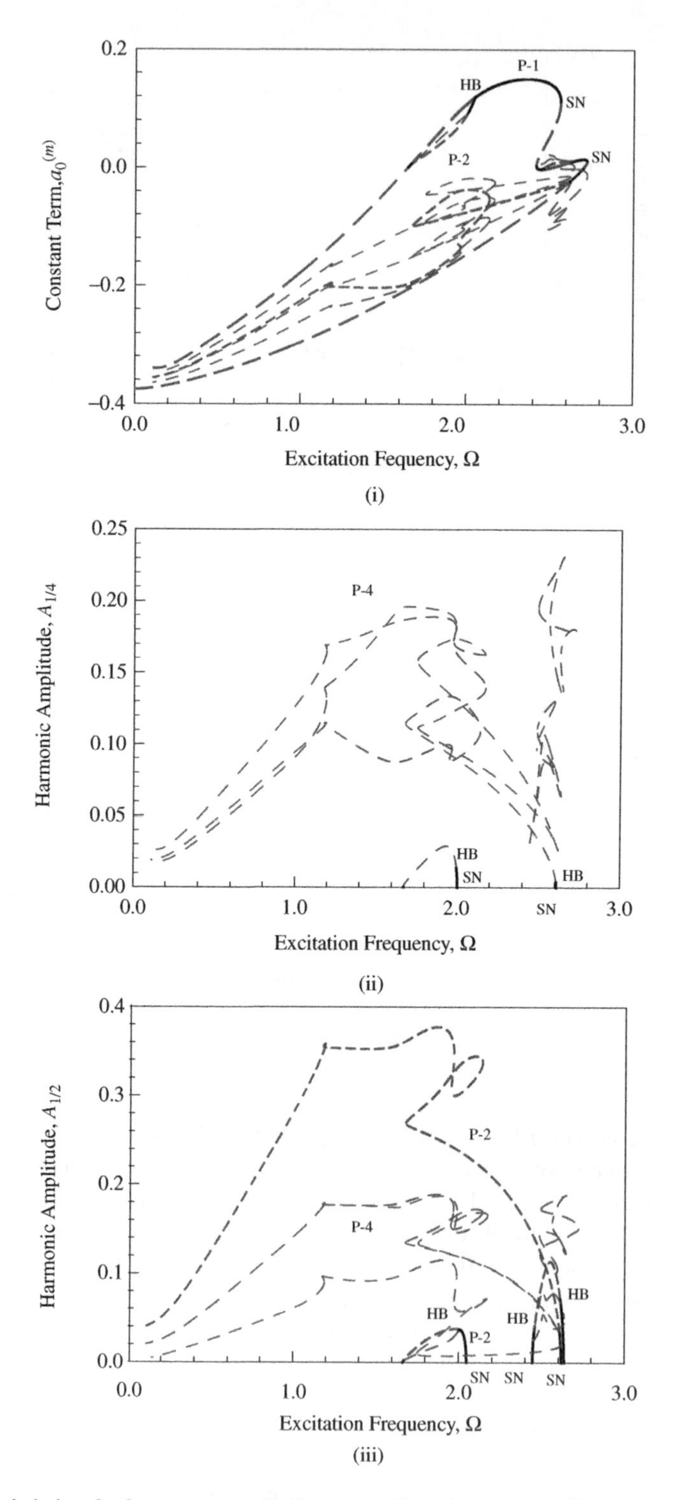

Figure 4.4 Global view for frequency-amplitude curves ($Q_0 = 15$) based on 80 harmonic terms (HB80) of period-1 motion to period-4 motion in the parametric, nonlinear quadratic oscillator: (i) constant term $a_0^{(m)}$, (ii)–(viii) harmonic amplitude $A_{k/m}$ ($k = 1, 2, \ldots, 4, 8, 12, 40, m = 4$). ($\delta = 0.5, \alpha = 5, \beta = 20$)

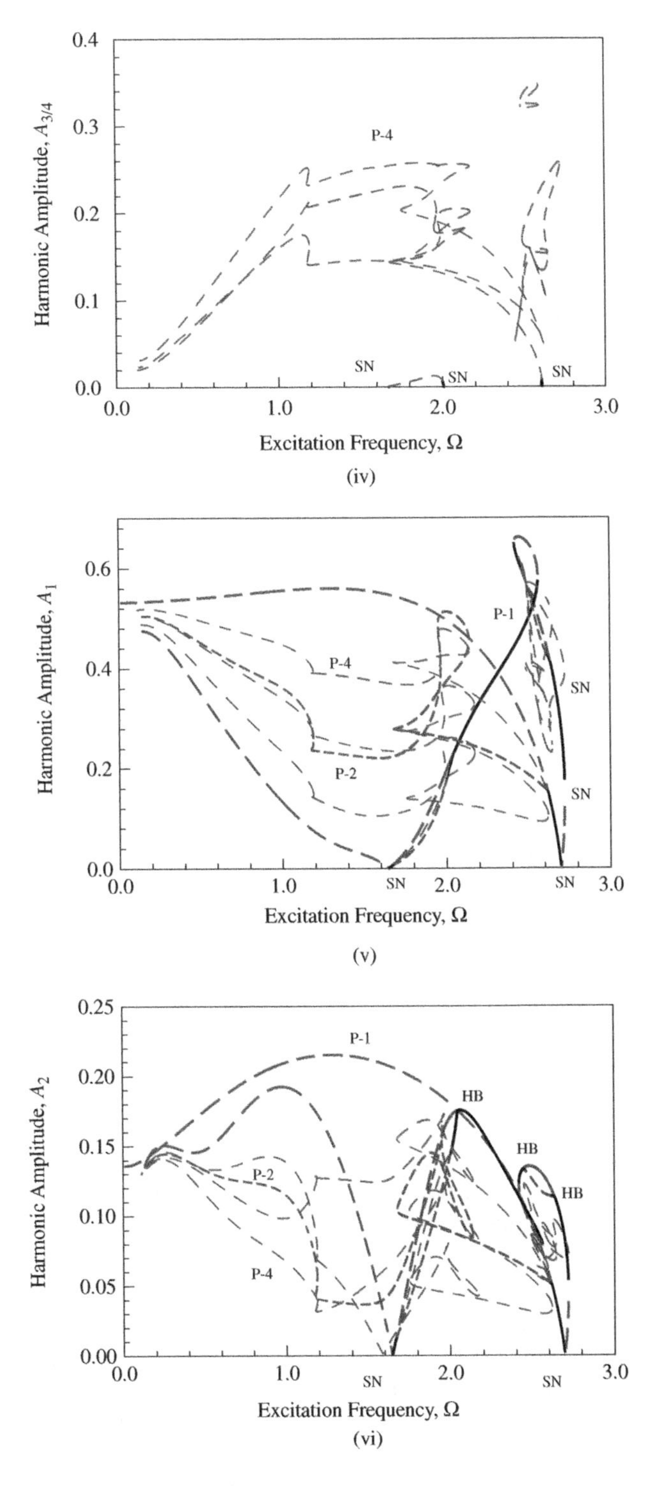

Figure 4.4 (*continued*)

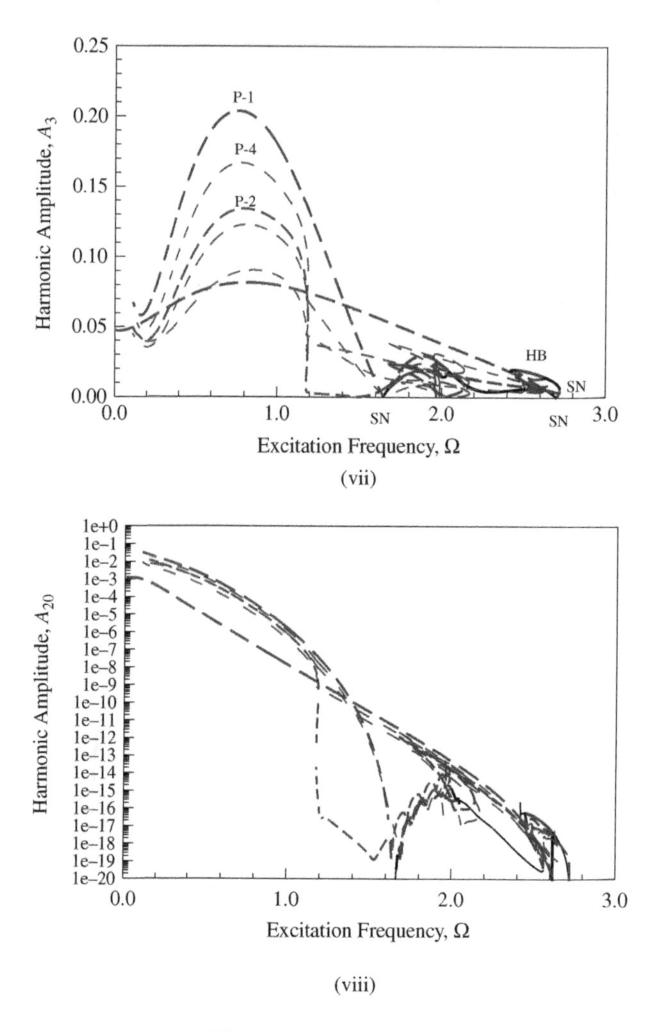

Figure 4.4 (*continued*)

For a linear parametric system, one can find period-2 motions instead of period-1 motion. However, for a nonlinear parametric system, period-1 motions can be found. The global view of the frequency-amplitude for bifurcation trees of period-1 motion to period-4 motion is presented for $Q_0 = 15$. Such illustrations of constants and harmonic amplitudes $(a_0^{(m)}, A_{1/4}, A_{1/2}, A_{3/4}, A_1, A_2, A_3, A_{20})$ are presented for the frequency range of $\Omega \in (0, 3.0)$ in Figure 4.4(i)–(viii), respectively. In Figure 4.4(i), the constant $a_0^{(m)} \in (-0.4, 0.2)$ versus excitation frequency is presented. There are a few branches of bifurcation trees and the stable and unstable solutions of period-1 to period-4 motions are crowded together. The saddle-node bifurcation of period-1 motion occurs between the stable and unstable period-1 motion without the onset of a new periodic motion. In addition to unstable period-1 motion, the Hopf bifurcation of stable period-1 motion will generate the onset of period-2 motion. Continually, the saddle-node bifurcation of period-2 motion is the Hopf bifurcation of the period-1 motion, and the Hopf bifurcation of period-2 motion is the onset of period-4 motion with a saddle-node bifurcation. For period-4 motion, the harmonic amplitude $A_{1/4} \sim 0.25$ is presented in Figure 4.4(ii). Most of solutions are unstable and the stable solutions are in a few short ranges. In addition, independent unstable period-4 motions are observed. The harmonic amplitude $A_{1/2} \sim 0.4$ is presented in Figure 4.4(iii). The onset of period-2 motion is at the HB of the period-1 motion. For the period-4 motion, the harmonic amplitude $A_{3/4} \sim 0.4$ is presented in Figure 4.4(iv), which is similar to $A_{1/4}$. The harmonic amplitude $A_1 \sim 0.7$ is presented in Figure 4.4(v). To show the quantity levels of harmonic amplitudes, the harmonic amplitudes $A_2 \sim 0.25$ is presented in Figure 4.4(vi). The harmonic amplitude $A_3 \sim 0.25$ is presented in Figure 4.4(vii), which is different from A_2. For lower frequency, the quantity levels of A_1 and A_2 are quite close, but for higher frequency, the quantity level of A_1 is much higher than the quantity level of A_2. To show the change of quantity levels, the harmonic amplitude A_{20} is presented in Figure 4.4(viii) with a common logarithmic scale. The quantity level of the harmonic amplitude A_{20} decreases with excitation frequency with a power law. For stable solutions, the harmonic amplitude $A_{20} \sim 10^{-13}$ is observed.

To show bifurcation trees of period-1 motion to period-4 motion, the harmonic amplitudes ($a_0^{(m)}$ and $A_{k/4}$, $k = 4l$, $l = 1, 2, 3, 4, 9$) are presented for the bifurcation tree-1 with $\Omega \in (2.55, 2.71)$ in Figure 4.5(i)–(vi), respectively. The bifurcation tree-2 for $\Omega \in (2.40, 2.7)$ is based on the period-1 motion to period-2 motion, and the harmonic amplitudes ($a_0^{(m)}$ and $A_{k/2}$, $k = 2l$, $l = 1, 2, 3, 4, 9$) are illustrated in Figure 4.6(i)–(vi). In Figure 4.7(i)–(vi), the harmonic amplitudes ($a_0^{(m)}$ and $A_{k/4}$, $k = 4l$, $l = 1, 2, 3, 4, 9$) for the bifurcation tree-3 are presented for $\Omega \in (1.98, 2.06)$. The bifurcation tree-4 for $\Omega \in (1.665, 1.685)$ is presented in Figure 4.8(i)–(vi) through the harmonic amplitudes ($a_0^{(m)}, A_1, A_2, \ldots, A_4, A_9$).

4.1.3 Numerical Simulations

For system parameters ($\delta = 0.5$, $\alpha = 5$, $\beta = 20$), a period-2 motion ($\Omega = 6.8$) is presented in Figure 4.9 for $Q_0 = 15$ and the initial condition ($x_0 \approx -0.174477$, $y_0 \approx 0.065932$) is computed from the analytical solution with 10 harmonic terms (HB10). In Figure 4.9(i), the trajectory of period-2 motion is presented. The asymmetry of periodic motion with one cycle is observed but not a simple cycle. The amplitude spectrum based on five harmonic terms is presented in Figure 4.9(ii). The main harmonic amplitudes with different harmonic orders are $a_0^{(2)} \approx -0.029046$, $A_{1/2} \approx 0.130685$, $A_1 \approx 5.850010\text{E-}3$, $A_{3/2} \approx 9.633124\text{E-}3$. The other harmonic amplitudes are $A_2 \sim 10^{-4}$, $A_{5/2} \sim 2.5 \times 10^{-4}$, $A_3 \sim 3 \times 10^{-6}$, $A_{7/2} \sim 3.2 \times 10^{-6}$, $A_4 \sim 10^{-7}$, $A_{9/2} \sim 2.6 \times 10^{-8}$, and $A_5 \sim 2 \times 10^{-9}$. For this periodic motion, the harmonic amplitude $A_{1/2} \approx 0.130685$ plays an important role in period-2 motion. For $k \geq 6$, the harmonic amplitudes $A_{k/2}$ can be ignored. For another branch of period-2 motion to period-4 motion, only phase trajectories and spectrums for period-2 motion will be presented in Figure 4.9(iii) and (vi) for $\Omega = 3.042402$ with ($x_0 \approx -0.427224$, $y_0 \approx 0.753339$,) and $\Omega = 2.89$ with ($x_0 \approx -0.069109$, $y_0 \approx 0.629068$). In Figure 4.9(iii), the trajectory of period-2 motion with $\Omega = 3.042402$ is presented. The analytical solutions possess 20 harmonic terms (HB24). The harmonic amplitudes in spectrum are shown in Figure 4.9(iv). The main harmonic amplitudes with different harmonic orders are $a_0 \approx -0.043146$, $A_{1/2} \approx 0.382101$, $A_1 \approx 0.299341$, $A_{3/2} \approx 0.057045$, and $A_2 \approx 0.054044$. The other harmonic amplitudes are $A_{5/2} \sim 6.2 \times 10^{-3}$, $A_3 \sim 1.1 \times 10^{-4}$, $A_{7/2} \sim 6.4 \times 10^{-4}$, $A_4 \sim 2.8 \times 10^{-4}$, $A_{9/2} \sim 4.5 \times 10^{-5}$, and $A_5 \sim 1 \times 10^{-5}$. $A_{k/2} \in (10^{-10}, 10^{-5})$ for $k = 11, 12, \ldots, 20$. For this periodic motion,

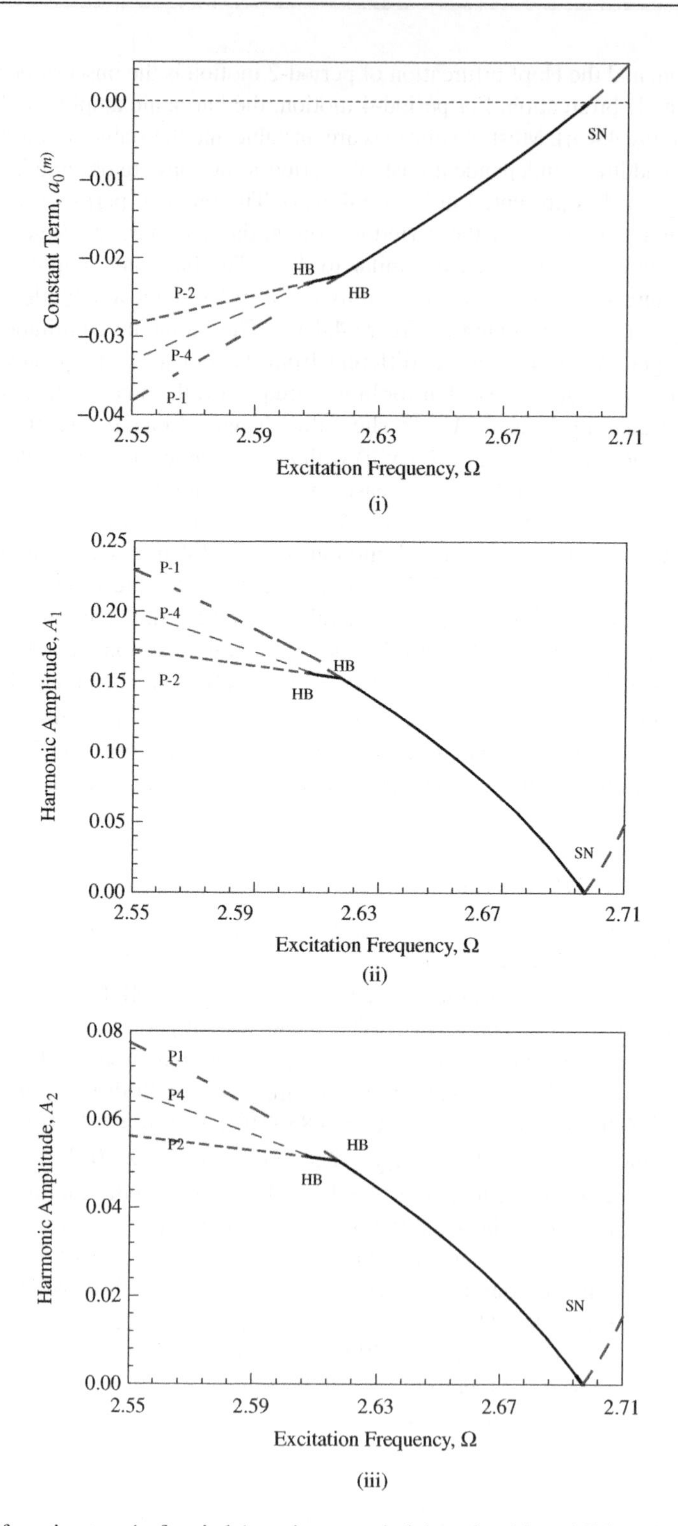

Figure 4.5 Bifurcation tree-1 of period-1 motion to period-4 motion ($Q_0 = 15$) based on 80 harmonic terms (HB80) in the parametric, nonlinear quadratic oscillator: (i) constant term $a_0^{(m)}$, (ii)–(viii) harmonic amplitude $A_{k/m}$ ($k = 4, 8, 12, 16, 36, m = 4$). ($\delta = 0.5, \alpha = 5, \beta = 20$)

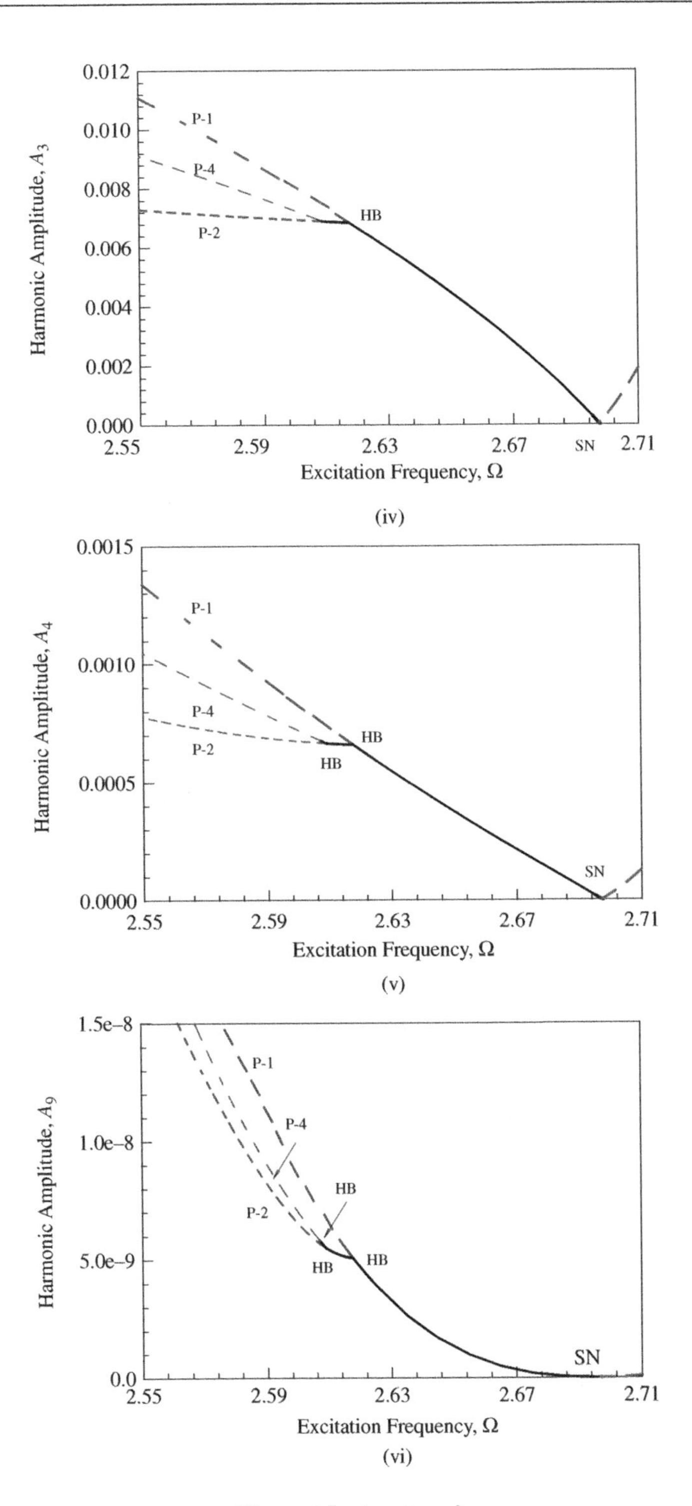

Figure 4.5 *(continued)*

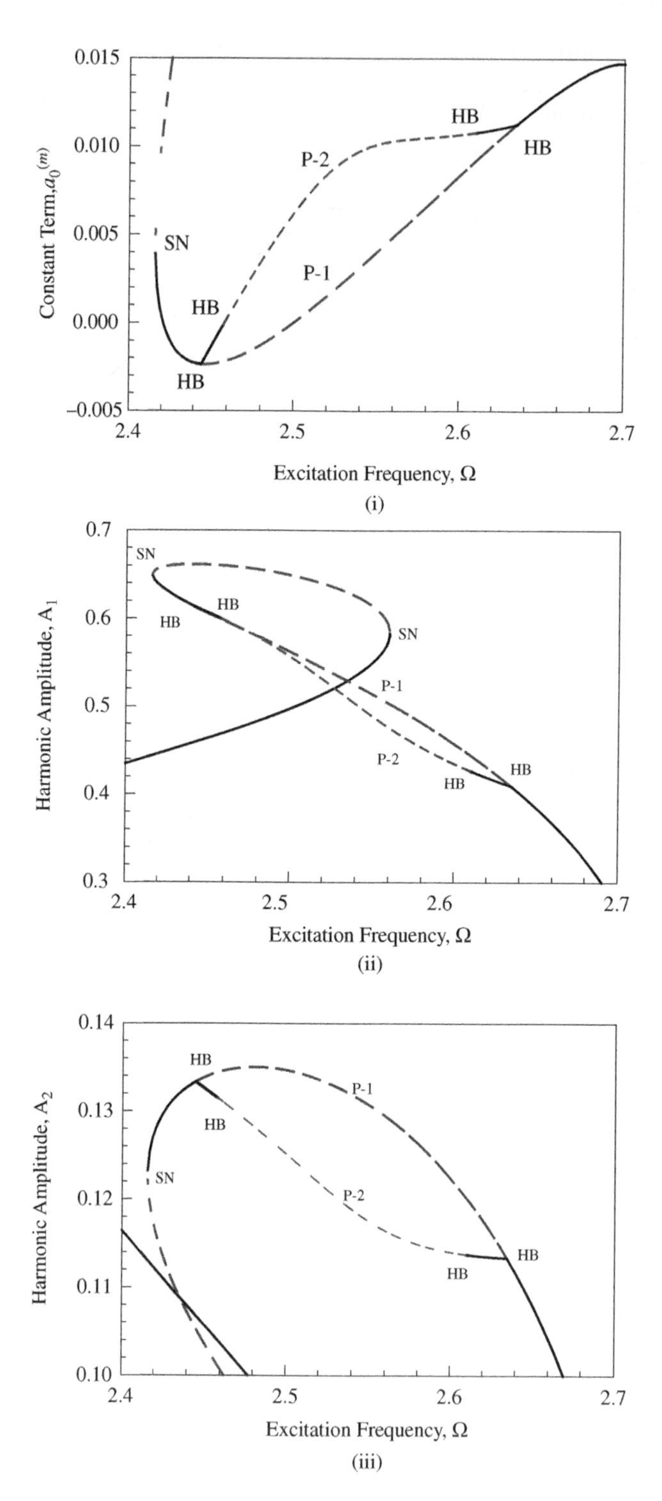

Figure 4.6 Bifurcation tree-2 of period-1 motion to period-2 motion ($Q_0 = 15$) based on 40 harmonic terms (HB40) in the parametric, nonlinear quadratic oscillator: (i) constant term $a_0^{(m)}$, (ii)–(viii) harmonic amplitude $A_{k/m}$ ($k = 2, 4, 6, 8, 18, m = 2$). ($\delta = 0.5, \alpha = 5, \beta = 20$)

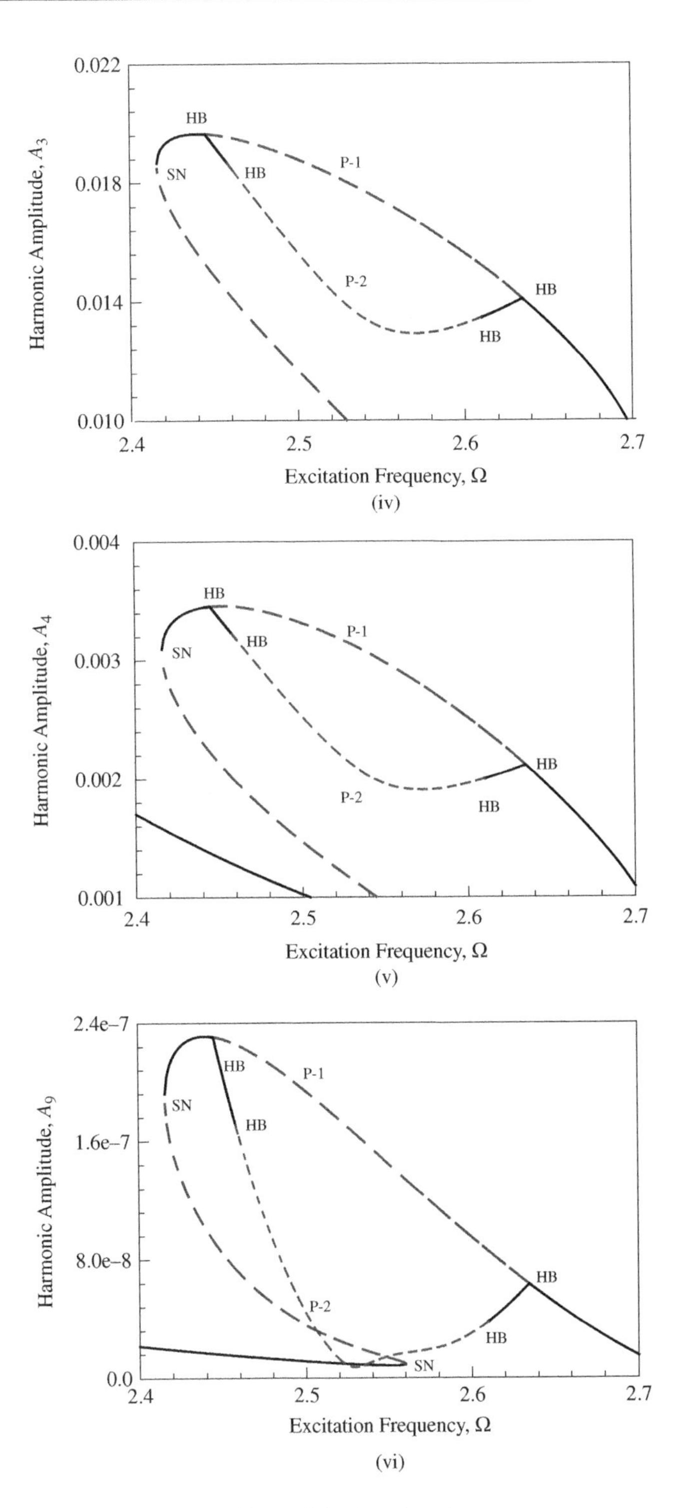

Figure 4.6 *(continued)*

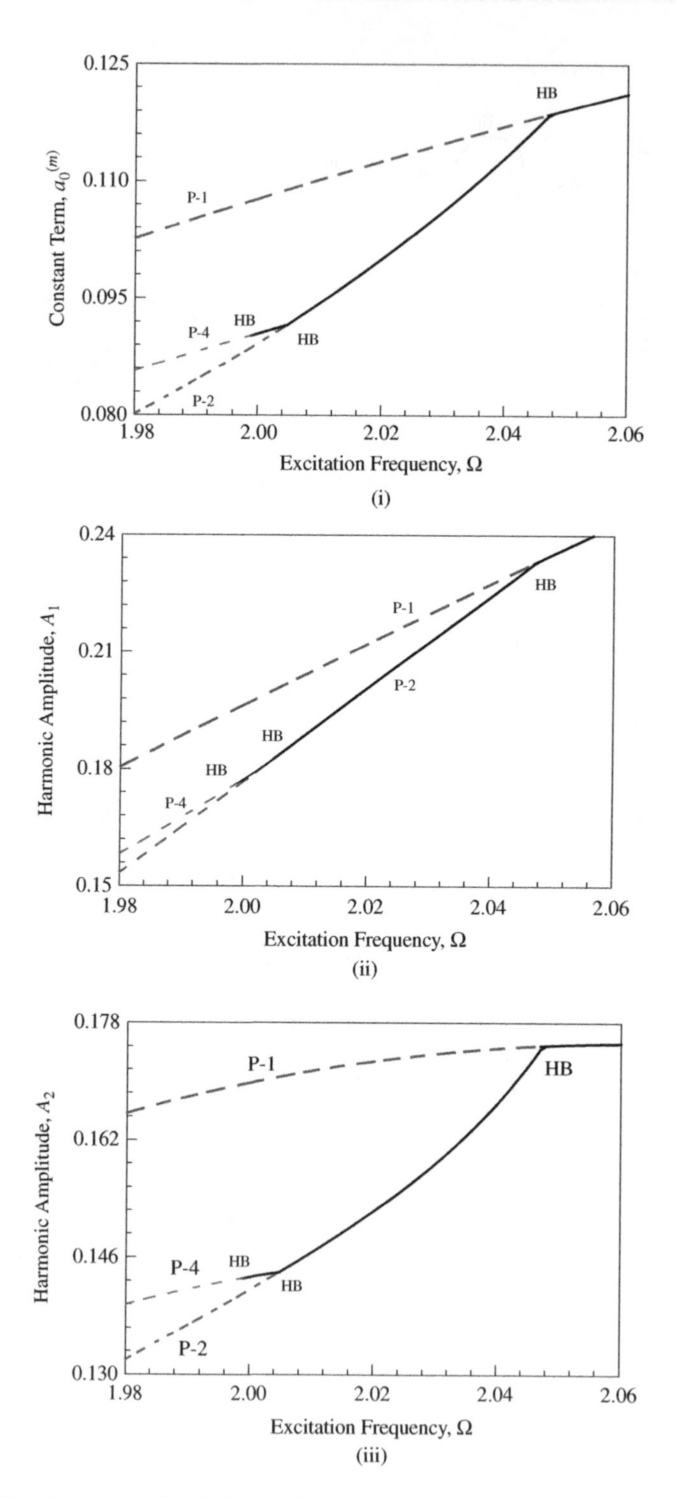

Figure 4.7 Bifurcation tree-3 of period-1 motion to period-4 motion ($Q_0 = 15$) based on 80 harmonic terms (HB80) in the parametric, nonlinear quadratic oscillator: (i) constant term $a_0^{(m)}$, (ii)–(vi) harmonic amplitude $A_{k/m}$ ($k = 4, 8, 12, 16, 36$, $m = 4$). ($\delta = 0.5$, $\alpha = 5$, $\beta = 20$)

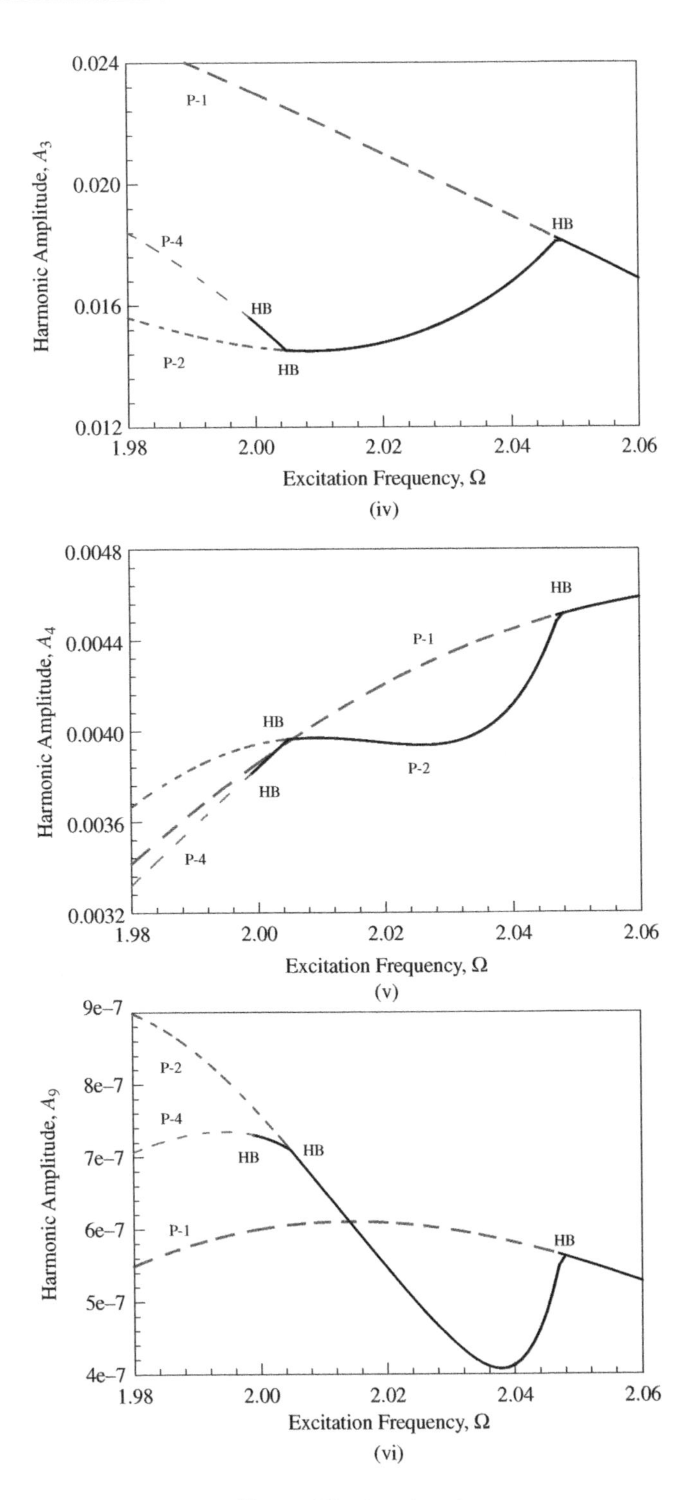

Figure 4.7 *(continued)*

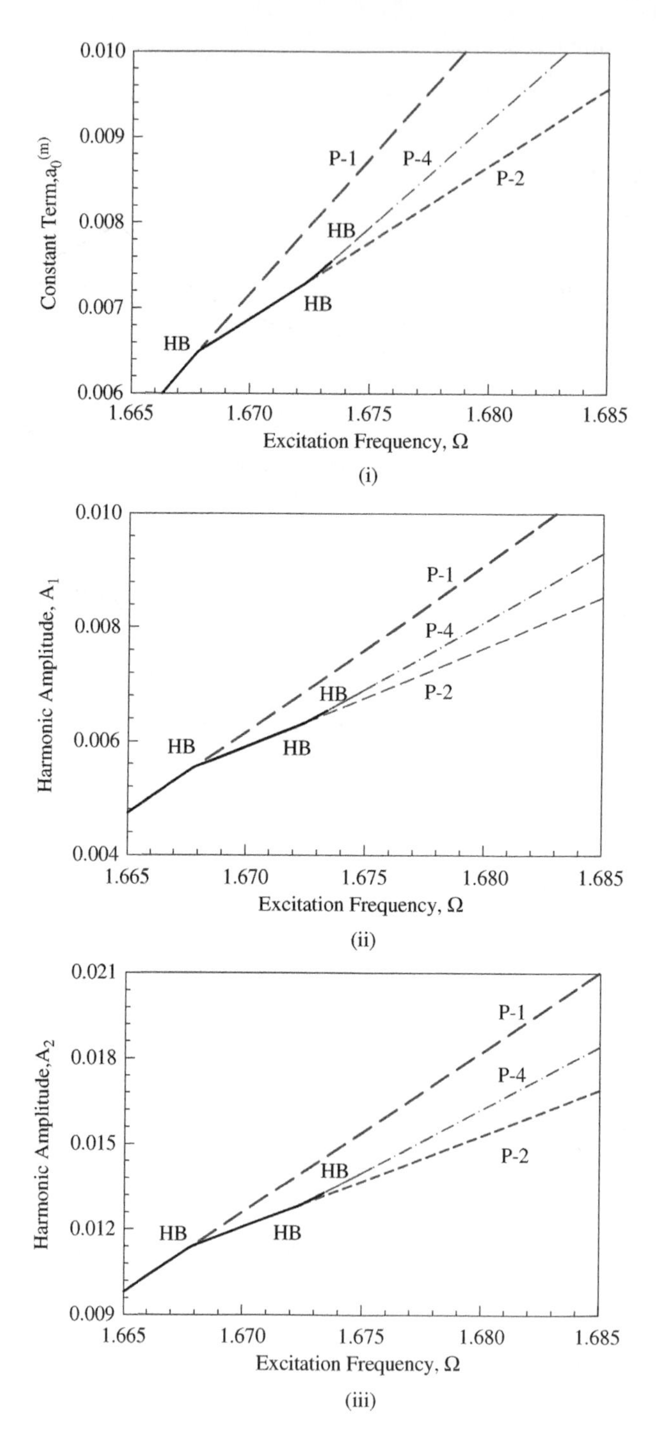

Figure 4.8 Bifurcation tree-4 of period-1 motion to period-4 motion ($Q_0 = 15$) based on 80 harmonic terms (HB80) in the parametric, nonlinear quadratic oscillator: (i) constant term $a_0^{(m)}$, (ii)–(vi) harmonic amplitude $A_{k/m}$ ($k = 4, 8, 12, 16, 36$, $m = 4$). ($\delta = 0.5$, $\alpha = 5$, $\beta = 20$)

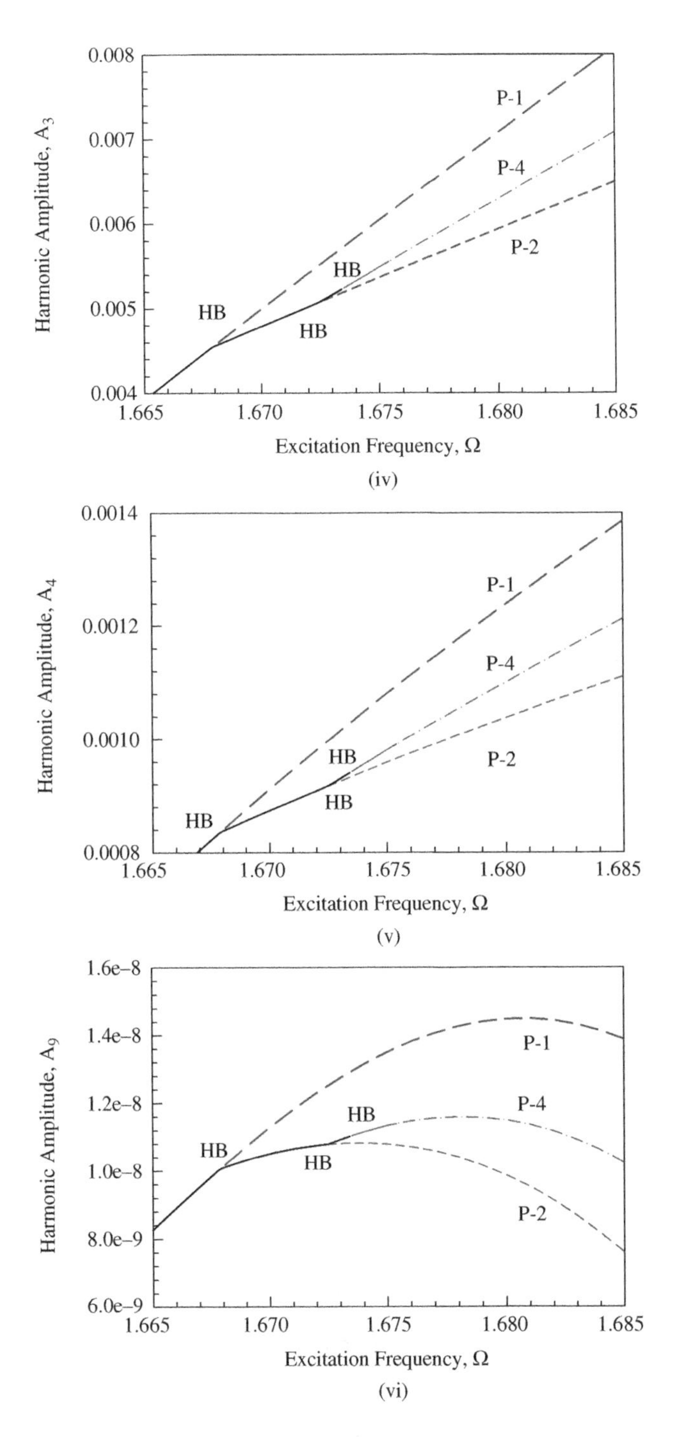

Figure 4.8 *(continued)*

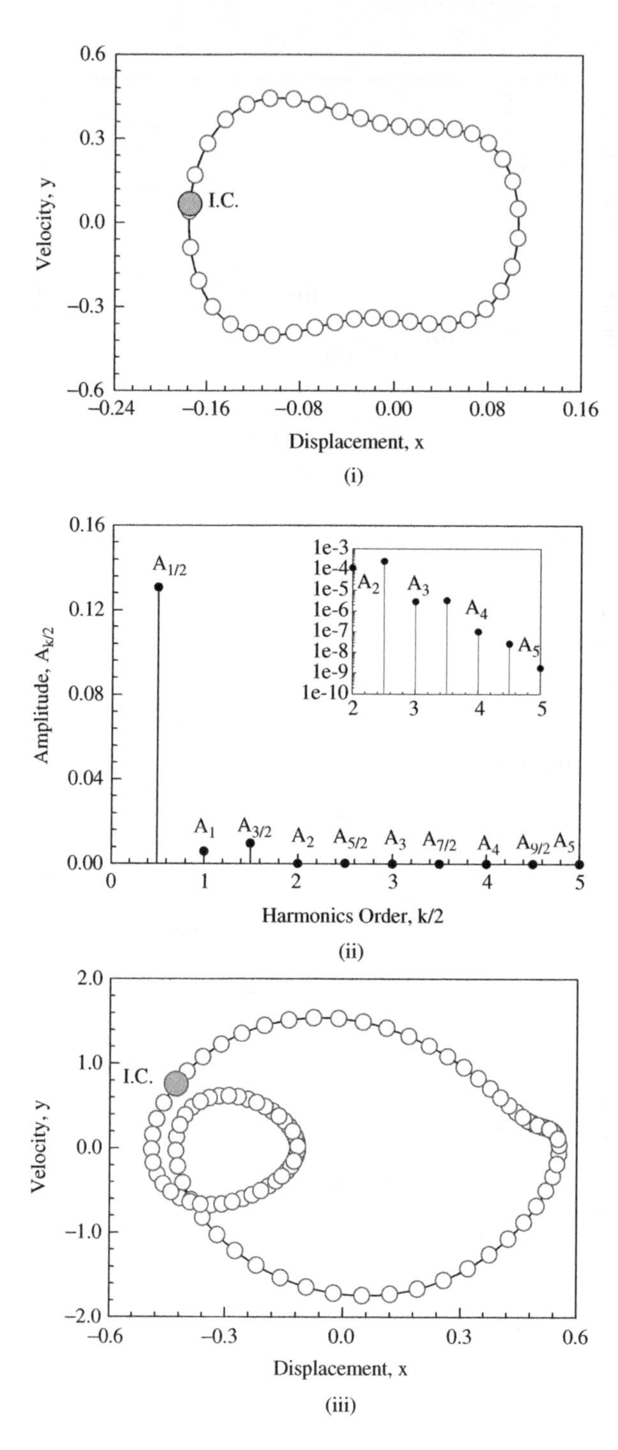

Figure 4.9 Period-2 motions: (i) trajectory and (ii) amplitude ($\Omega = 6.8$, $x_0 \approx -0.174477$, $y_0 \approx 0.065932$), (iii) trajectory and (iv) amplitude ($\Omega = 3.042402$, $x_0 \approx -0.427224$, $y_0 \approx 0.753339$), (v) trajectory and (vi) amplitude ($\Omega = 2.89$, $x_0 \approx -0.069109$, $y_0 \approx 0.629068$). ($\delta = 0.5$, $\alpha = 5$, $\beta = 20$, $Q_0 = 15$)

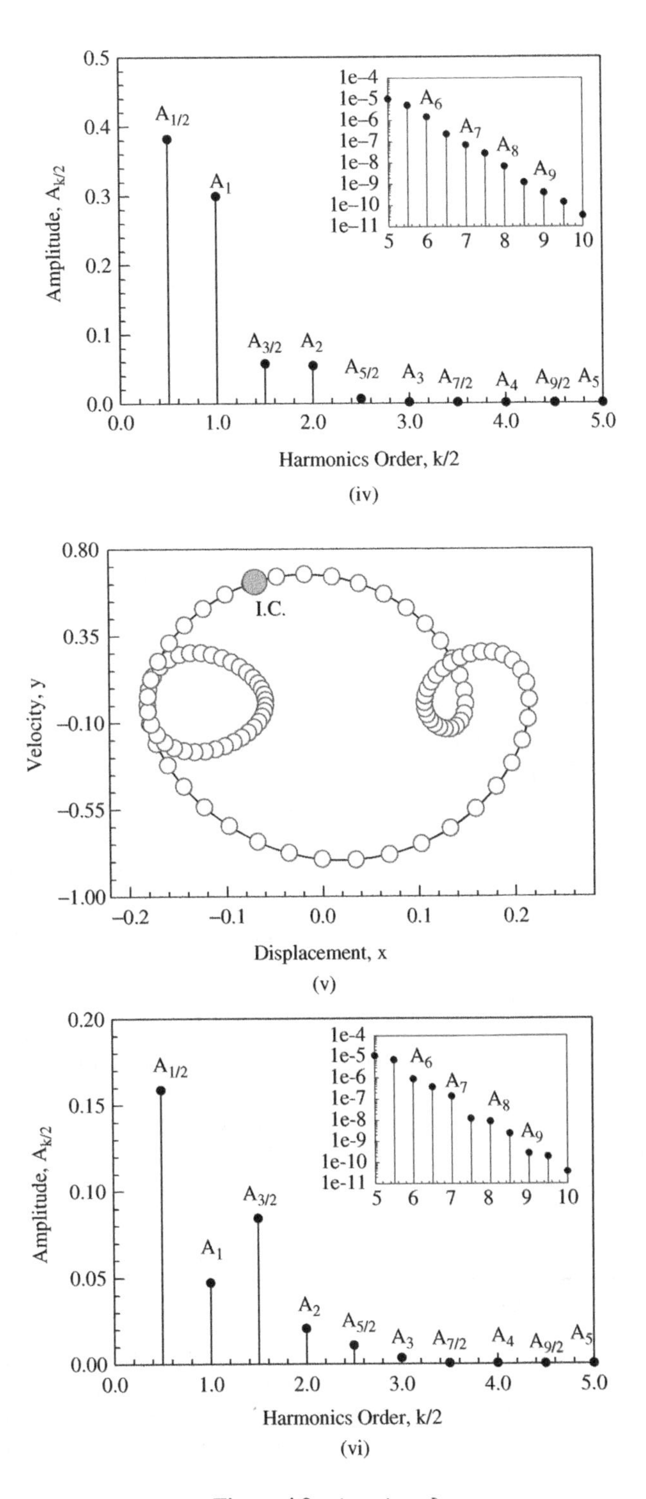

Figure 4.9 *(continued)*

the harmonic amplitudes $A_{1/2}$ to A_2 make significant contribution on period-2 motion rather than majorly from $A_{1/2}$. $A_{1/2}$ and A_1 are two most important terms. For $k \geq 10$, the harmonic amplitudes $A_{k/2}$ can be ignored. In Figure 4.9(v), the phase trajectory of period-2 motion $\Omega = 2.89$ is illustrated, which is different from the one in Figure 4.9(iii). The corresponding spectrum of period-2 motion is presented in Figure 4.9(vi). The main harmonic amplitudes with different harmonic orders are $a_0^{(2)} \approx -2.292999\text{e-}3$, $A_{1/2} \approx 0.158649$, $A_1 \approx 0.047177$, $A_{3/2} \approx 0.084358$, $A_2 \approx 0.020375$, and $A_{5/2} \approx 0.010720$. The other harmonic amplitudes are $A_3 \sim 3.4 \times 10^{-3}$, $A_{7/2} \sim 3.0 \times 10^{-4}$, $A_4 \sim 2.9 \times 10^{-4}$, $A_{9/2} \sim 5.7 \times 10^{-5}$, and $A_5 \sim 1.1 \times 10^{-5}$. $A_{k/2} \in (10^{-11}, 10^{-5})$ for $k = 11, 12, \ldots, 20$. For this period-2 motion, the harmonic amplitudes ($A_{1/2}$ to $A_{5/2}$) make significant contribution on period-2 motion. The harmonic amplitude $A_{3/2}$ has more contribution than A_1 on period-2 motion (i.e., $A_{3/2} \approx 2A_1$).

For this kind of periodic motion, once the Hopf bifurcation occurs, the period-4 motion can be observed in such a bifurcation tree. Thus, with system parameters ($\delta = 0.5$, $\alpha = 5$, $\beta = 20$), a period-4 motion ($\Omega = 6.78$) is illustrated in Figure 4.10 for $Q_0 = 15$ and the initial condition ($x_0 \approx -0.159366$, $y_0 \approx 0.065306$) is computed from the analytical solution with 24 harmonic terms (HB24). In Figure 4.10(a), the trajectory of period-4 motion is presented, and there are two cycles instead of one cycle in period-2 motion. The amplitude spectrum based on 24 harmonic terms is presented in Figure 4.10(b). The main harmonic amplitudes are $a_0^{(4)} \approx -0.030139$, $A_{1/4} \approx 0.013961$, $A_{1/2} \approx 0.132269$, $A_{3/4} \approx 2.824103\text{E-}3$, $A_1 \approx 6.130666\text{E-}3$, $A_{5/4} \approx 1.371135\text{E-}3$, $A_{3/2} \approx 9.793038\text{E-}3$, $A_{5/4} \approx 1.048286\text{E-}4$. The other harmonic amplitudes are $A_2 \sim 1.3 \times 10^{-4}$, $A_{9/4} \sim 4 \times 10^{-5}$, $A_{5/2} \sim 2.5 \times 10^{-4}$, $A_{11/4} \sim 9.6 \times 10^{-7}$, $A_3 \sim 3 \times 10^{-6}$, $A_{13/4} \sim 5.4 \times 10^{-7}$, $A_{7/2} \sim 3.3 \times 10^{-6}$, $A_{15/4} \sim 1.8 \times 10^{-8}$, $A_4 \sim 10^{-7}$, $A_{17/4} \sim 4 \times 10^{-9}$, $A_{9/2} \sim 2.6 \times 10^{-8}$, $A_{19/4} \sim 5.2 \times 10^{-9}$, $A_5 \sim 1.8 \times 10^{-9}$. $A_{21/4} \sim 2.1 \times 10^{-11}$, $A_{11/2} \sim 1.6 \times 10^{-10}$, $A_{19/4} \sim 7 \times 10^{-12}$, and $A_6 \sim 2.2 \times 10^{-11}$. For this periodic motion, the harmonic amplitudes of $A_{1/4} \approx 0.013961$ and $A_{1/2} \approx 0.132269$ make important contribution on the period-4 motion. For $k \geq 12$, the harmonic amplitudes $A_{k/4}$ can be ignored.

Next illustrations are trajectories for period-1 motion to period-4 motions. In linear parametric oscillators, no such period-1 motions can be observed. However, the nonlinear parametric oscillator possesses bifurcation tree of period-1 motion to chaos. For different branches, nonlinear dynamical behaviors are different, as shown in Figures 4.11 and 4.12. To save space, only phase trajectories and spectrums are illustrated, and the initial conditions are listed in Table 4.1.

On the bifurcation tree relative to period-1 motion with $\Omega = 1.665$, the trajectories and amplitude spectrums of the period-1 motion ($\Omega = 1.665$), period-2 motion ($\Omega = 1.67$), and period-4 motion ($\Omega = 1.73$) are presented in Figure 4.11(i)–(vi), respectively.

Table 4.1 Input data for numerical simulations ($\delta = 0.5, \alpha = 5, \beta = 20$)

	Frequency Ω	Initial conditions (x_0, y_0)	Type and stability
Figure 4.11(i),(ii)	1.665	(−0.012807, 2.851630E-3)	Period-1 (stable)
Figure 4.11(iii),(iv)	1.670	(−0.019510, 5.134178E-3)	Period-2 (stable)
Figure 4.11(v),(vi)	1.673	(−0.021881, 6.075298E-3)	Period-4 (stable)
Figure 4.12(i),(ii)	2.625	(−0.050625,0.659436)	Period-1 (stable)
Figure 4.12(iii),(iv)	2.613	(−0.040570,0.606281)	Period-2 (stable)
Figure 4.12(v),(vi)	2.608	(−0.098209, 0.849246)	Period-4 (stable)

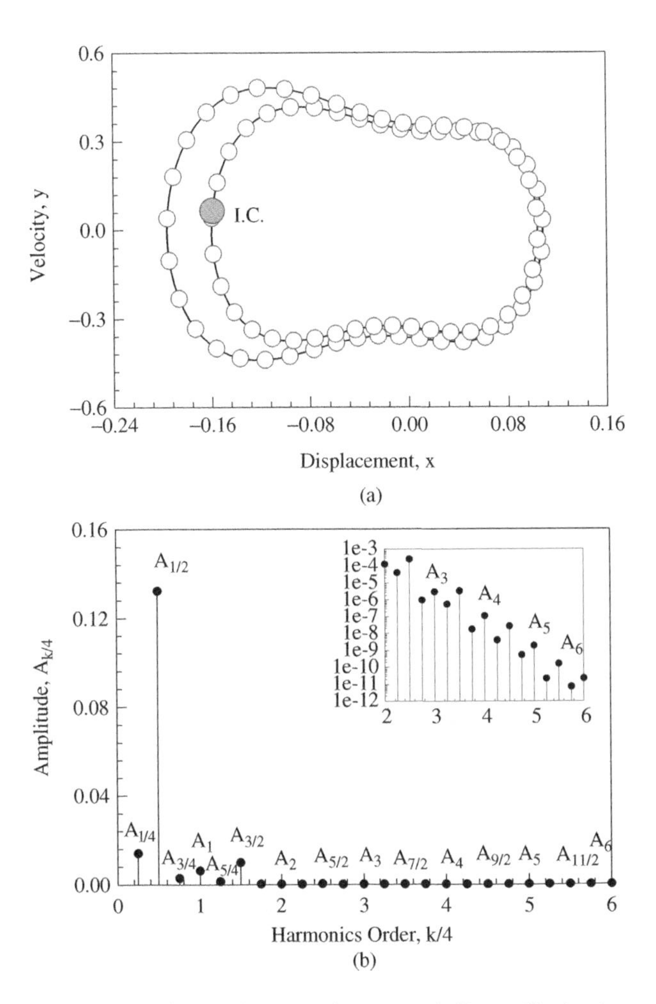

Figure 4.10 Period-4 motion ($\Omega = 6.78$): (a) trajectory and (b) amplitude. ($x_0 \approx -0.159366$, $y_0 \approx 0.065306$). ($\delta = 0.5$, $\alpha = 5$, $\beta = 20$, $Q_0 = 15$)

In Figure 4.11(i), the trajectory of period-1 motion for $\Omega = 1.665$ possesses two cycles because the second harmonic term (A_2) plays an important role on this period-1 motion. In Figure 4.11(ii), the main harmonic amplitudes are $a_0 = 5.583469\text{E-}3$, $A_1 = 4.734440\text{E-}3$, $A_2 = 9.802613\text{E-}3$, $A_3 = 3.923209\text{E-}3$, $A_4 = 7.304942\text{E-}4$, $A_5 = 7.274665\text{E-}5$. The other harmonic amplitudes are $A_6 \sim 2.5 \times 10^{-6}$, $A_7 \sim 4.1 \times 10^{-7}$, $A_8 \sim 8.5 \times 10^{-8}$, $A_9 \sim 8.3 \times 10^{-9}$, $A_{10} \sim 3.8 \times 10^{-10}$. With increasing excitation frequency, period-2 motion can be observed. In Figure 4.11(iii), the trajectory of period-2 motion for $\Omega = 1.67$ possesses four cycles. The distribution of harmonic amplitudes presented in Figure 4.11(iv) can give us an important clue. The main harmonic amplitudes for this period-2 motion are $a_0^{(2)} \approx 6.872412\text{E-}3$, $A_{1/2} \approx 1.656075\text{E-}3$, $A_1 \approx 5.901387\text{E-}3$, $A_{3/2} \approx 2.555508\text{E-}3$, $A_2 \approx 0.012082$, $A_{5/2} \approx 1.834976\text{E-}3$, $A_3 \approx 4.788544\text{E-}3$, $A_{7/2} \approx 4.661318\text{E-}4$, $A_4 \approx 8.737198\text{E-}4$, $A_{9/2} \approx 5.460212\text{E-}5$, $A_5 \approx 8.153977\text{E-}5$. The other harmonic amplitudes are $A_{k/2} \in (10^{-10}, 10^{-5})$ ($k = 11, 12, \ldots, 20$). For period-4 motion ($\Omega = 1.673$), the corresponding trajectory with eight cycles is

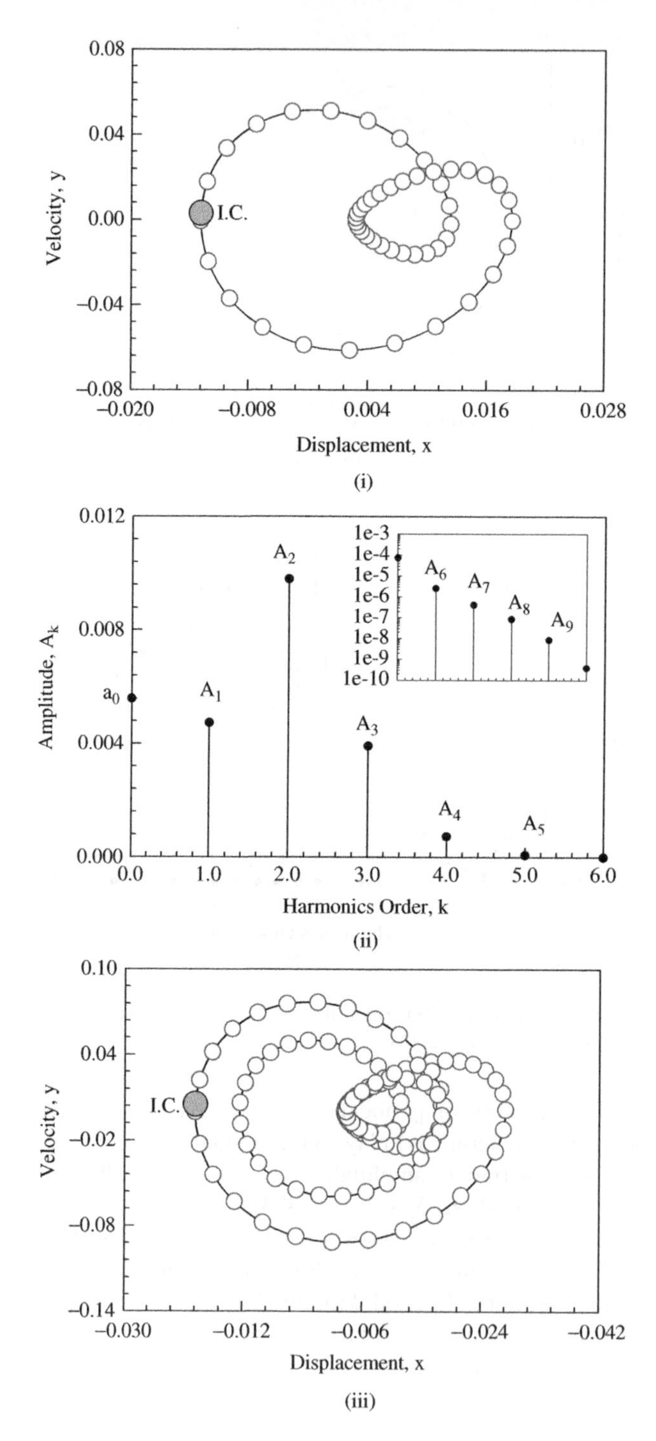

Figure 4.11 Period-1 motion ($\Omega = 1.665$) with ($x_0 \approx -0.012807, y_0 \approx 2.851630\text{E-3}$): (i) trajectory and (ii) amplitude. Period-2 motion ($\Omega = 1.67$) with ($x_0 \approx -0.019150$, $y_0 \approx 5.134178\text{E-3}$): (iii) trajectory and (iv) amplitude. Period-4 motion ($\Omega = 1.673$) with ($x_0 \approx -0.021881$, $y_0 \approx 6.075298\text{E-3}$): (v) trajectory and (vi) amplitude. ($\delta = 0.5$, $\alpha = 5$, $\beta = 20$, $Q_0 = 15$)

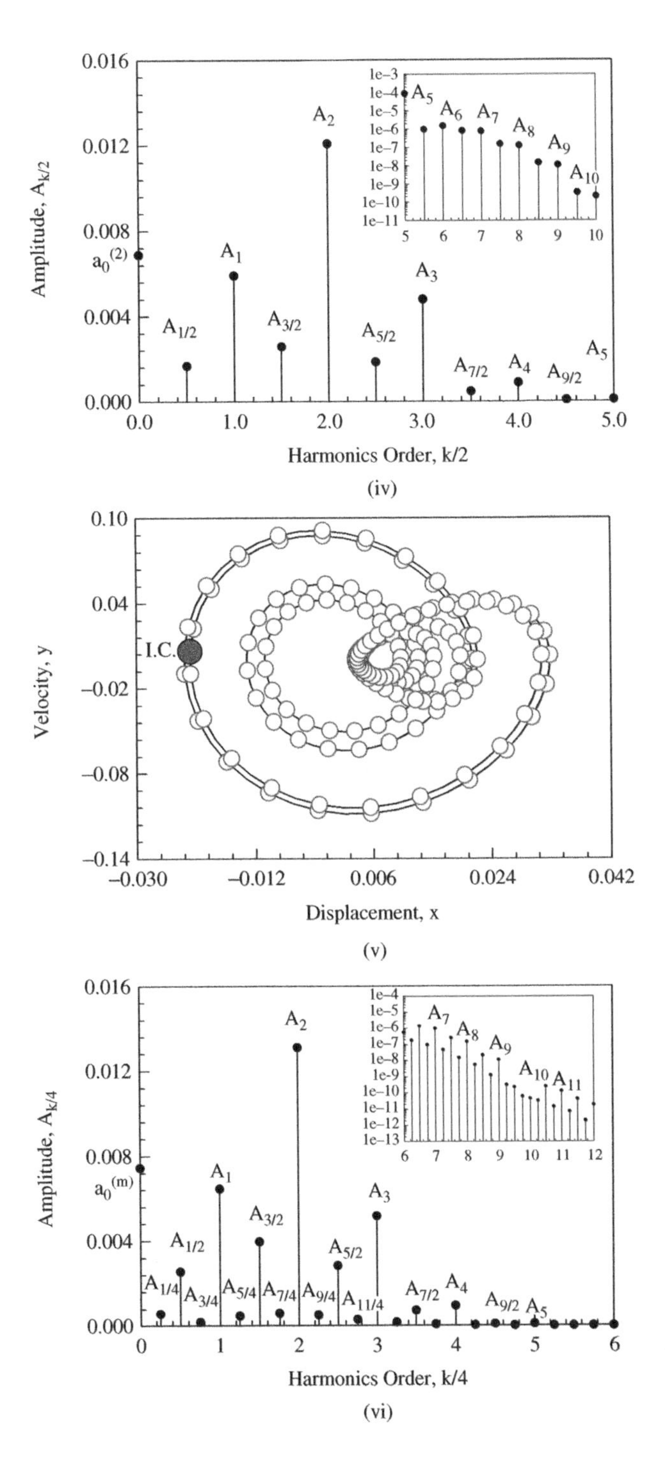

Figure 4.11 *(continued)*

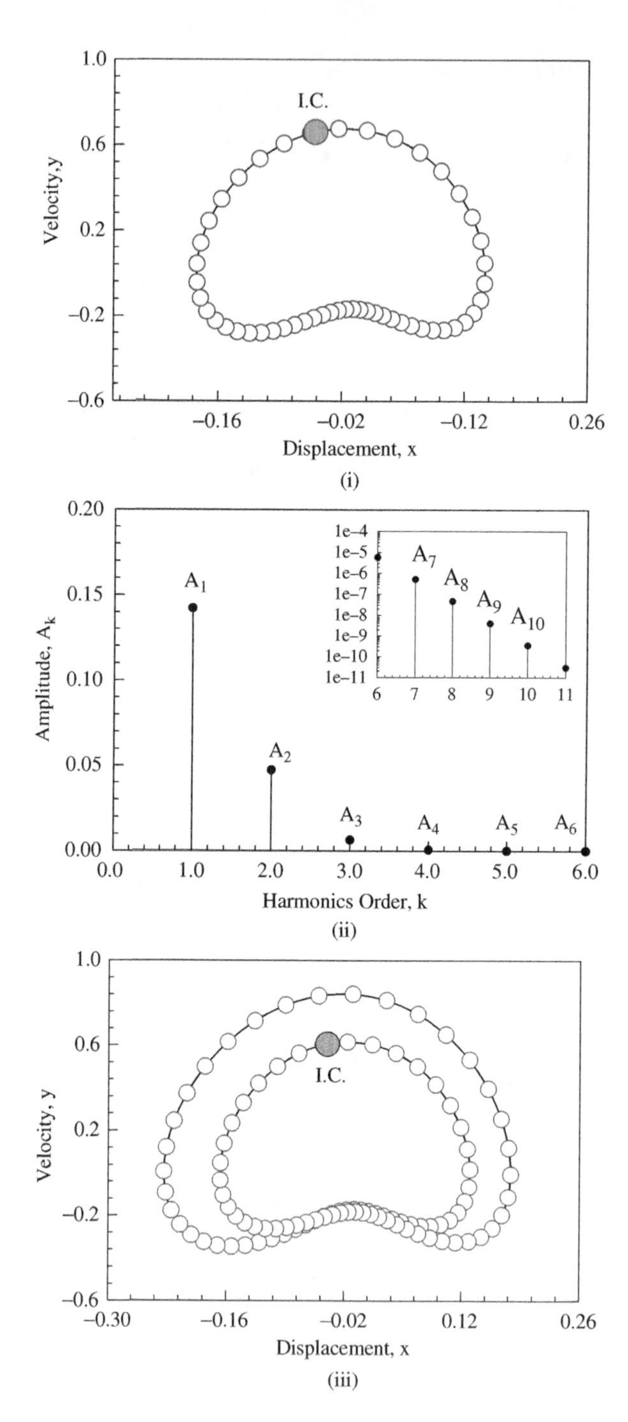

Figure 4.12 Period-1 motion ($\Omega = 2.625$) with ($x_0 \approx -0.050625$, $y_0 \approx 0.659436$): (i) trajectory and (ii) amplitude. Period-2 motion ($\Omega = 2.613$) with ($x_0 \approx -0.040570$, $y_0 \approx 0.606281$): (iii) trajectory and (iv) amplitude. Period-4 motion ($\Omega = 2.608$) with ($x_0 \approx -0.098209$, $y_0 \approx 0.849246$): (v) trajectory and (vi) amplitude. ($\delta = 0.5$, $\alpha = 5$, $\beta = 20$, $Q_0 = 15$)

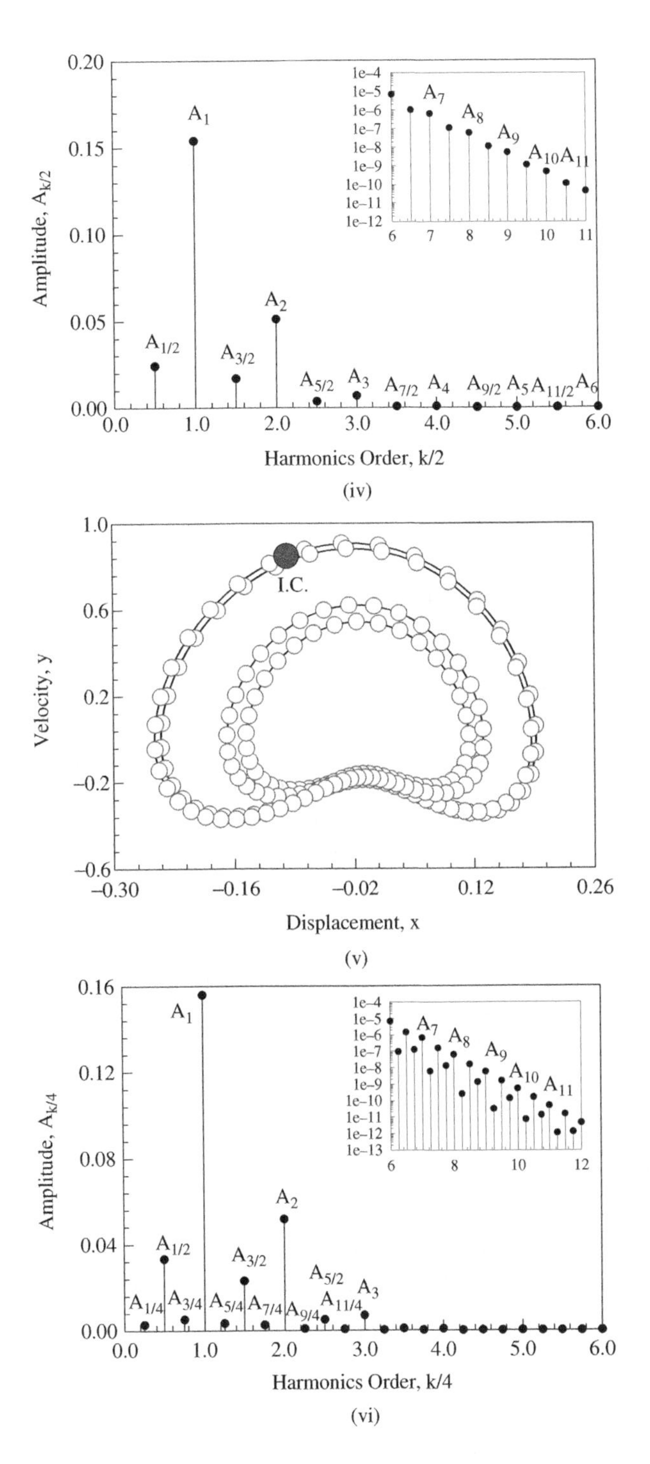

Figure 4.12 *(continued)*

presented in Figure 4.11(v), and the harmonic amplitude distribution in spectrum is presented in Figure 4.11(vi). The main harmonic amplitudes for the period-4 motion are $a_0^{(4)} \approx$ 7.445405E-3, $A_{1/4} \approx$ 5.545468E-4, $A_{1/2} \approx$ 2.551518E-3, $A_{3/4} \approx$ 1.699559E-4, $A_1 \approx$ 6.442443E-3, $A_{5/4} \approx$ 4.533820E-4, $A_{3/2} \approx$ 3.962114E-3, $A_{7/4} \approx$ 5.715145E-4, $A_2 \approx 0.013100$, $A_{9/4} \approx$ 4.976451E-4, $A_{5/2} \approx$ 2.830418E-3, $A_{11/4} \approx$ 2.921089E-4, $A_3 \approx$ 5.161562E-3, $A_{13/4} \approx$ 1.562348E-4, $A_{7/2} \approx$ 7.127523E-4, $A_{15/4} \approx$ 6.135480E-5, $A_4 \approx$ 9.300047E-4, $A_{17/4} \approx$ 2.306539E-5, $A_{9/2} \approx$ 8.124188E-5, $A_{19/4} \approx$ 6.126134E-6, and $A_5 \approx$ 8.316787E-5. The other harmonic amplitudes are $A_{k/4} \in (10^{-11}, 10^{-5})$ $(k = 11, 12, \ldots, 48)$.

For the bifurcation tree relative to period-1 motion with $\Omega = 2.625$, the trajectories and amplitude spectrums of the period-1 motion ($\Omega = 2.625$), period-2 motion ($\Omega = 2.613$), and period-4 motion ($\Omega = 2.608$) are presented in Figure 4.12(i)–(vi), respectively. In Figure 4.12(i), the phase trajectory of period-1 motion for $\Omega = 2.625$ possesses one cycle because the first harmonic term (A_1) plays an important role on the period-1 motion. In Figure 4.12(ii), the main harmonic amplitudes are $a_0 \approx -0.020284$, $A_1 \approx 0.142279$, $A_2 \approx 0.047170$, $A_3 \approx$ 6.332490E-3, $A_4 \approx$ 5.912902E-4, $A_5 \approx$ 5.863546E-5, and $A_6 \approx$ 5.590259E-6. The other harmonic amplitudes are $A_7 \sim 5.0 \times 10^{-7}$, $A_8 \sim 4.5 \times 10^{-8}$, $A_9 \sim 3.9 \times 10^{-9}$, $A_{10} \sim 2.4 \times 10^{-10}$, $A_{11} \sim 2.9 \times 10^{-11}$. With decreasing excitation frequency, period-2 motion can be observed. In Figure 4.12(iii), the trajectory of period-2 motion for $\Omega = 2.613$ possesses two cycles. The distribution of harmonic amplitudes is presented in Figure 4.12(iv). The main harmonic amplitudes for the period-2 motion are $a_0^{(2)} \approx -0.022574$, $A_{1/2} \approx 0.024025$, $A_1 \approx 0.153918$, $A_{3/2} \approx 0.016735$, $A_2 \approx 0.051052$, $A_{5/2} \approx$ 3.589764E-3, $A_3 \approx$ 6.870026E-3, $A_{7/2} \approx$ 5.394541E-4 $A_4 \approx$ 6.614716E-4, $A_{9/2} \approx$ 7.394803E-5, $A_5 \approx$ 6.848459E-5, $A_{11/2} \approx$ 8.867560E-6, and $A_6 \approx$ 6.699010E-6. The other harmonic amplitudes are $A_{k/2} \in (10^{-9}, 10^{-5})$ $(k = 13, 14, \ldots, 24)$. For period-4 motion ($\Omega = 2.608$), the corresponding trajectory with four cycles is presented in Figure 4.12(v), and the harmonic amplitude distribution in spectrum is presented in Figure 4.12(vi). The main harmonic amplitudes for the period-4 motion are $a_0^{(4)} \approx -0.023123$, $A_{1/4} \approx$ 2.682261E-3, $A_{1/2} \approx 0.033327$, $A_{3/4} \approx$ 5.144045E-3, $A_1 \approx 0.155966$, $A_{5/4} \approx$ 3.338796E-3, $A_{3/2} \approx 0.023219$, $A_{7/4} \approx$ 2.529025E-3, $A_2 \approx 0.051662$, $A_{9/4} \approx$ 7.907204E-4, $A_{5/2} \approx$ 4.983564E-3, $A_{11/4} \approx$ 4.856891E-4, $A_3 \approx$ 6.929255E-3, $A_{13/4} \approx$ 1.001560E-4, $A_{7/2} \approx$ 7.514113E-4, $A_{15/4} \approx$ 7.000160E-5, $A_4 \approx$ 6.697876E-4, $A_{17/4} \approx$ 1.153707E-5, $A_{9/2} \approx$ 1.033680E-4, $A_{19/4} \approx$ 9.237744E-6, $A_5 \approx$ 7.010572E-5, $A_{21/4} \approx$ 1.138380E-6, $A_{11/2} \approx$ 1.242875E-5, $A_{23/4} \approx$ 1.087214E-6, $A_6 \approx$ 6.901329E-6. The other harmonic amplitudes are $A_{k/4} \in (10^{-11}, 10^{-5})$ $(k = 13, 14, \ldots, 48)$.

4.2 Parametric Duffing Oscillators

In this section, periodic motions in a parametric Duffing oscillator will be discussed. Period-*m* motions in a parametrically forced, Duffing oscillator will be presented based on the prescribed accuracy of harmonic amplitudes. Period-1 and period-2 motions in such a parametric Duffing oscillator will be presented.

4.2.1 Formulations

Consider a parametrically excited, Duffing oscillator as

$$\ddot{x} + \delta\dot{x} + (\alpha + Q_0 \cos \Omega t)x + \beta x^3 = 0 \tag{4.37}$$

where $\dot{x} = dx/dt$ is velocity, Q_0 and Ω are parametric excitation amplitude and frequency, respectively. The damping coefficient δ, linear and nonlinear terms (α and β) are for the parametrically excited Duffing oscillator. Equation (4.37) can be expressed in a standard form of

$$\ddot{x} = F(x, \dot{x}, t) \tag{4.38}$$

where

$$F(x, \dot{x}, t) = -\delta\dot{x} - (\alpha + Q_0 \cos \Omega t)x - \beta x^3. \tag{4.39}$$

In Luo (2012a), the analytical solution of period-m motion with $\theta = \Omega t$ can be written as

$$x^{(m)*}(t) = a_0^{(m)}(t) + \sum_{k=1}^{N} b_{k/m}(t) \cos\left(\frac{k}{m}\theta\right) + c_{k/m}(t) \sin\left(\frac{k}{m}\theta\right). \tag{4.40}$$

Taking the first and second order derivatives of Equation (4.41) with respect to time generates

$$\begin{aligned}\dot{x}^{(m)*}(t) = \dot{a}_0^{(m)} + \sum_{k=1}^{N} \left[\left(\dot{b}_{k/m} + \frac{k\Omega}{m} c_{k/m}\right) \cos\left(\frac{k}{m}\theta\right)\right.\\ \left. + \left(\dot{c}_{k/m} - \frac{k\Omega}{m} b_{k/m}\right) \sin\left(\frac{k}{m}\theta\right)\right]\end{aligned} \tag{4.41}$$

$$\begin{aligned}\ddot{x}^{(m)*}(t) = \ddot{a}_0^{(m)} + \sum_{k=1}^{N} \left[\left(\ddot{b}_{k/m} + 2\frac{k\Omega}{m} \dot{c}_{k/m} - \left(\frac{k\Omega}{m}\right)^2 b_{k/m}\right) \cos\left(\frac{k}{m}\theta\right)\right.\\ \left. + \left(\ddot{c}_{k/m} - 2\frac{k\Omega}{m} \dot{b}_{k/m} - \left(\frac{k\Omega}{m}\right)^2 c_{k/m}\right) \sin\left(\frac{k}{m}\theta\right)\right].\end{aligned} \tag{4.42}$$

Substitution of Equations (4.40)–(4.42) to Equation (4.38) and averaging all terms of $\cos(k\theta/m)$ and $\sin(k\theta/m)$ during m-periods mT ($T = 2\pi/\Omega$) gives

$$\begin{aligned}&\ddot{a}_0^{(m)} = F_0^{(m)}(a_0^{(m)}, \mathbf{b}^{(m)}, \mathbf{c}^{(m)}, \dot{a}_0^{(m)}, \dot{\mathbf{b}}^{(m)}, \dot{\mathbf{c}}^{(m)}),\\ &\ddot{b}_{k/m} + 2\frac{k\Omega}{m} \dot{c}_{k/m} - \left(\frac{k\Omega}{m}\right)^2 b_{k/m}\\ &\quad = F_{1k}^{(m)}(a_0^{(m)}, \mathbf{b}^{(m)}, \mathbf{c}^{(m)}, \dot{a}_0^{(m)}, \dot{\mathbf{b}}^{(m)}, \dot{\mathbf{c}}^{(m)}),\\ &\ddot{c}_{k/m} - 2\frac{k\Omega}{m} \dot{b}_{k/m} - \left(\frac{k\Omega}{m}\right)^2 c_{k/m}\\ &\quad = F_{2k}^{(m)}(a_0^{(m)}, \mathbf{b}^{(m)}, \mathbf{c}^{(m)}, \dot{a}_0^{(m)}, \dot{\mathbf{b}}^{(m)}, \dot{\mathbf{c}}^{(m)})\\ &\text{for } k = 1, 2, \ldots, N\end{aligned} \tag{4.43}$$

The coefficients of constant, $\cos(k\theta/m)$ and $\sin(k\theta/m)$ for the function of $F(x, \dot{x}, t)$ in the Fourier series are

$$\begin{aligned}&F_0^{(m)}(a_0^{(m)}, \mathbf{b}^{(m)}, \mathbf{c}^{(m)}, \dot{a}_0^{(m)}, \dot{\mathbf{b}}^{(m)}, \dot{\mathbf{c}}^{(m)})\\ &\quad = \frac{1}{mT} \int_0^{mT} F(x^{(m)*}, \dot{x}^{(m)*}, t) dt\\ &\quad = -\delta\dot{a}_0^{(m)} - \alpha a_0^{(m)} - \frac{1}{2} Q_0 b_{k/m} \delta_m^k - \beta f^{(0)},\end{aligned}$$

$$\begin{aligned}
&F_{1k}^{(m)}(a_0^{(m)}, \mathbf{b}^{(m)}, \mathbf{c}^{(m)}, \dot{a}_0^{(m)}, \dot{\mathbf{b}}^{(m)}, \dot{\mathbf{c}}^{(m)}) \\
&\quad = \frac{2}{mT}\int_0^{mT} F(x^{(m)*}, \dot{x}^{(m)*}, t)\cos\left(\frac{k}{m}\Omega t\right) dt \\
&\quad = -\delta\left(\dot{b}_{k/m} + \frac{k\Omega}{m}c_{k/m}\right) - \alpha b_{k/m} - a_0^{(m)}Q_0\delta_m^k \\
&\qquad - \frac{1}{2}Q_0\sum_{i=1}^{N} a_{i/m}(\delta_{i+m}^k + \delta_{m-i}^k + \delta_{i-m}^k) - \beta f^{(c)}, \\
&F_{2k}^{(m)}(a_0^{(m)}, \mathbf{b}^{(m)}, \mathbf{c}^{(m)}, \dot{a}_0^{(m)}, \dot{\mathbf{b}}^{(m)}, \dot{\mathbf{c}}^{(m)}) \\
&\quad = \frac{2}{mT}\int_0^{mT} F(x^{(m)*}, \dot{x}^{(m)*}, t)\sin\left(\frac{k}{m}\Omega t\right) dt \\
&\quad = -\delta\left(\dot{c}_{k/m} - \frac{k\Omega}{m}b_{k/m}\right) - \alpha c_{k/m} \\
&\qquad - \frac{1}{2}Q_0\sum_{i=1}^{N} b_{i/m}(\delta_{i+m}^k + \delta_{i-m}^k - \delta_{m-i}^k) - \beta f^{(s)}
\end{aligned} \tag{4.44}$$

where

$$\begin{aligned}
f^{(0)} = (a_0^{(m)})^3 + \sum_{l=1}^{N}\sum_{j=1}^{N}\sum_{i=1}^{N}\Bigg[&\frac{3a_0^{(m)}}{2N}(b_{i/m}b_{j/m}\delta_{i-j}^0 + c_{i/m}c_{j/m}\delta_{i-j}^0) \\
&+ \frac{1}{4}b_{i/m}b_{j/m}b_{l/m}(\delta_{i-j-l}^0 + \delta_{i-j+l}^0 + \delta_{i+j-l}^0) \\
&+ \frac{3}{4}b_{i/m}c_{j/m}c_{l/m}\left(\delta_{i+j-l}^0 + \delta_{i-j+l}^0 - \delta_{i-j-l}^0\right)\Bigg],
\end{aligned} \tag{4.45}$$

$$\begin{aligned}
f^{(c)} = \sum_{l=1}^{N}\sum_{j=1}^{N}\sum_{i=1}^{N}\Bigg[&3\left(\frac{a_0^{(m)}}{N}\right)^2 b_{l/m}\delta_l^k + \frac{3a_0^{(m)}}{2N}b_{l/m}b_{j/m}(\delta_{|l-j|}^k + \delta_{l+j}^k) \\
&+ \frac{3a_0^{(m)}}{2N}c_{l/m}c_{j/m}(\delta_{|l-j|}^k - \delta_{l+j}^k) \\
&+ \frac{1}{4}b_{l/m}b_{j/m}b_{i/m}(\delta_{|l-j-i|}^k + \delta_{l+j+i}^k + \delta_{|l-j+i|}^k + \delta_{|l+j-i|}^k) \\
&+ \frac{3}{4}b_{l/m}c_{j/m}c_{i/m}\left(\delta_{|l+j-i|}^k - \delta_{l+j+i}^k + \delta_{|l-j+i|}^k - \delta_{|l-j-i|}^k\right)\Bigg],
\end{aligned} \tag{4.46}$$

$$\begin{aligned}
f^{(s)} = \sum_{l=1}^{N}\sum_{j=1}^{N}\sum_{i=1}^{N}&3\left(\frac{a_0^{(m)}}{N}\right)^2 c_{l/m}\delta_l^k + \frac{3a_0^{(m)}}{N}b_{l/m}c_{j/m}[\delta_{l+j}^k - \mathrm{sgn}(l-j)\delta_{|l-j|}^k] \\
&+ \frac{1}{4}c_{l/m}c_{j/m}c_{i/m}[\mathrm{sgn}(l-j+i)\delta_{|l-j+i|}^k - \delta_{l+j+i}^k
\end{aligned}$$

$$
\begin{aligned}
&+ \operatorname{sgn}(l+j-i)\delta^k_{|l+j-i|} - \operatorname{sgn}(l-j-i)\delta^k_{|l-j-i|}] \\
&+ \frac{3}{4} b_{l/m} b_{j/m} c_{i/m} [\operatorname{sgn}(l-j+i)\delta^k_{|l-j+i|} + \delta^k_{l+j+i} \\
&- \operatorname{sgn}(l+j-i)\delta^k_{|l+j-i|} - \operatorname{sgn}(l-j-i)\delta^k_{|l-j-i|})].
\end{aligned}
\tag{4.47}
$$

Introduce vectors to express the unknown time-varying coefficients as

$$
\begin{aligned}
\mathbf{z}^{(m)} &\triangleq (a_0^{(m)}, \mathbf{b}^{(m)}, \mathbf{c}^{(m)})^{\mathrm{T}} \\
&= (a_0^{(m)}, b_{1/m}, \ldots, b_{N/m}, c_{1/m}, \ldots, c_{N/m})^{\mathrm{T}} \\
&\equiv (z_0^{(m)}, z_1^{(m)}, \ldots, z_{2N}^{(m)})^{\mathrm{T}}, \\
\mathbf{z}_1 &= \dot{\mathbf{z}} = (\dot{a}_0^{(m)}, \dot{\mathbf{b}}^{(m)}, \dot{\mathbf{c}}^{(m)})^{\mathrm{T}} \\
&= (\dot{a}_0^{(m)}, \dot{b}_{1/m}, \ldots, \dot{b}_{N/m}, \dot{c}_{1/m}, \ldots, \dot{c}_{N/m})^{\mathrm{T}} \\
&\equiv (\dot{z}_0^{(m)}, \dot{z}_1^{(m)}, \ldots, \dot{z}_{2N}^{(m)})^{\mathrm{T}}
\end{aligned}
\tag{4.48}
$$

where

$$
\begin{aligned}
\mathbf{b}^{(m)} &= (b_{1/m}, b_{2/m}, \ldots, b_{N/m})^{\mathrm{T}}, \\
\mathbf{c}^{(m)} &= (c_{1/m}, c_{2/m}, \ldots, c_{N/m})^{\mathrm{T}}.
\end{aligned}
\tag{4.49}
$$

Equation (4.43) can be expressed in the form of vector field as

$$
\dot{\mathbf{z}}^{(m)} = \mathbf{z}_1^{(m)} \text{ and } \dot{\mathbf{z}}_1^{(m)} = \mathbf{g}^{(m)}(\mathbf{z}^{(m)}, \mathbf{z}_1^{(m)}),
\tag{4.50}
$$

where

$$
\mathbf{g}^{(m)}(\mathbf{z}^{(m)}, \mathbf{z}_1^{(m)}) = \begin{pmatrix}
F_0^{(m)}(\mathbf{z}^{(m)}, \mathbf{z}_1^{(m)}) \\
\mathbf{F}_1^{(m)}(\mathbf{z}^{(m)}, \mathbf{z}_1^{(m)}) - 2\mathbf{k}_1 \left(\frac{\Omega}{m}\right) \dot{\mathbf{c}}^{(m)} + \mathbf{k}_2 \left(\frac{\Omega}{m}\right)^2 \mathbf{b}^{(m)} \\
\mathbf{F}_2^{(m)}(\mathbf{z}^{(m)}, \mathbf{z}_1^{(m)}) + 2\mathbf{k}_1 \left(\frac{\Omega}{m}\right) \dot{\mathbf{b}}^{(m)} + \mathbf{k}_2 \left(\frac{\Omega}{m}\right)^2 \mathbf{c}^{(m)}
\end{pmatrix},
\tag{4.51}
$$

$$
\begin{aligned}
&\mathbf{k}_1 = diag(1, 2, \ldots, N), \\
&\mathbf{k}_2 = diag(1, 2^2, \ldots, N^2), \\
&\mathbf{F}_1^{(m)} = (F_{11}^{(m)}, F_{12}^{(m)}, \ldots, F_{1N}^{(m)})^{\mathrm{T}}, \\
&\mathbf{F}_2^{(m)} = (F_{21}^{(m)}, F_{22}^{(m)}, \ldots, F_{2N}^{(m)})^{\mathrm{T}} \\
&\text{for } N = 1, 2, \ldots, \infty;
\end{aligned}
\tag{4.52}
$$

and

$$
\mathbf{y}^{(m)} \equiv (\mathbf{z}^{(m)}, \mathbf{z}_1^{(m)}) \text{ and } \mathbf{f}^{(m)} = (\mathbf{z}_1^{(m)}, \mathbf{g}^{(m)})^{\mathrm{T}}.
\tag{4.53}
$$

Thus, Equation (4.50) becomes

$$
\dot{\mathbf{y}}^{(m)} = \mathbf{f}^{(m)}(\mathbf{y}^{(m)}).
\tag{4.54}
$$

The solutions of steady-state periodic motion can be obtained by setting $\dot{\mathbf{y}}^{(m)} = \mathbf{0}$, that is,

$$\begin{aligned} &F_0^{(m)}(a_0^{(m)*}, \mathbf{b}^{(m)*}, \mathbf{c}^{(m)*}, 0, \mathbf{0}, \mathbf{0}) = 0, \\ &\mathbf{F}_1^{(m)}(a_0^{(m)*}, \mathbf{b}^{(m)*}, \mathbf{c}^{(m)*}, 0, \mathbf{0}, \mathbf{0}) + \frac{\Omega^2}{m^2}\mathbf{k}_2\mathbf{b}^{(m)*} = \mathbf{0}, \\ &\mathbf{F}_2^{(m)}(a_0^{(m)*}, \mathbf{b}^{(m)*}, \mathbf{c}^{(m)*}, 0, \mathbf{0}, \mathbf{0}) + \frac{\Omega^2}{m^2}\mathbf{k}_2\mathbf{c}^{(m)*} = \mathbf{0}. \end{aligned} \tag{4.55}$$

The solutions of the $(2N+1)$ nonlinear equations in Equation (4.55) are computed from the Newton–Raphson method. The linearized equation at the equilibrium point $\mathbf{y}^{(m)*} = (\mathbf{z}^{(m)*}, \mathbf{0})^{\mathrm{T}}$ is

$$\Delta\dot{\mathbf{y}}^{(m)} = D\mathbf{f}^{(m)}(\mathbf{y}^{*(m)})\Delta\mathbf{y}^{(m)} \tag{4.56}$$

where

$$D\mathbf{f}^{(m)}(\mathbf{y}^{*(m)}) = \partial\mathbf{f}^{(m)}(\mathbf{y}^{(m)})/\partial\mathbf{y}^{(m)}|_{\mathbf{y}^{(m)*}}. \tag{4.57}$$

The Jacobian matrix is

$$D\mathbf{f}^{(m)}(\mathbf{y}^{(m)}) = \begin{bmatrix} \mathbf{0}_{(2N+1)\times(2N+1)} & \mathbf{I}_{(2N+1)\times(2N+1)} \\ \mathbf{G}_{(2N+1)\times(2N+1)} & \mathbf{H}_{(2N+1)\times(2N+1)} \end{bmatrix} \tag{4.58}$$

and

$$\mathbf{G} = \frac{\partial\mathbf{g}^{(m)}}{\partial\mathbf{z}^{(m)}} = (\mathbf{G}^{(0)}, \mathbf{G}^{(c)}, \mathbf{G}^{(s)})^{\mathrm{T}} \tag{4.59}$$

with

$$\begin{aligned} \mathbf{G}^{(0)} &= (G_0^{(0)}, G_1^{(0)}, \ldots, G_{2N}^{(0)}), \\ \mathbf{G}^{(c)} &= (\mathbf{G}_1^{(c)}, \mathbf{G}_2^{(c)}, \ldots, \mathbf{G}_N^{(c)})^{\mathrm{T}}, \\ \mathbf{G}^{(s)} &= (\mathbf{G}_1^{(s)}, \mathbf{G}_2^{(s)}, \ldots, \mathbf{G}_N^{(s)})^{\mathrm{T}} \end{aligned} \tag{4.60}$$

for $N = 1, 2, \ldots \infty$ with

$$\begin{aligned} \mathbf{G}_k^{(c)} &= (G_{k0}^{(c)}, G_{k1}^{(c)}, \ldots, G_{k(2N)}^{(c)}), \\ \mathbf{G}_k^{(s)} &= (G_{k0}^{(s)}, G_{k1}^{(s)}, \ldots, G_{k(2N)}^{(s)}) \end{aligned} \tag{4.61}$$

for $k = 1, 2, \ldots N$. The corresponding components are

$$\begin{aligned} G_r^{(0)} &= -\alpha\delta_0^r - \frac{1}{2}Q_0\delta_m^r - \beta g_{2r}^{(0)}, \\ G_{kr}^{(c)} &= \left(\frac{k\Omega}{m}\right)^2\delta_k^r - \alpha\delta_k^r - \delta\frac{k\Omega}{m}\delta_{k+N}^r - \frac{1}{2}Q_0\delta_m^k\delta_0^r \\ &\quad - \frac{1}{4}Q_0\sum_{i=1}^{N}(\delta_{i+m}^k + \delta_{m-i}^k + \delta_{i-m}^k)\delta_i^r - \beta g_{2kr}^{(c)}, \\ G_{kr}^{(s)} &= \left(\frac{k\Omega}{m}\right)^2\delta_{k+N}^r + \delta\frac{k\Omega}{m}\delta_k^r - \alpha\delta_{k+N}^r \\ &\quad - \frac{1}{4}Q_0\sum_{i=1}^{N}(\delta_{i+m}^k + \delta_{i-m}^k - \delta_{m-i}^k)\delta_{i+N}^r - \beta g_{2kr}^{(s)} \end{aligned} \tag{4.62}$$

where

$$
\begin{aligned}
g_{2r}^{(0)} = {} & 3(a_0^{(m)})^2\delta_r^0 + \sum_{l=1}^{N}\sum_{j=1}^{N}\sum_{i=1}^{N}\frac{3}{2N}(b_{i/m}b_{j/m}\delta_r^0 + 2a_0^{(m)}b_{i/m}\delta_j^r)\delta_{i-j}^0 \\
& + \frac{3}{2N}(c_{i/m}c_{j/m}\delta_r^0 + 2a_0^{(m)}c_{i/m}\delta_{j+N}^r)\delta_{i-j}^0 \\
& + \frac{3}{4}b_{i/m}b_{j/m}\delta_l^r(\delta_{i-j-l}^0 + \delta_{i-j+l}^0 + \delta_{i+j-l}^0) \\
& + \frac{3}{4}(c_{j/m}c_{l/m}\delta_i^r + 2b_{i/m}c_{j/m}\delta_{l+N}^r)(\delta_{i+j-l}^0 + \delta_{i-j+l}^0 - \delta_{i-j-l}^0),
\end{aligned}
\tag{4.63}
$$

$$
\begin{aligned}
g_{2kr}^{(c)} = {} & \sum_{l=1}^{N}\sum_{j=1}^{N}\sum_{i=1}^{N}3\frac{a_0^{(m)}}{N^2}\left[a_0^{(m)}\delta_l^r + 2b_{l/m}\delta_0^r\right]\delta_l^k \\
& + \frac{3}{2N}(b_{l/m}b_{j/m}\delta_0^r + 2a_0^{(m)}b_{j/m}\delta_l^r)(\delta_{|l-j|}^k + \delta_{l+j}^k) \\
& + \frac{3}{2N}(c_{l/m}c_{j/m}\delta_0^r + a_0^{(m)}c_{j/m}\delta_{l+N}^r)(\delta_{|l-j|}^k - \delta_{l+j}^k) \\
& + \frac{3}{4}b_{j/m}b_{i/m}\delta_l^r(\delta_{|l-j-i|}^k + \delta_{l+j+i}^k + \delta_{|l-j+i|}^k + \delta_{|l+j-i|}^k) \\
& + \frac{3}{4}(c_{j/m}c_{i/m}\delta_l^r + 2b_{l/m}c_{i/m}\delta_{j+N}^r)(\delta_{|l+j-i|}^k - \delta_{l+j+i}^k + \delta_{|l-j+i|}^k - \delta_{|l-j-i|}^k),
\end{aligned}
\tag{4.64}
$$

$$
\begin{aligned}
g_{2kr}^{(s)} = {} & \sum_{l=1}^{N}\sum_{j=1}^{N}\sum_{i=1}^{N}3\frac{a_0^{(m)}}{N^2}\left[a_0^{(m)}\delta_{l+N}^r + 2c_{l/m}\delta_0^r\right]\delta_l^k \\
& + \frac{3}{N}(a_0^{(m)}c_{j/m}\delta_l^r + a_0^{(m)}b_{l/m}\delta_{j+N}^r + b_{l/m}c_{j/m}\delta_0^r)\left[\delta_{l+j}^k - \operatorname{sgn}(l-j)\,\delta_{|l-j|}^k\right] \\
& + \frac{3}{4}c_{j/m}c_{i/m}\delta_{l+N}^r\left[\operatorname{sgn}(l-j+i)\,\delta_{|l-j+i|}^k - \delta_{l+j+i}^k\right. \\
& \left.+\operatorname{sgn}(l+j-i)\,\delta_{|l+j-i|}^k - \operatorname{sgn}(l-j-i)\delta_{|l-j-i|}^k\right] \\
& + \frac{3}{4}(b_{l/m}b_{j/m}\delta_{i+N}^r + 2b_{j/m}c_{i/m}\delta_l^r)\left[\operatorname{sgn}(l-j+i)\,\delta_{|l-j+i|}^k + \delta_{l+j+i}^k\right. \\
& \left.-\operatorname{sgn}(l+j-i)\delta_{|l+j-i|}^k - \operatorname{sgn}(l-j-i)\delta_{|l-j-i|}^k)\right]
\end{aligned}
\tag{4.65}
$$

for $r = 0, 1, \ldots 2N$.

$$
\mathbf{H} = \frac{\partial \mathbf{g}^{(m)}}{\partial \mathbf{z}_1{}^{(m)}} = (\mathbf{H}^{(0)}, \mathbf{H}^{(c)}, \mathbf{H}^{(s)})^{\mathrm{T}}
\tag{4.66}
$$

where

$$\begin{aligned}\mathbf{H}^{(0)} &= (H_0^{(0)}, H_1^{(0)}, \ldots, H_{2N}^{(0)}),\\ \mathbf{H}^{(c)} &= (\mathbf{H}_1^{(c)}, \mathbf{H}_2^{(c)}, \ldots, \mathbf{H}_N^{(c)})^{\mathrm{T}},\\ \mathbf{H}^{(s)} &= (\mathbf{H}_1^{(s)}, \mathbf{H}_2^{(s)}, \ldots, \mathbf{H}_N^{(s)})^{\mathrm{T}}\end{aligned} \tag{4.67}$$

for $N = 1, 2, \ldots \infty$, with

$$\begin{aligned}\mathbf{H}_k^{(c)} &= (H_{k0}^{(c)}, H_{k1}^{(c)}, \ldots, H_{k(2N)}^{(c)}),\\ \mathbf{H}_k^{(s)} &= (H_{k0}^{(s)}, H_{k1}^{(s)}, \ldots, H_{k(2N)}^{(s)})\end{aligned} \tag{4.68}$$

for $k = 1, 2, \ldots N$. The corresponding components are

$$\begin{aligned}H_r^{(0)} &= -\delta\delta_0^r,\\ H_{kr}^{(c)} &= -2\frac{k\Omega}{m}\delta_{k+N}^r - \delta\delta_k^r,\\ H_{kr}^{(s)} &= 2\frac{k\Omega}{m}\delta_k^r - \delta\delta_{k+N}^r\end{aligned} \tag{4.69}$$

for $r = 0, 1, \ldots, 2N$.

The corresponding eigenvalues are given by

$$|D\mathbf{f}^{(m)}(\mathbf{y}^{*(m)}) - \lambda\mathbf{I}_{2(2N+1)\times 2(2N+1)}| = 0. \tag{4.70}$$

The boundary between the stable and unstable solutions is given by the bifurcation conditions.

For the asymmetric period-1 motion, $a_0^{(m)} \neq 0$ is required. For one harmonic term balance, setting $m = k = 1$, Equation (4.44) becomes

$$\begin{aligned}F_0^{(1)}(a_0, b_1, c_1, \dot{a}_0, \dot{b}_1, \dot{c}_1) &= -\delta\dot{a}_0 - \alpha a_0 - \frac{1}{2}Q_0 b_1 - \beta f_0^{(1)}\\ F_{11}^{(1)}(a_0, b_1, c_1, \dot{a}_0, \dot{b}_1, \dot{c}_1) &= -\delta(\dot{b}_1 + \Omega c_1) - \alpha b_1 - a_0 Q_0 - \beta f_1^{(c)};\\ F_{21}^{(1)}(a_0, b_1, c_1, \dot{a}_0, \dot{b}_1, \dot{c}_1) &= -\delta(\dot{c}_1 - \Omega b_1) - \alpha c_1 - \beta f_1^{(s)}.\end{aligned} \tag{4.71}$$

where for $i = j = l = 1$ Equations (4.45)–(4.47) gives

$$\begin{aligned}f_0^{(1)} &= a_0\left[a_0^2 + \frac{3}{2}\left(b_1^2 + c_1^2\right)\right],\\ f_1^{(c)} &= 3b_1\left[a_0^2 + \frac{1}{4}\left(b_1^2 + c_1^2\right)\right],\\ f_1^{(s)} &= 3c_1\left[a_0^2 + \frac{1}{4}\left(b_1^2 + c_1^2\right)\right].\end{aligned} \tag{4.72}$$

Thus for $m = k = 1$, Equation (4.43) becomes

$$\begin{aligned}\ddot{a}_0 &= -\delta\dot{a}_0 - \alpha a_0 - \frac{1}{2}Q_0 b_1 - \beta a_0\left[a_0^2 + \frac{3}{2}\left(b_1^2 + c_1^2\right)\right],\\ \ddot{b}_1 + 2\Omega\dot{c}_1 - \Omega^2 b_1 &= -\delta(\dot{b}_1 + \Omega c_1) - \alpha b_1 - a_0 Q_0 - 3\beta b_1\left[a_0^2 + \frac{1}{4}\left(b_1^2 + c_1^2\right)\right],\\ \ddot{c}_1 - 2\Omega\dot{b}_1 - \Omega^2 c_1 &= -\delta(\dot{c}_1 - \Omega b_1) - \alpha c_1 - 3\beta c_1\left[a_0^2 + \frac{1}{4}\left(b_1^2 + c_1^2\right)\right].\end{aligned} \tag{4.73}$$

The algebraic equations for the traditional harmonic balance with one term is given by the equilibrium point of Equation (4.73), that is,

$$
\begin{aligned}
0 &= -\alpha a_0^* - \frac{1}{2}Q_0 b_1^* - \beta a_0^* \left[a_0^{*2} + \frac{3}{2}\left(b_1^{*2} + c_1^{*2}\right)\right], \\
-\Omega^2 b_1^* &= -\delta\Omega c_1^* - \alpha b_1^* - a_0^* Q_0 - 3\beta b_1^* \left[a_0^{*2} + \frac{1}{4}\left(b_1^{*2} + c_1^{*2}\right)\right], \\
-\Omega^2 c_1^* &= \delta\Omega b_1^* - \alpha c_1^* - 3\beta c_1^* \left[a_0^{*2} + \frac{1}{4}\left(b_1^{*2} + c_1^{*2}\right)\right].
\end{aligned}
\tag{4.74}
$$

Setting $A_1^2 = c_1^{*2} + b_1^{*2}$, and deformation of Equation (4.74) produces

$$
\begin{aligned}
&- a_0^* \left[\alpha + \beta \left[a_0^{*2} + \frac{3}{2}A_1^2\right]\right] = \frac{1}{2}Q_0 b_1^*, \\
&(\delta\Omega)^2 A_1^2 + A_1^2\left[\left(\alpha - \Omega^2\right) + 3\beta\left(a_0^{*2} + \frac{1}{4}A_1^2\right)\right]^2 = (a_0^* Q_0)^2.
\end{aligned}
\tag{4.75}
$$

From Equation (4.75), the first harmonic amplitude A_1 and a_0^* can be determined. Further, from Equation (4.74), the coefficient b_1^* and c_1^* are determined. The corresponding stability and bifurcations are from the eigenvalue analysis of the linearized equation of Equation (4.73). At equilibrium point (a_0^*, b_1^*, c_1^*), the linearized equation is

$$
\ddot{\mathbf{u}} + \mathbf{C}\dot{\mathbf{u}} + \mathbf{K}\mathbf{u} = \mathbf{0},
\tag{4.76}
$$

where

$$
\mathbf{u} = (\Delta a_0, \Delta b_1, \Delta c_1)^{\mathrm{T}}, \dot{\mathbf{u}} = (\Delta \dot{a}_0, \Delta \dot{b}_1, \Delta \dot{c}_1)^{\mathrm{T}}, \ddot{\mathbf{u}} = (\Delta \ddot{a}_0, \Delta \ddot{b}_1, \Delta \ddot{c}_1)^{\mathrm{T}}
$$

$$
\begin{aligned}
&\mathbf{C} = \begin{bmatrix} \delta & 0 & 0 \\ 0 & \delta & 2\Omega \\ 0 & -2\Omega & \delta \end{bmatrix}, \mathbf{K} = \begin{bmatrix} K_{00} & K_{01} & K_{02} \\ K_{10} & K_{11} & K_{12} \\ K_{20} & K_{21} & K_{22} \end{bmatrix}; \\
&K_{00} = \alpha + 3\beta\left[a_0^2 + \frac{1}{2}\left(b_1^2 + c_1^2\right)\right], \\
&K_{01} = \frac{1}{2}Q_0 + 3\beta a_0 b_1, K_{02} = 3\beta a_0 c_1, \\
&K_{10} = Q_0 + 6\beta b_1 a_0, \\
&K_{11} = \alpha - \Omega^2 + \beta\left(3a_0^{*2} + \frac{9}{4}b_1^{*2} + \frac{3}{4}c_1^{*2}\right), \\
&K_{12} = \delta\Omega + \frac{3}{2}\beta b_1^* c_1^*, \\
&K_{20} = 6\beta c_1 a_0, K_{21} = -\delta\Omega + \frac{3}{2}\beta b_1^* c_1^*, \\
&K_{22} = \alpha - \Omega^2 + \beta\left(3a_0^{*2} + \frac{3}{4}b_1^{*2} + \frac{9}{4}c_1^{*2}\right).
\end{aligned}
\tag{4.77}
$$

The eigenvalues of the linearized equation is determined by

$$|\lambda^2\mathbf{I} + \lambda\mathbf{C} + \mathbf{K}| = \mathbf{0}. \tag{4.78}$$

In other words,

$$\begin{vmatrix} \lambda^2 + \delta\lambda + K_{00} & K_{01} & K_{02} \\ K_{10} & \lambda^2 + \delta\lambda + K_{11} & 2\Omega\lambda + K_{12} \\ K_{20} & -2\Omega\lambda + K_{21} & \lambda^2 + \delta\lambda + K_{22} \end{vmatrix} = 0. \tag{4.79}$$

From the eigenvalues, the stability and bifurcation of approximate asymmetric period-1 motion can be determined. For one harmonic term, the asymmetric period-1 motion cannot be approximated well.

The harmonic amplitude varying with excitation frequency Ω is presented through the kth order harmonic amplitude and phase as

$$A_{k/m} \equiv \sqrt{b_{k/m}^2 + c_{k/m}^2}, \varphi_{k/m} = \arctan \frac{c_{k/m}}{b_{k/m}} \tag{4.80}$$

and the corresponding solution in Equation (4.40) is

$$x^*(t) = a_0^{(m)} + \sum_{k=1}^{N} A_{k/m} \cos\left(\frac{k}{m}\Omega t - \varphi_{k/m}\right). \tag{4.81}$$

4.2.2 *Parametric Hardening Duffing Oscillators*

In this section, asymmetric period-1 and symmetric period-2 motions in the parametric hardening Duffing oscillator will be presented herein. One used thc perturbation analysis, symmetric period-2 motions in such a parametric oscillator can be determined. It is very difficult to get the asymmetric period-1 motion. Through this discussion, a complete picture for periodic motion in the parametric hardening Duffing oscillator can be provided.

As in Luo and O'Connor (2014), consider a set of system parameters as

$$\delta = 0.2, \alpha = 20, \beta = 50. \tag{4.82}$$

In all frequency-amplitude curves, the acronym "SN" represents the saddle-node bifurcation. Solid and dashed curves represent stable and unstable period-m motions.

The first branch of frequency-amplitude curves for asymmetric period-1 motion based on 12 harmonic terms are presented in Figure 4.13 for $Q_0 = 16, 17.5, 20$. The asymmetric period-1 motion lies in the range of $\Omega \in (1.8, 2.4)$ for such excitation amplitudes. For the given parameters in Equation (4.82), the minimum value of parametric excitation is $Q_{0\min} \approx 15.79$ to generate the asymmetric period-1 motion. The constant term $a_0 \equiv a_0^{(m)}$ $(m = 1)$ is presented for $Q_0 = 16, 17.5, 20$ in Figure 4.13(i), and the asymmetric period-1 motion on the right hand of $x = 0$ is presented. The values lie in the range of $a_0 \in (0, 0.06)$. For the asymmetric period-1 motion on the left hand, the constant term is opposite to the results on the right hand, that is, ${}^La_0 = -{}^Ra_0$. Other harmonic amplitudes are the same, but the harmonic phases satisfy ${}^L\varphi_k = \text{mod}({}^R\varphi_k + (k+1)\pi, 2\pi)$. Thus, the first harmonic amplitude A_1 is presented for $Q_0 = 16, 17.5, 20$ in Figure 4.13(ii). The corresponding values are in the range of $A_1 \in (0, 0.2)$. The second and third order amplitudes are presented in Figure 4.13(iii) and (iv), respectively.

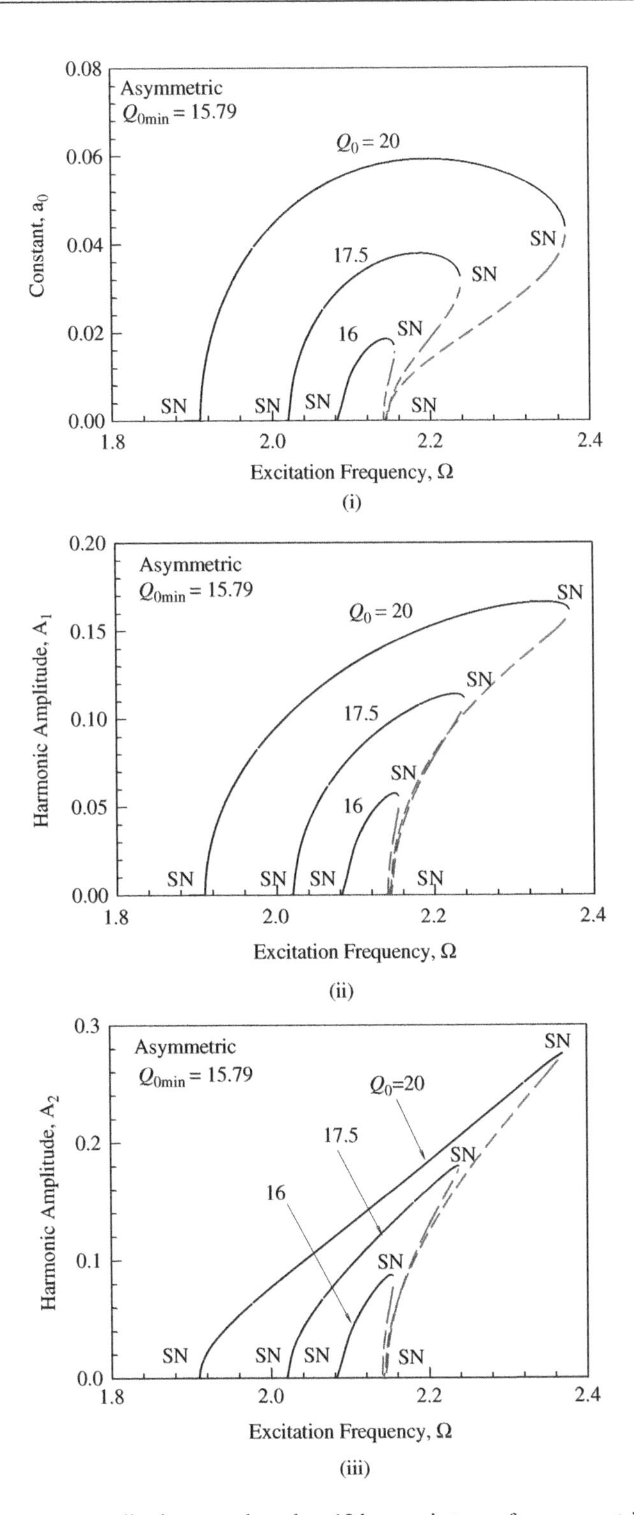

Figure 4.13 Frequency-amplitude curves based on 12 harmonic terms for asymmetric period-1 motion in the parametric Duffing oscillator: (i) constant term $Q_0 = 15, 17.5, 20$, (ii)–(vi) harmonic amplitude $\Omega \in (3.0, 8.0)$ ($Q_{0\,\min} \approx 8.5$). ($a_0 \equiv a_0^{(m)}$ $m = 1$ $Q_0 = 15, 17.5, 20$ $x = 0$)

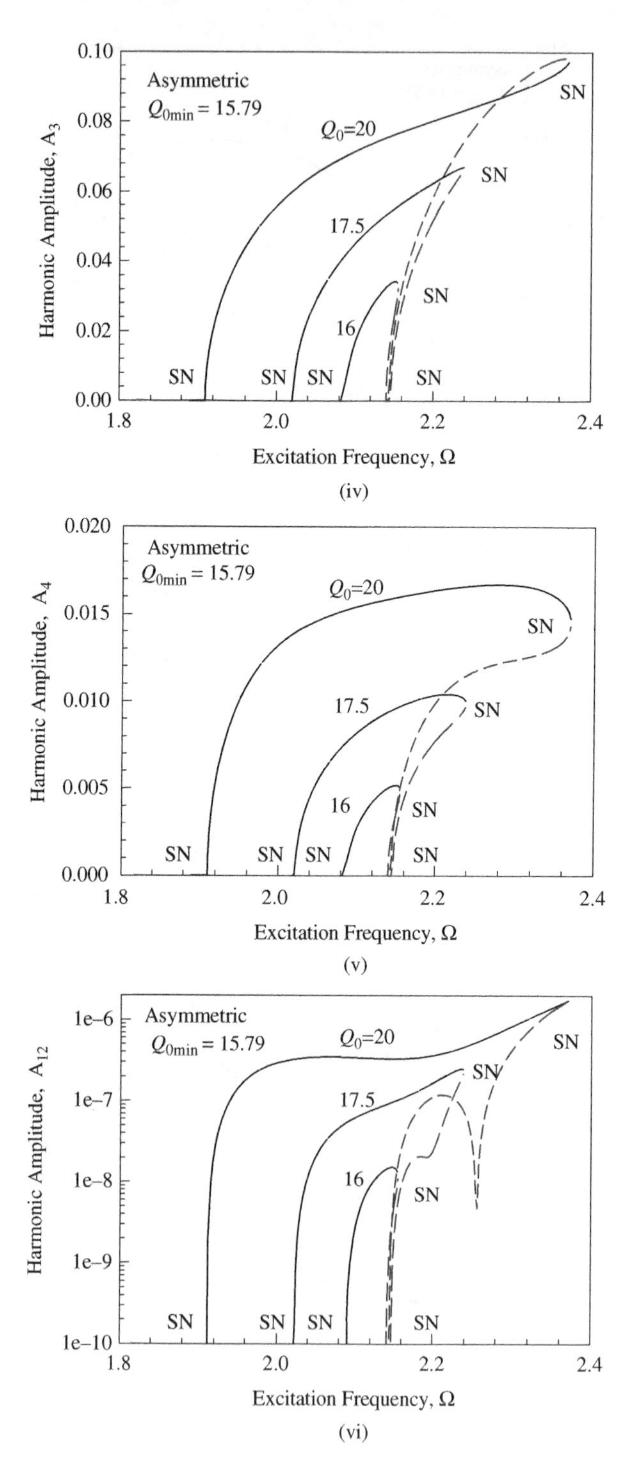

Figure 4.13 (*continued*)

One can obtain $A_2 \in (0, 0.3)$ and $A_3 \in (0, 0.1)$, which are the same quantity levels as the first harmonic amplitude. The fourth order harmonic amplitude $A_4 \in (0, 0.02)$ is plotted in Figure 4.13(v). However, the fifth order harmonic amplitude is $A_5 \in (0, 3 \times 10^{-3})$. The quantity levels of the harmonic amplitudes are dropped. To avoid abundant illustrations, other harmonic amplitudes and harmonic phases will not be presented. The 12th order harmonic amplitude $A_{12} \in (0, 1 \times 10^{-6})$ is shown in Figure 4.13(vi).

The second branch of frequency-amplitude curves for asymmetric period-1 motion is also based on 12 harmonic terms, which are presented in Figure 4.14 for $Q_0 = 15, 17.5, 20$. The asymmetric period-1 motion lies in $\Omega \in (3.0, 8.0)$ for such excitation amplitude. For the given parameters in Equation (4.82), the minimum value of parametric excitation is $Q_{0\min} = 8.5$ to generate the asymmetric period-1 motion. The constant term $a_0 = a_0^{(m)}$ $(m = 1)$ is presented for $Q_0 = 15, 17.5, 20$, in Figure 4.14(i), and the asymmetric period-1 motion on the right hand of $x = 0$ is presented. The values lie in the range of $Q_0 \in (0, 0.15)$ which is much bigger than the first branch. Thus, the first harmonic amplitude a_0 is presented for $Q_0 = 15, 17.5, 20$, in Figure 4.14(ii). The corresponding values are in the range of $A_1 \in (0, 1.0)$ The second and third order amplitudes are presented in Figure 4.14(iii) and (iv), respectively. One can obtain $A_2 \in (0, 0.08)$ and $A_3 \in (0, 0.03)$, which dramatically drop compared to the first harmonic amplitude. Compared to the first branch, the frequency-amplitude curves are distinguishing. The fourth order harmonic amplitude $A_4 \in (0, 3 \times 10^{-3})$ is plotted in Figure 4.14(v). To avoid abundant illustrations, other harmonic amplitudes and harmonic phases will not be presented. The 12th order harmonic amplitude $A_{12} \in (0, 3 \times 10^{-8})$ is shown in Figure 4.14(vi).

The frequency-amplitude curves for the first branch of symmetric period-2 motions based on 24 harmonic terms are presented in Figure 4.15 for $Q_0 = 15, 17.5, 20$. The symmetric period-2 motion lies in $\Omega \in (2.4, 3.6)$. For the specific parameters, the minimum value of parametric excitation is $Q_{0\min} \approx 13.25$ to generate the symmetric period-2 motion. For the symmetric period-2 motion, $a_0^{(2)} = 0$ and $A_{k/2} = 0$ for $k = 2l$ $(l = 1, 2, \ldots)$. In Figure 4.15(i), the first order harmonic amplitude $A_{1/2} \in (0, 0.2)$ is presented for $Q_0 = 15, 17.5, 20$. The third order harmonic amplitude in Figure 4.15(ii) lies in $A_{3/2} \in (0, 0.5)$, which plays an important role on the symmetric period-2 motion. The fifth order harmonic amplitude $A_{5/2} \in (0, 0.1)$ is plotted in Figure 4.15(iii). However, the seventh order harmonic amplitude lies in the range of $A_{7/2} \in (0, 1.2 \times 10^{-2})$, as shown in Figure 4.15(iv). The ninth order harmonic amplitude $A_{9/2} \in (0, 4 \times 10^{-3})$ is presented in Figure 4.15(v). To avoid abundant illustrations, other harmonic amplitudes and harmonic phases for this symmetric period-2 motion will not be presented. The 23th order harmonic amplitude $A_{23/2} \in (0, 5 \times 10^{-6})$ is shown in Figure 4.15(vi).

The frequency-amplitude curves for the second branch of symmetric period-2 motions based on 12 harmonic terms are presented in Figure 4.16 for $Q_0 = 15, 17.5, 20$. The symmetric period-2 motion lies in $\Omega \in (5.0, 110.0)$. For the specific parameters, the minimum value of parametric excitation is $Q_{0\min} \approx 1.81$ to generate the symmetric period-2 motion. Such the minimum excitation amplitude is close to zero compared to the other branches of solutions. In Figure 4.16(i), the first order harmonic amplitude $A_{1/2} \in (0, 9.0)$ is presented for $Q_0 = 15, 17.5, 20$. Such a harmonic term plays an important role on the symmetric period-2 motion. The third order harmonic amplitude $A_{3/2} \in (0, 0.4)$ is presented in Figure 4.16(ii). The third one contributes to such a symmetric period-2 motion much less than the first order harmonics. The fifth and seventh order harmonic amplitudes are plotted in Figure 4.16(iii) and (iv). $A_{5/2} \in (0, 2 \times 10^{-2})$ and $A_{7/2} \in (0, 8 \times 10^{-4})$. The 9th and 11th order harmonic amplitudes are

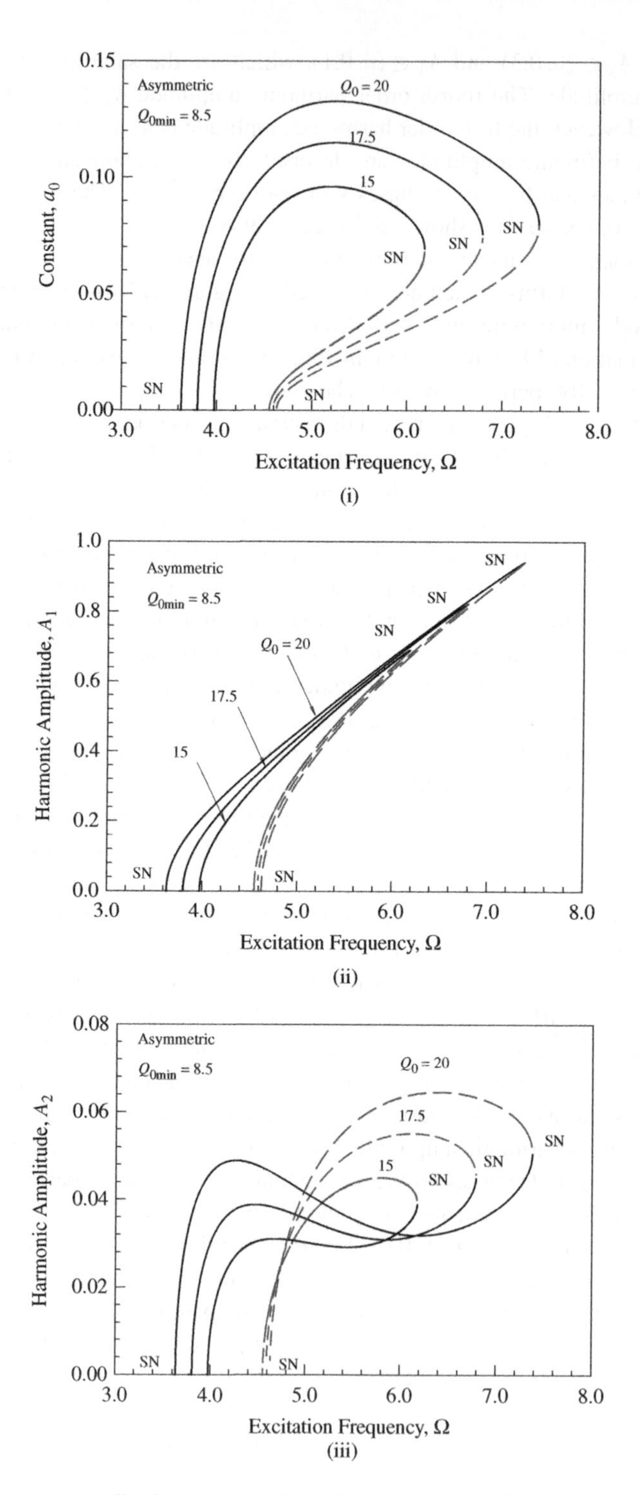

Figure 4.14 Frequency-amplitude curves based on 12 harmonic terms for asymmetric period-1 motion in the parametric Duffing oscillator: (i) constant term $a_0 \in (0, 0.15)$, (ii)–(vi) harmonic amplitude A_1 ($Q_0 = 15, 17.5, 20$). ($A_1 \in (0, 1.0)$. $A_2 \in (0, 0.08)$ $A_3 \in (0, 0.03)$ $A_4 \in (0, 3 \times 10^{-3})$)

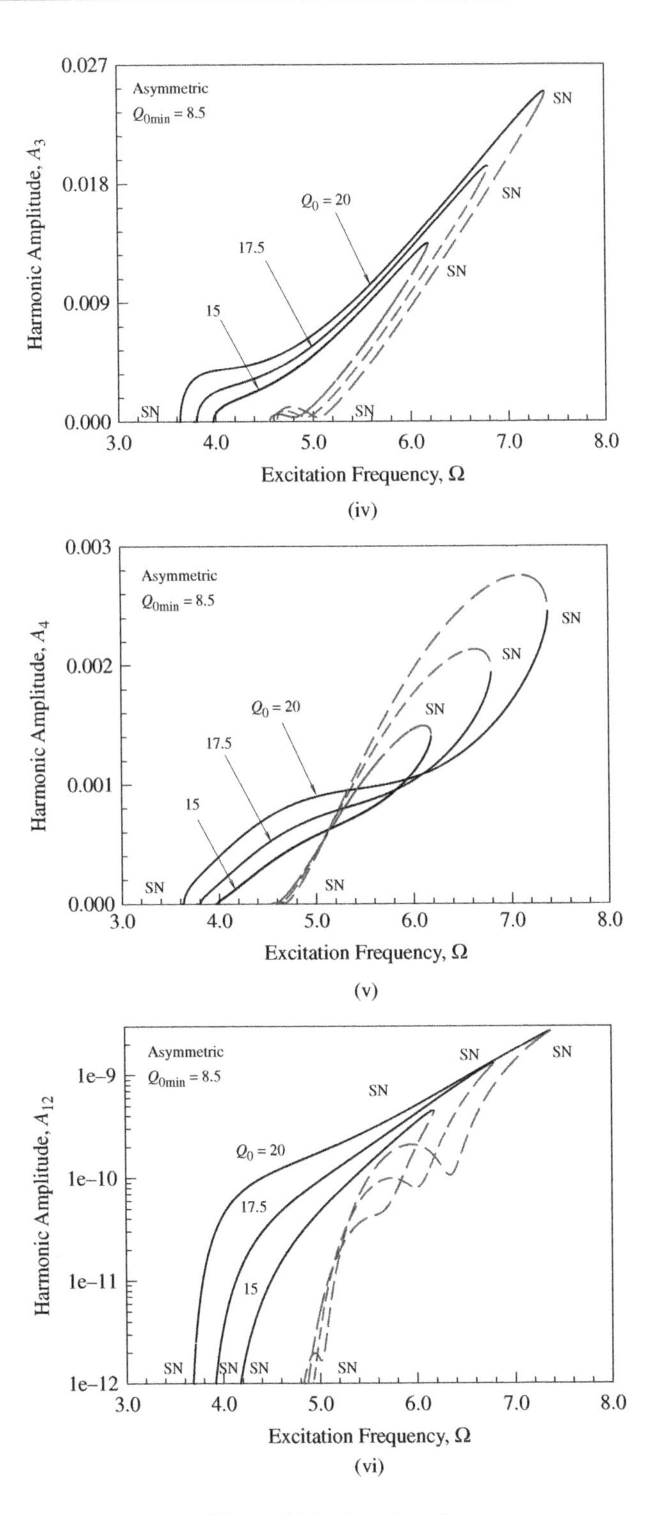

Figure 4.14 *(continued)*

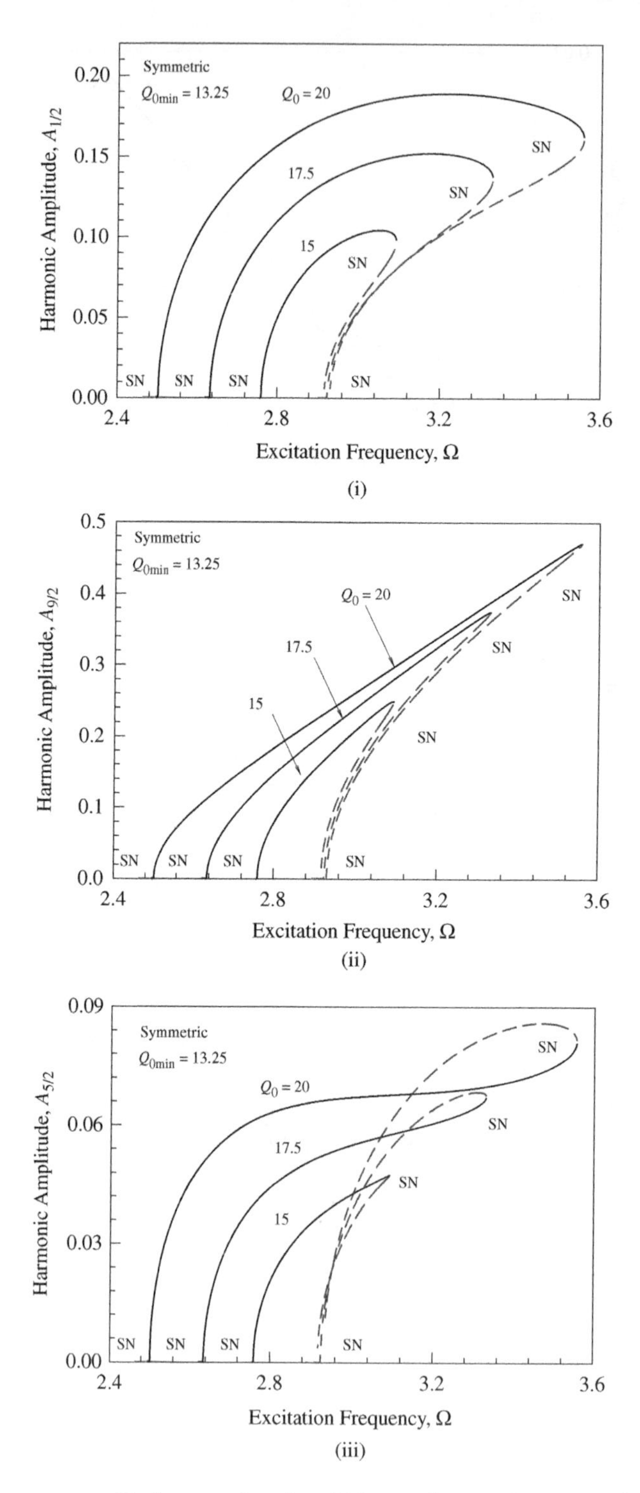

Figure 4.15 Frequency-amplitude curves based on 24 harmonic terms for symmetric period-2 motion in the parametric Duffing oscillator: (i)–(vi) harmonic term $A_{k/2}$ ($k = 2l - 1, l = 1, 2, \ldots, 5, 12$). ($\delta = 0.2$, $\alpha = 20$, $\beta = 50$, $Q_0 = 15, 17.5, 20$)

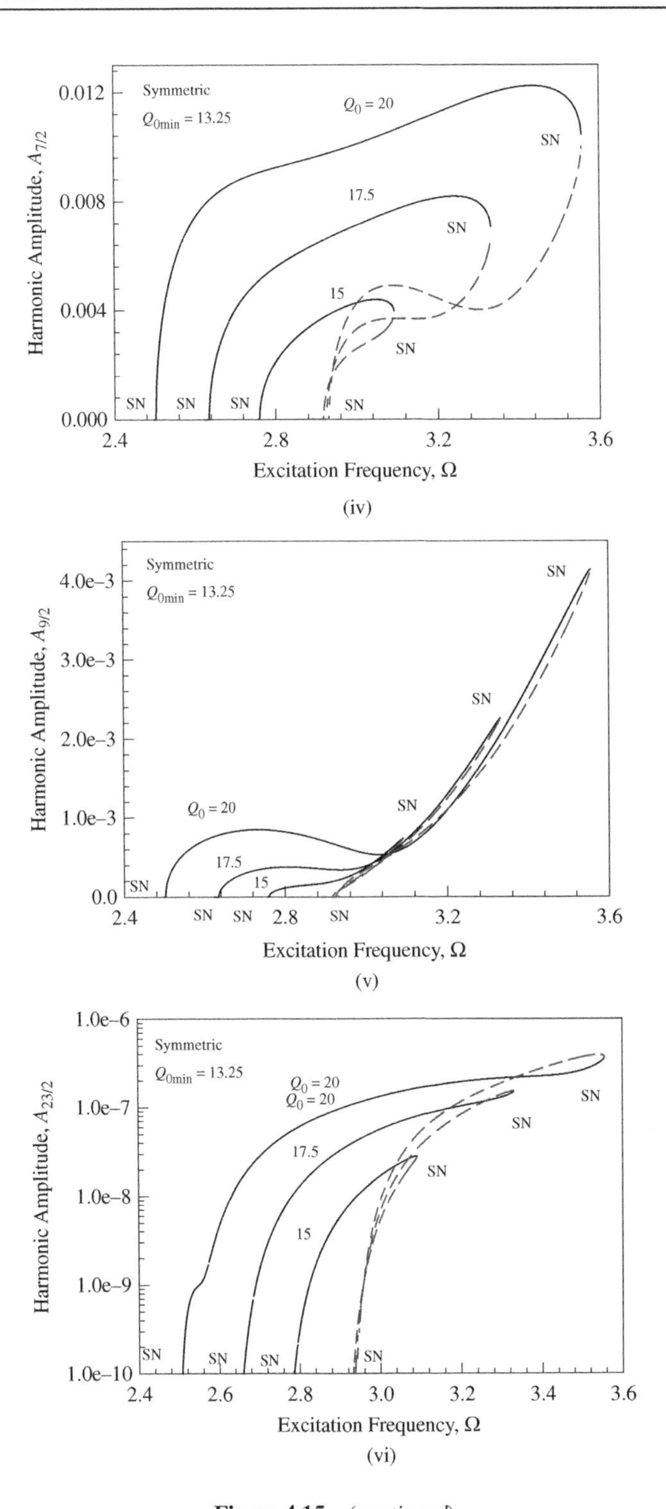

Figure 4.15 *(continued)*

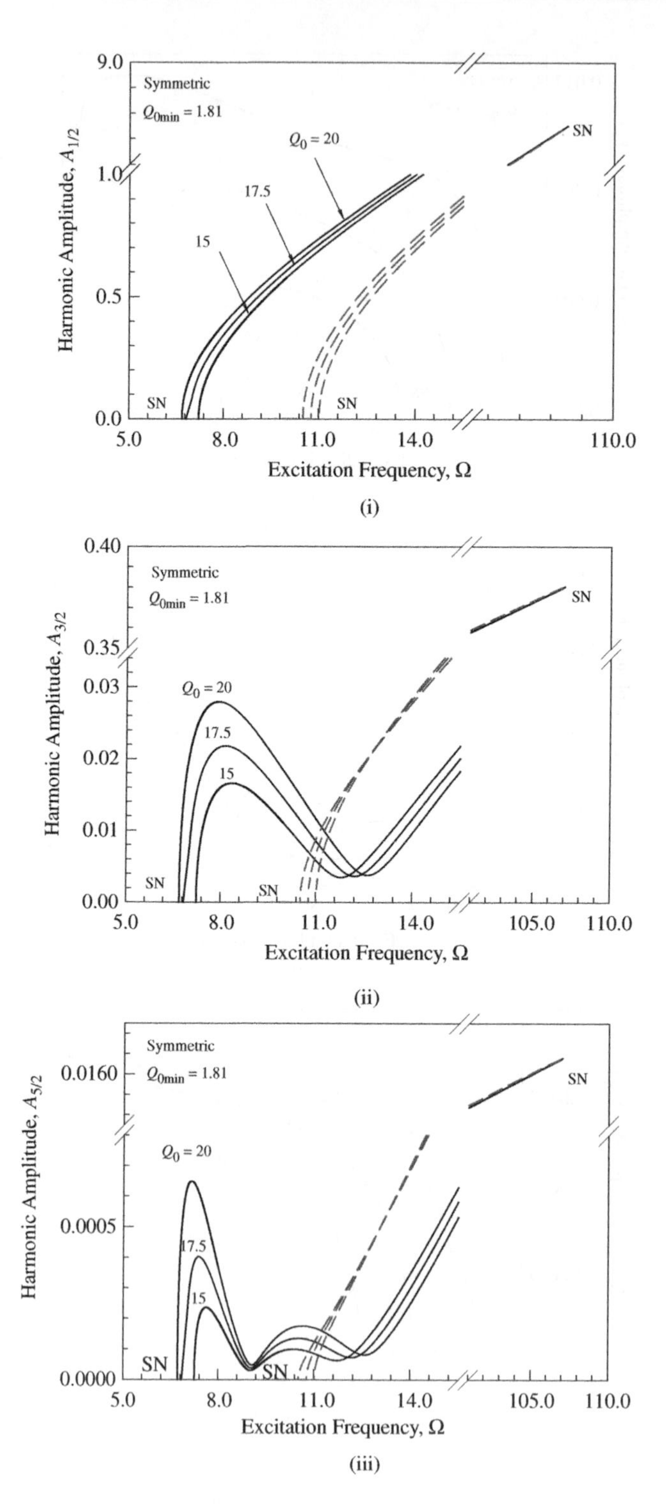

Figure 4.16 Frequency-amplitude curves based on 12 harmonic terms for symmetric period-2 motion in the parametric Duffing oscillator: (i)–(vi) harmonic term $A_{k/2}$ $(k = 2l - 1,\ l = 1, 2, \ldots, 6)$. $(\delta = 0.2,\ \alpha = 20,\ \beta = 50,\ Q_0 = 15, 17.5, 20)$

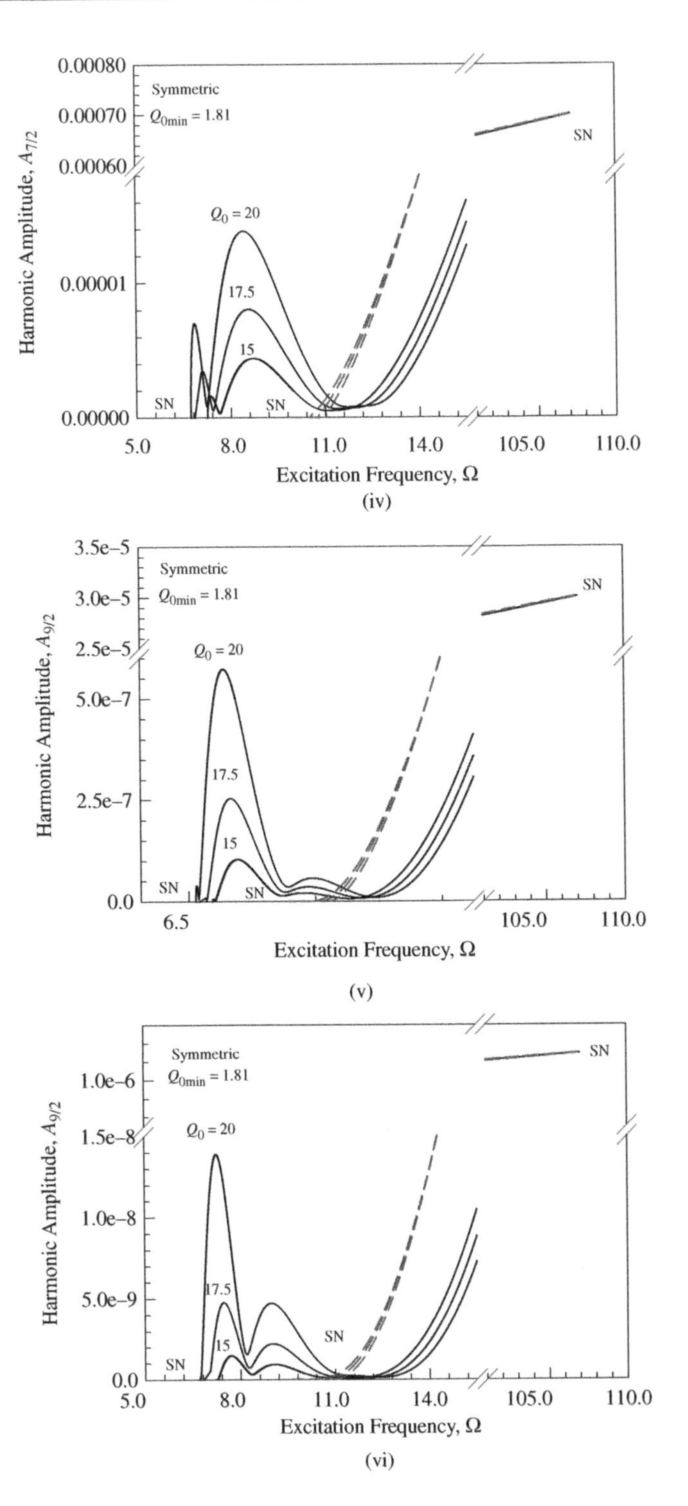

Figure 4.16 *(continued)*

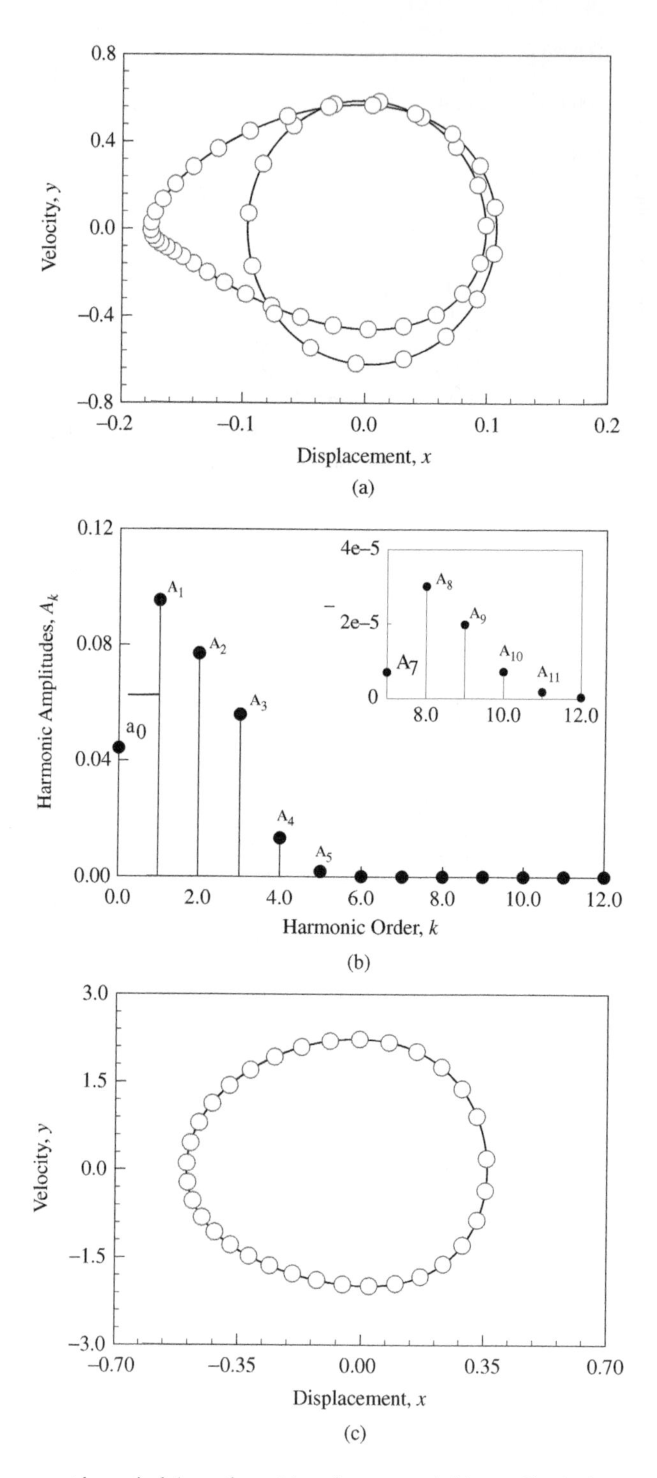

Figure 4.17 Asymmetric period-1 motion: (a) trajectory and (b) amplitude for ($\Omega = 2.0$, $Q_0 = 20$) and ($x_0 \approx -0.094591$, $y_0 = -0.113990$); (c) trajectory and (d) amplitude for ($\Omega = 5.0$, $Q_0 = 15$) and ($x_0 \approx 0.343711$, $y_0 = 0.648798$). System parameters ($\delta = 0.2$, $\alpha = 20$, $\beta = 50$)

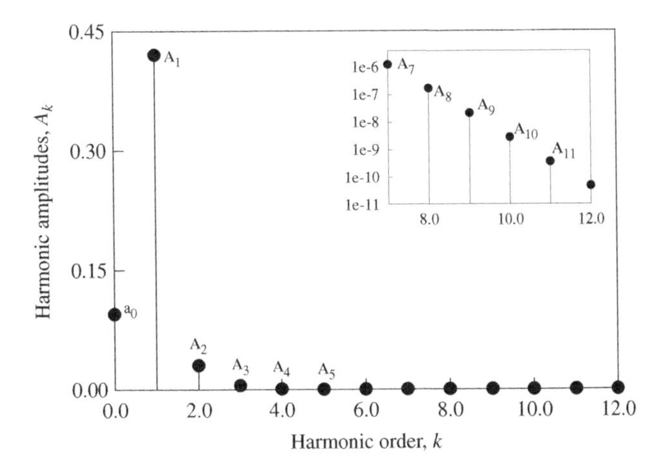

Figure 4.17 *(continued)*

in the ranges of $A_{9/2} \in (0, 3 \times 10^{-5})$ and $A_{11/2} \in (0, 1 \times 10^{-6})$, as presented in Figure 4.16(v) and (vi).

From the frequency-amplitude analysis, there are four independent periodic motions, including two symmetric period-2 motions and two asymmetric period-1 motions. For each of them, stable periodic motions are presented. The numerical solutions of periodic motions are generated via the midpoint discrete scheme. In all plots, circular symbols give approximate analytical solutions, and solid curves give numerical results.

The asymmetric period-1 motion ($\Omega = 2.0$) on the first branch of asymmetric period-1 motion is presented in Figure 4.17 for $Q_0 = 20$. For system parameters in Equation (4.82), the initial condition ($x_0 \approx -0.094591, y_0 \approx -0.113990$) is computed from the analytical solution. In Figure 4.17(a), the trajectory of the asymmetric period-1 motion is presented in phase plane. The asymmetry of periodic motion is observed, and such a period-1 motion is not a simple cycle. Such asymmetric periodic motion has two cycles. The amplitude spectrum based on 12 harmonic terms is presented in Figure 4.17(b). The harmonic amplitudes with different harmonic orders are $a_0 \approx 0.0442$, $A_1 \approx 0.0953$, $A_2 \approx 0.0769$, $A_3 \approx 0.0559$, $A_4 \approx 0.0133$, $A_5 \sim 1.8 \times 10^{-3}$, $A_6 \sim 8.5 \times 10^{-3}$, $A_7 \sim 6.8 \times 10^{-6}$, $A_8 \sim 3 \times 10^{-5}$, $A_9 \sim 2.0 \times 10^{-5}$, $A_{10} \sim 7.0 \times 10^{-6}$, $A_{11} \sim 1.7 \times 10^{-6}$, and $A_{12} \sim 2.9 \times 10^{-7}$. The asymmetric period-1 motion ($\Omega = 5.0$) for the second branch of asymmetric period-1 motion is also presented in Figure 4.17 for $Q_0 = 15$. The initial condition ($x_0 \approx 0.343711, y_0 \approx 0.648798$) is computed from the analytical solution. In Figure 4.17(c), the trajectory of the asymmetric period-1 motion is presented in phase plane. The asymmetry of periodic motion is observed with a little shift on the left hand side, and such a period-1 motion is a simple cycle. The amplitude spectrum based on 12 harmonic terms is presented in Figure 4.17(d). The harmonic amplitudes with different harmonic orders are $a_0 \approx 0.095$, $A_1 \approx 0.420$, $A_2 \approx 0.030$, $A_3 \approx 4.97 \times 10^{-3}$, $A_4 \approx 5.70 \times 10^{-4}$, $A_5 \sim 7.3 \times 10^{-5}$, $A_6 \sim 1.0 \times 10^{-5}$, $A_7 \sim 1.2 \times 10^{-6}$, $A_8 \sim 1.7 \times 10^{-7}$, $A_9 \sim 2.1 \times 10^{-8}$, $A_{10} \sim 2.8 \times 10^{-9}$, $A_{11} \sim 3.6 \times 10^{-10}$, and $A_{12} \sim 4.7 \times 10^{-11}$. Compared to the first branch of asymmetric periodic motion, the primary harmonic term plays an important role in such a simple asymmetric period-1 motion.

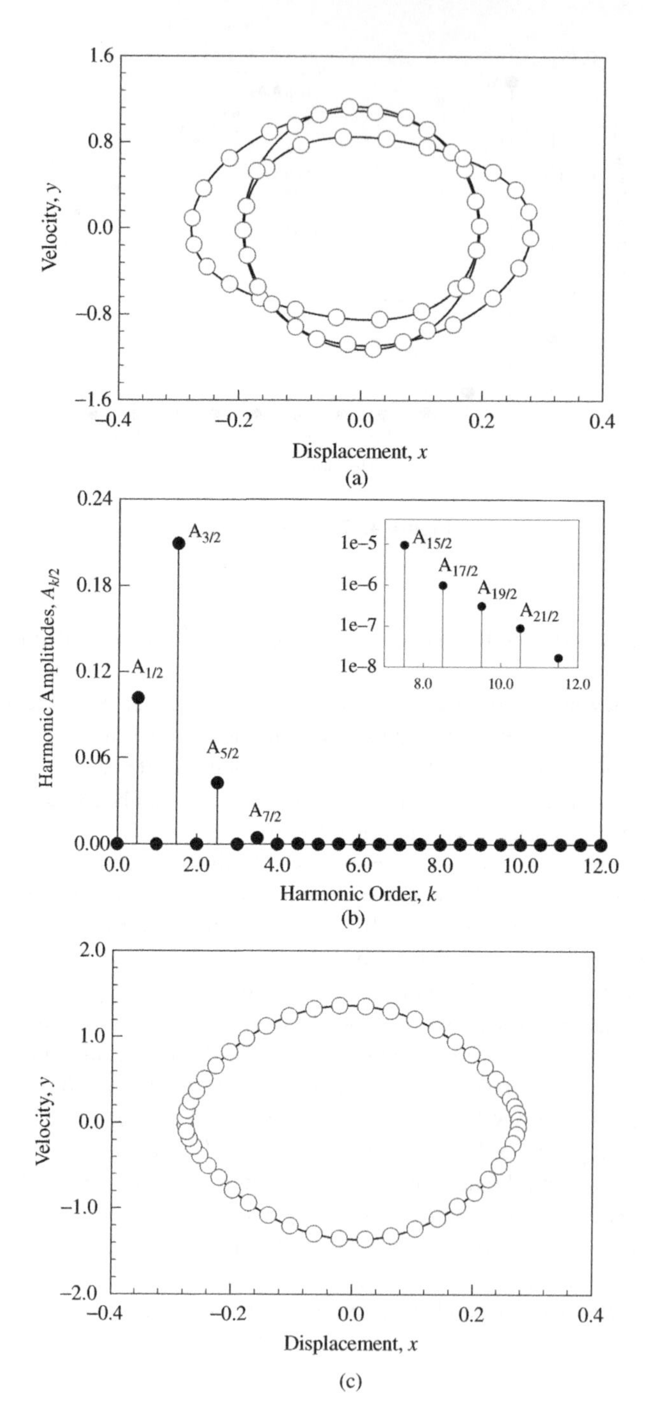

Figure 4.18 Symmetric period-2 motions: (a) trajectory and (b) amplitude for ($\Omega = 3.0$, $Q_0 = 15$) and ($x_0 \approx 0.098997$, $y_0 = -0.984744$); (c) trajectory and (d) amplitude for ($\Omega = 5.0$, $Q_0 = 15$) and ($x_0 \approx 0.015645$, $y_0 = -1.363112$). System parameters ($\delta = 0.2$, $\alpha = 20$, $\beta = 50$)

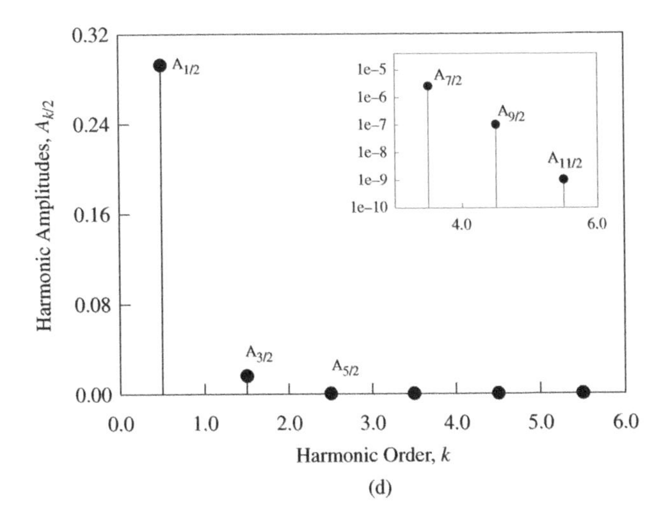

(d)

Figure 4.18 (*continued*)

A symmetric period-2 motion on the first branch of symmetric period-2 motion are presented in Figure 4.18 for $\Omega = 3.0$ and $Q_0 = 15$ with system parameters in Equation (4.82). The initial condition is ($x_0 \approx 0.098997$, $y_0 \approx -0.984744$). In Figure 4.18(a), the trajectory of symmetric period-2 motion with multiple cycles is presented in phase plane. The amplitude spectrum based on 24 harmonic terms is presented in Figure 4.17(b). The harmonic amplitudes are $A_{1/2} \approx 0.1017, A_{3/2} \approx 0.2091, A_{5/2} \approx 0.0425, A_{7/2} \sim 4.3 \times 10^{-3}, A_{9/2} \sim 3.7 \times 10^{-4}, A_{11/2} \sim 2.4 \times 10^{-4}$, $A_{13/2} \sim 5.9 \times 10^{-5}$, $A_{15/2} \sim 9.2 \times 10^{-6}$, $A_{17/2} \sim 9.7 \times 10^{-7}$, $A_{19/2} \sim 3.6 \times 10^{-7}$, $A_{21/2} \sim 8.9 \times 10^{-8}$, $A_{21/2} \sim 1.7 \times 10^{-8}$. This period-2 motion possesses the harmonic terms with $A_{2m/2} = 0$ $(m = 1, 2, 3, \ldots)$. In the traditional perturbation analysis, one can find the simple period-2 motion in the parametric Duffing oscillator. A simple symmetric period-2 motion is presented in Figure 4.18 for $\Omega = 8.0$ and $Q_0 = 15$ with the initial condition ($x_0 \approx 0.015645$, $y_0 \approx -1.363112$). In Figure 4.18(c), the trajectory of symmetric period-2 motion with a single cycle is presented. The amplitude spectrum based on 12 harmonic terms is presented in Figure 4.18(d). $A_{1/2} \approx 0.2922$, $A_{3/2} \approx 0.0160$, $A_{5/2} \approx 1.9 \times 10^{-4}$, $A_{7/2} \sim 2.6 \times 10^{-6}$, $A_{9/2} \sim 1.0 \times 10^{-7}, A_{11/2} \sim 1.0 \times 10^{-9}$.

5

Nonlinear Jeffcott Rotor Systems

In this chapter, analytical solutions for period-m motions in a nonlinear Jeffcott rotor system will be discussed. This rotor system with two-degrees of freedom is a simple rotor dynamical system and periodic excitations are from the rotor eccentricity. The analytical expressions of periodic solutions will be developed. The corresponding stability and bifurcation analyses of period-m motions will be carried out. Analytical bifurcation trees of period-1 motions to chaos will be presented. The Hopf bifurcations of periodic motions can cause not only the bifurcation trees but quasi-periodic motions. Displacement orbits of periodic motions in nonlinear rotor systems show motion complexity, and harmonic amplitude spectrums gives harmonic effects on periodic motions of the nonlinear rotor.

5.1 Analytical Periodic Motions

In this section, the appropriate analytical solutions will also be presented with finite harmonic terms based on the prescribed accuracy of harmonic amplitudes. Rotors in mechanical engineering are extensively used as an important element. The dynamical behaviors of rotors will directly effect the performance of the entire machine. It is significant to investigate the nonlinear behaviors of rotor. As in Huang and Luo (2014), consider a nonlinear Jeffcott rotor dynamical system as

$$\ddot{x} + \delta\dot{x} + (1+\gamma)[\alpha x + \beta x(x^2+y^2)] = e\Omega^2\cos\Omega t,$$
$$\ddot{y} + \delta\dot{y} + \alpha y + \beta y(x^2+y^2) = e\Omega^2\sin\Omega t \tag{5.1}$$

where δ is the linear damping coefficient. α and β are linear and nonlinear spring coefficients, respectively. e and Ω are the eccentric distance and rotation frequency of rotor, respectively. γ is the asymmetric coefficient. The standard form of Equation (5.1) is

$$\ddot{x} = F_1(x,y,\dot{x},\dot{y},t), \ddot{y} = F_2(x,y,\dot{x},\dot{y},t) \tag{5.2}$$

where

$$F_1(x,y,\dot{x},\dot{y},t) = -\delta\dot{x} - (1+\gamma)[\alpha x + \beta x(x^2+y^2)] + e\Omega^2\cos\Omega t,$$
$$F_2(x,y,\dot{x},\dot{y},t) = -\delta\dot{y} - \alpha y - \beta y(x^2+y^2) + e\Omega^2\sin\Omega t. \tag{5.3}$$

In Luo (2012a), the analytical solution of period-m motion in Equation (5.1) is

$$x^{(m)*}(t) = a_{10}^{(m)}(t) + \sum_{k=1}^{N} b_{1k/m}(t)\cos\left(\frac{k}{m}\theta\right) + c_{1k/m}(t)\sin\left(\frac{k}{m}\theta\right),$$
$$y^{(m)*}(t) = a_{20}^{(m)}(t) + \sum_{k=1}^{N} b_{2k/m}(t)\cos\left(\frac{k}{m}\theta\right) + c_{2k/m}(t)\sin\left(\frac{k}{m}\theta\right). \tag{5.4}$$

where $a_{i0}^{(m)}(t)$, $b_{ik/m}(t)$, and $c_{ik/m}(t)$ $(i = 1, 2)$ vary with time and $\theta = \Omega t$. The first and second order of derivatives of $x^*(t)$ and $y^*(t)$ are

$$\dot{x}^{(m)*}(t) = \dot{a}_{10}^{(m)} + \sum_{k=1}^{N}\left[\left(\dot{b}_{1k/m} + \frac{k\Omega}{m}c_{1k/m}\right)\cos\left(\frac{k}{m}\theta\right)\right.$$
$$\left. + \left(\dot{c}_{1k/m} - \frac{\Omega w}{m}b_{1k/m}\right)\sin\left(\frac{k}{m}\theta\right)\right],$$
$$\dot{y}^{(m)*}(t) = \dot{a}_{20}^{(m)} + \sum_{k=1}^{N}\left[\left(\dot{b}_{2k/m} + \frac{k\Omega}{m}c_{yk/m}\right)\cos\left(\frac{k}{m}\theta\right)\right.$$
$$\left. + \left(\dot{c}_{yk/m} - \frac{k\Omega}{m}b_{yk/m}\right)\sin\left(\frac{k}{m}\theta\right)\right]. \tag{5.5}$$

$$\ddot{x}^{(m)*}(t) = \ddot{a}_{10}^{(m)} + \sum_{k=1}^{N}\left[\ddot{b}_{1k/m} + 2\frac{k\Omega}{m}\dot{c}_{1k/m} - \left(\frac{k\Omega}{m}\right)^2 b_{1k/m}\right]\cos\left(\frac{k}{m}\theta\right)$$
$$+ \left[\ddot{c}_{1k/m} - 2\frac{k\Omega}{m}\dot{b}_{1k/m} - \left(\frac{k\Omega}{m}\right)^2 c_{1k/m}\right]\sin\left(\frac{k}{m}\theta\right),$$
$$\ddot{y}^{(m)*}(t) = \ddot{a}_{20}^{(m)} + \sum_{k=1}^{N}\left[\ddot{b}_{2k/m} + 2\frac{k\Omega}{m}\dot{c}_{2k/m} - \left(\frac{k\Omega}{m}\right)^2 b_{2k/m}\right]\cos\left(\frac{k}{m}\theta\right)$$
$$+ \left[\ddot{c}_{2k/m} - 2\frac{k\Omega}{m}\dot{b}_{2k/m} - \left(\frac{k\Omega}{m}\right)^2 c_{2k/m}\right]\sin\left(\frac{k}{m}\theta\right). \tag{5.6}$$

Define

$$\mathbf{a}_0^{(m)} = (a_{10}^{(m)}, a_{20}^{(m)})^{\mathrm{T}},$$
$$\mathbf{b}^{(m)} = (b_{11/m}, \dots, b_{1N/m}; b_{21/m}, \dots, b_{2N/m})^{\mathrm{T}} = (\mathbf{b}_1^{(m)}, \mathbf{b}_2^{(m)})^{\mathrm{T}},$$
$$\mathbf{c}^{(m)} = (c_{11/m}, \dots, c_{1N/m}; c_{21/m}, \dots, c_{2N/m})^{\mathrm{T}} = (\mathbf{c}_1^{(m)}; \mathbf{c}_2^{(m)})^{\mathrm{T}}. \tag{5.7}$$

Substitution of Equations (5.4)–(5.6) into Equation (5.1) and averaging for the harmonic terms of $\cos(k\theta/m)$ and $\sin(k\theta/m)$ $(k = 0, 1, 2, \dots)$ gives

$$\ddot{a}_{10}^{(m)} = F_{10}^{(m)}(\mathbf{a}_0^{(m)}, \mathbf{b}^{(m)}, \mathbf{c}^{(m)}, \dot{\mathbf{a}}_0^{(m)}, \dot{\mathbf{b}}^{(m)}, \dot{\mathbf{c}}^{(m)}),$$
$$\ddot{b}_{1k/m} + 2\frac{k\Omega}{m}\dot{c}_{1k/m} - \left(\frac{k\Omega}{m}\right)^2 b_{1k/m}$$
$$= F_{1k/m}^{(c)}(\mathbf{a}_0^{(m)}, \mathbf{b}^{(m)}, \mathbf{c}^{(m)}, \dot{\mathbf{a}}_0^{(m)}, \dot{\mathbf{b}}^{(m)}, \dot{\mathbf{c}}^{(m)}),$$

$$\ddot{c}_{1k/m} - 2\frac{k\Omega}{m}\dot{b}_{1k/m} - \left(\frac{k\Omega}{m}\right)^2 c_{1k/m}$$
$$= F^{(s)}_{1k/m}(\mathbf{a}^{(m)}_0, \mathbf{b}^{(m)}, \mathbf{c}^{(m)}, \dot{\mathbf{a}}^{(m)}_0, \dot{\mathbf{b}}^{(m)}, \dot{\mathbf{c}}^{(m)}); \tag{5.8}$$

$$\ddot{a}^{(m)}_{20} = F^{(m)}_{20}(\mathbf{a}^{(m)}_0, \mathbf{b}^{(m)}, \mathbf{c}^{(m)}, \dot{\mathbf{a}}^{(m)}_0, \dot{\mathbf{b}}^{(m)}, \dot{\mathbf{c}}^{(m)}),$$

$$\ddot{b}_{2k/m} + 2\frac{k\Omega}{m}\dot{c}_{2k/m} - \left(\frac{k\Omega}{m}\right)^2 b_{2k/m}$$
$$= F^{(c)}_{2k/m}(\mathbf{a}^{(m)}_0, \mathbf{b}^{(m)}, \mathbf{c}^{(m)}, \dot{\mathbf{a}}^{(m)}_0, \dot{\mathbf{b}}^{(m)}, \dot{\mathbf{c}}^{(m)}),$$

$$\ddot{c}_{2k/m} - 2\frac{k\Omega}{m}\dot{b}_{2k/m} - \left(\frac{k\Omega}{m}\right)^2 c_{2k/m}$$
$$= F^{(s)}_{2k/m}(\mathbf{a}^{(m)}_0, \mathbf{b}^{(m)}, \mathbf{c}^{(m)}, \dot{\mathbf{a}}^{(m)}_0, \dot{\mathbf{b}}^{(m)}, \dot{\mathbf{c}}^{(m)}) \tag{5.9}$$

where

$$F^{(m)}_{10}(\mathbf{a}^{(m)}_0, \mathbf{b}^{(m)}, \mathbf{c}^{(m)}, \dot{\mathbf{a}}^{(m)}_0, \dot{\mathbf{b}}^{(m)}, \dot{\mathbf{c}}^{(m)})$$
$$= \frac{1}{mT}\int_0^{mT} F_1(x^{(m)*}, \dot{x}^{(m)*}, y^{(m)*}, \dot{y}^{(m)*}, t)dt$$
$$= -\delta\dot{a}^{(m)}_{10} - (1+\gamma)(\alpha a^{(m)}_{10} + \beta f^{(m)}_{10}),$$

$$F^{(c)}_{1k}(\mathbf{a}^{(m)}_0, \mathbf{b}^{(m)}, \mathbf{c}^{(m)}, \dot{\mathbf{a}}^{(m)}_0, \dot{\mathbf{b}}^{(m)}, \dot{\mathbf{c}}^{(m)})$$
$$= \frac{2}{mT}\int_0^{mT} F_1(x^{(m)*}, \dot{x}^{(m)*}, y^{(m)*}, \dot{y}^{(m)*}, t)\cos\left(\frac{k}{m}\Omega t\right)dt$$
$$= -\delta\left(\dot{b}_{1k/m} + \frac{k\Omega}{m}c_{1k/m}\right) - (1+\gamma)(\alpha b_{1k/m} + \beta f^{(c)}_{1k/m}) + e\Omega^2\delta^m_k,$$

$$F^{(s)}_{1k}(\mathbf{a}^{(m)}_0, \mathbf{b}^{(m)}, \mathbf{c}^{(m)}, \dot{\mathbf{a}}^{(m)}_0, \dot{\mathbf{b}}^{(m)}, \dot{\mathbf{c}}^{(m)})$$
$$= \frac{2}{mT}\int_0^{mT} F_1(x^{(m)*}, \dot{x}^{(m)*}, y^{(m)*}, \dot{y}^{(m)*}, t)\sin\left(\frac{k}{m}\Omega t\right)dt$$
$$= -\delta\left(\dot{c}_{1k/m} - \frac{k\Omega}{m}b_{1k/m}\right) - (1+\gamma)(\alpha c_{1k/m} + \beta f^{(s)}_{1k/m}); \tag{5.10}$$

and

$$F^{(m)}_{20}(\mathbf{a}^{(m)}_0, \mathbf{b}^{(m)}, \mathbf{c}^{(m)}, \dot{\mathbf{a}}^{(m)}_0, \dot{\mathbf{b}}^{(m)}, \dot{\mathbf{c}}^{(m)})$$
$$= \frac{1}{mT}\int_0^{mT} F_2(x^{(m)*}, \dot{x}^{(m)*}, y^{(m)*}, \dot{y}^{(m)*}, t)dt$$
$$= -\delta\dot{a}^{(m)}_{20} - (\alpha a^{(m)}_{20} + \beta f^{(m)}_{10}),$$

$$
\begin{aligned}
&F_{2k}^{(c)}(\mathbf{a}_0^{(m)}, \mathbf{b}^{(m)}, \mathbf{c}^{(m)}, \dot{\mathbf{a}}_0^{(m)}, \dot{\mathbf{b}}^{(m)}, \dot{\mathbf{c}}^{(m)}) \\
&\quad = \frac{2}{mT}\int_0^{mT} F_2(x^{(m)*}, \dot{x}^{(m)*}, y^{(m)*}, \dot{y}^{(m)*}, t)\cos\left(\frac{k}{m}\Omega t\right) dt \\
&\quad = -\delta\left(\dot{b}_{2k/m} + \frac{k\Omega}{m}c_{2k/m}\right) - (\alpha b_{2k/m} + \beta f_{2k/m}^{(c)}), \\
&F_{2k}^{(s)}(\mathbf{a}_0^{(m)}, \mathbf{b}^{(m)}, \mathbf{c}^{(m)}, \dot{\mathbf{a}}_0^{(m)}, \dot{\mathbf{b}}^{(m)}, \dot{\mathbf{c}}^{(m)}) \\
&\quad = \frac{2}{mT}\int_0^{mT} F_2(x^{(m)*}, \dot{x}^{(m)*}, y^{(m)*}, \dot{y}^{(m)*}, t)\sin\left(\frac{k}{m}\Omega t\right) dt \\
&\quad = -\delta\left(\dot{c}_{2k/m} - \frac{k\Omega}{m}b_{2k/m}\right) - (\alpha c_{2k/m} + \beta f_{2k/m}^{(s)}) + e\Omega^2\delta_k^m.
\end{aligned} \tag{5.11}
$$

The functions for the first oscillator are

$$
\begin{aligned}
f_{10}^{(m)} &= (a_{10}^{(m)})^3 + a_{10}^{(m)}(a_{20}^{(m)})^2 + \sum_{q=1}^{11}\sum_{l=1}^{N}\sum_{j=1}^{N}\sum_{i=1}^{N} f_{10}^{(m)}(i,j,l,q), \\
f_{1k/m}^{(c)} &= 3(a_{10}^{(m)})^2 b_{1k/m} + 2a_{10}^{(m)}a_{20}^{(m)}b_{2k/m} + (a_{20}^{(m)})^2 b_{1k/m} \\
&\quad + \sum_{q=1}^{11}\sum_{l=1}^{N}\sum_{j=1}^{N}\sum_{i=1}^{N} f_{1k/m}^{(c)}(i,j,l,q), \\
f_{1k/m}^{(s)} &= 3(a_{10}^{(m)})^2 c_{1k/m} + 2a_{10}^{(m)}a_{20}^{(m)}c_{2k/m} + (a_{20}^{(m)})^2 c_{1k/m} \\
&\quad + \sum_{q=1}^{9}\sum_{l=1}^{N}\sum_{j=1}^{N}\sum_{i=1}^{N} f_{1k/m}^{(s)}(i,j,l,q),
\end{aligned} \tag{5.12}
$$

where

$$
\begin{aligned}
f_{10}^{(m)}(i,j,l,1) &= \frac{3a_{10}^{(m)}}{2N} b_{1i/m}b_{1j/m}\delta_{i-j}^0, \\
f_{10}^{(m)}(i,j,l,2) &= \frac{3a_{10}^{(m)}}{2N} c_{1i/m}c_{1j/m}\delta_{i-j}^0, \\
f_{10}^{(m)}(i,j,l,3) &= \frac{1}{4} b_{1i/m}b_{1j/m}b_{1l/m}(\delta_{i-j-l}^0 + \delta_{i-j+l}^0 + \delta_{i+j-l}^0), \\
f_{10}^{(m)}(i,j,l,4) &= \frac{3}{4} b_{1i/m}c_{1j/m}c_{1l/m}(\delta_{i+j-l}^0 + \delta_{i-j+l}^0 - \delta_{i-j-l}^0), \\
f_{10}^{(m)}(i,j,l,5) &= \frac{a_{10}^{(m)}}{2N} b_{2i/m}b_{2j/m}\delta_{i-j}^0, \\
f_{10}^{(m)}(i,j,l,6) &= \frac{a_{10}^{(m)}}{2N} c_{2i/m}c_{2j/m}\delta_{i-j}^0, \\
f_{10}^{(m)}(i,j,l,7) &= \frac{a_{20}^{(m)}}{N} b_{1i/m}b_{2j/m}\delta_{i-j}^0,
\end{aligned}
$$

$$f_{10}^{(m)}(i,j,l,8) = \frac{a_{20}^{(m)}}{N} c_{1i/m} c_{2j/m} \delta_{i-j}^{0},$$

$$f_{10}^{(m)}(i,j,l,9) = \frac{1}{4} b_{1i/m} b_{2j/m} b_{2l/m} (\delta_{i-j-l}^{0} + \delta_{i-j+l}^{0} + \delta_{i+j-l}^{0}),$$

$$f_{10}^{(m)}(i,j,l,10) = \frac{1}{4} b_{1i/m} c_{2j/m} c_{2l/m} (\delta_{i+j-l}^{0} + \delta_{i-j+l}^{0} - \delta_{i-j-l}^{0}),$$

$$f_{10}^{(m)}(i,j,l,11) = \frac{1}{2} c_{1i/m} c_{2j/m} b_{2l/m} (\delta_{i-j-l}^{0} + \delta_{i-j+l}^{0} - \delta_{i+j-l}^{0}); \tag{5.13}$$

and

$$f_{1k/m}^{(c)}(i,j,l,1) = \frac{3a_{10}^{(m)}}{2N} b_{1i/m} b_{1j/m} (\delta_{|i-j|}^{k} + \delta_{i+j}^{k}),$$

$$f_{1k/m}^{(c)}(i,j,l,2) = \frac{3a_{10}^{(m)}}{2N} c_{1i/m} c_{1j/m} (\delta_{|i-j|}^{k} - \delta_{i+j}^{k}),$$

$$f_{1k/m}^{(c)}(i,j,l,3) = \frac{1}{4} b_{1i/m} b_{1j/m} b_{1l/m} (\delta_{|i-j-l|}^{k} + \delta_{|i-j+l|}^{k} + \delta_{|i+j-l|}^{k} + \delta_{i+j+l}^{k}),$$

$$f_{1k/m}^{(c)}(i,j,l,4) = \frac{3}{4} b_{1i/m} c_{1j/m} c_{1l/m} (\delta_{|i-j+l|}^{k} + \delta_{|i+j-l|}^{k} - \delta_{|i-j-l|}^{k} - \delta_{i+j+l}^{k}),$$

$$f_{1k/m}^{(c)}(i,j,l,5) = \frac{a_{10}^{(m)}}{2N} b_{2i/m} b_{2j/m} (\delta_{|i-j|}^{k} + \delta_{i+j}^{k}),$$

$$f_{1k/m}^{(c)}(i,j,l,6) = \frac{a_{10}^{(m)}}{2N} c_{2i/m} c_{2j/m} (\delta_{|i-j|}^{k} - \delta_{i+j}^{k}),$$

$$f_{1k/m}^{(c)}(i,j,l,7) = \frac{a_{20}^{(m)}}{N} b_{1i/m} b_{2j/m} (\delta_{|i-j|}^{k} + \delta_{i+j}^{k}),$$

$$f_{1k/m}^{(c)}(i,j,l,8) = \frac{a_{20}^{(m)}}{N} c_{1i/m} c_{2j/m} (\delta_{|i-j|}^{k} - \delta_{i+j}^{k}),$$

$$f_{1k/m}^{(c)}(i,j,l,9) = \frac{1}{4} b_{1i/m} b_{2j/m} b_{2l/m} (\delta_{|i-j-l|}^{k} + \delta_{|i-j+l|}^{k} + \delta_{|i+j-l|}^{k} + \delta_{i+j+l}^{k}),$$

$$f_{1k/m}^{(c)}(i,j,l,10) = \frac{1}{4} b_{1i/m} c_{2j/m} c_{2l/m} (\delta_{|i-j+l|}^{k} + \delta_{|i+j-l|}^{k} - \delta_{|i-j-l|}^{k} - \delta_{i+j+l}^{k}),$$

$$f_{1k/m}^{(c)}(i,j,l,11) = \frac{1}{2} b_{2i/m} c_{2j/m} c_{1l/m} (\delta_{|i-j+l|}^{k} + \delta_{|i+j-l|}^{k} - \delta_{|i-j-l|}^{k} - \delta_{i+j+l}^{k}); \tag{5.14}$$

and

$$f_{1k/m}^{(s)}(i,j,l,1) = \frac{3a_{10}^{(m)}}{N} b_{1i/m} c_{1j/m} [\delta_{i+j}^{k} - \mathrm{sgn}(i-j)\delta_{|i-j|}^{k}],$$

$$f_{1k/m}^{(s)}(i,j,l,2) = \frac{1}{4} c_{1i/m} c_{1j/m} c_{1l/m} [\mathrm{sgn}(i-j+l)\delta_{|i-j+l|}^{k} + \mathrm{sgn}(i+j-l)\delta_{|i+j-l|}^{k} - \mathrm{sgn}(i-j-l)\delta_{|i-j-l|}^{k} - \delta_{i+j+l}^{k}],$$

$$
\begin{aligned}
f_{1k/m}^{(s)}(i,j,l,3) &= \frac{3}{4} b_{1i/m} b_{1j/m} c_{1l/m} [\operatorname{sgn}(i-j+l)\delta_{|i-j+l|}^{k} + \delta_{i+j+l}^{k} \\
&\quad - \operatorname{sgn}(i+j-l)\delta_{|i+j-l|}^{k} - \operatorname{sgn}(i-j-l)\delta_{|i-j-l|}^{k}], \\
f_{1k/m}^{(s)}(i,j,l,4) &= \frac{a_{20}^{(m)}}{N} b_{1i/m} c_{2j/m} [\delta_{i+j}^{k} - \operatorname{sgn}(i-j)\delta_{|i-j|}^{k}], \\
f_{1k/m}^{(s)}(i,j,l,5) &= \frac{a_{20}^{(m)}}{N} c_{1i/m} b_{2j/m} [\delta_{i+j}^{k} + \operatorname{sgn}(i-j)\delta_{|i-j|}^{k}], \\
f_{1k/m}^{(s)}(i,j,l,6) &= \frac{a_{10}^{(m)}}{N} b_{2i/m} c_{2j/m} [\delta_{i+j}^{k} - \operatorname{sgn}(i-j)\delta_{|i-j|}^{k}], \\
f_{1k/m}^{(s)}(i,j,l,7) &= \frac{1}{4} c_{1i/m} c_{2j/m} c_{2l/m} [\operatorname{sgn}(i-j+l)\delta_{|i-j+l|}^{k} + \operatorname{sgn}(i+j-l)\delta_{|i+j-l|}^{k} \\
&\quad - \operatorname{sgn}(i-j-l)\delta_{|i-j-l|}^{k} - \delta_{i+j+l}^{k}], \\
f_{1k/m}^{(s)}(i,j,l,8) &= \frac{1}{2} b_{1i/m} b_{2j/m} c_{2l/m} [\operatorname{sgn}(i-j+l)\delta_{|i-j+l|}^{k} + \delta_{i+j+l}^{k} \\
&\quad - \operatorname{sgn}(i+j-l)\delta_{|i+j-l|}^{k} - \operatorname{sgn}(i-j-l)\delta_{|i-j-l|}^{k}], \\
f_{1k/m}^{(s)}(i,j,l,9) &= \frac{1}{4} b_{2i/m} b_{2j/m} c_{1l/m} [\operatorname{sgn}(i-j+l)\delta_{|i-j+l|}^{k} + \delta_{i+j+l}^{k} \\
&\quad - \operatorname{sgn}(i+j-l)\delta_{|i+j-l|}^{k} - \operatorname{sgn}(i-j-l)\delta_{|i-j-l|}^{k}].
\end{aligned} \tag{5.15}
$$

The functions for the second oscillator are

$$
\begin{aligned}
f_{20}^{(m)} &= (a_{20}^{(m)})^3 + (a_{10}^{(m)})^2 a_{20}^{(m)} + \sum_{q=1}^{11}\sum_{l=1}^{N}\sum_{j=1}^{N}\sum_{i=1}^{N} f_{20}^{(m)}(i,j,l,q), \\
f_{2k/m}^{(c)} &= 3(a_{20}^{(m)})^2 b_{2k/m} + 2a_{10}^{(m)} a_{20}^{(m)} b_{1k/m} + (a_{10}^{(m)})^2 b_{2k/m} \\
&\quad + \sum_{q=1}^{11}\sum_{l=1}^{N}\sum_{j=1}^{N}\sum_{i=1}^{N} f_{2k/m}^{(c)}(i,j,l,q), \\
f_{2k/m}^{(s)} &= 3(a_{20}^{(m)})^2 c_{2k/m} + 2a_{10}^{(m)} a_{20}^{(m)} c_{1k/m} + (a_{10}^{(m)})^2 c_{2k/m} \\
&\quad + \sum_{q=1}^{9}\sum_{l=1}^{N}\sum_{j=1}^{N}\sum_{i=1}^{N} f_{2k/m}^{(s)}(i,j,l,q),
\end{aligned} \tag{5.16}
$$

where

$$
\begin{aligned}
f_{20}^{(m)}(i,j,l,1) &= \frac{3a_{20}^{(m)}}{2N} b_{2i/m} b_{2j/m} \delta_{i-j}^{0}, \\
f_{20}^{(m)}(i,j,l,2) &= \frac{3a_{20}^{(m)}}{2N} c_{2i/m} c_{2j/m} \delta_{i-j}^{0},
\end{aligned}
$$

$$
\begin{aligned}
f_{20}^{(m)}(i,j,l,3) &= \frac{1}{4}b_{2i/m}b_{2j/m}b_{2l/m}(\delta_{i-j-l}^{0} + \delta_{i-j+l}^{0} + \delta_{i+j-l}^{0}),\\
f_{20}^{(m)}(i,j,l,4) &= \frac{3}{4}b_{2i/m}c_{2j/m}c_{2l/m}(\delta_{i+j-l}^{0} + \delta_{i-j+l}^{0} - \delta_{i-j-l}^{0}),\\
f_{20}^{(m)}(i,j,l,5) &= \frac{a_{20}^{(m)}}{2N}b_{2i/m}b_{2j/m}\delta_{i-j}^{0},\\
f_{20}^{(m)}(i,j,l,6) &= \frac{a_{20}^{(m)}}{2N}c_{1i/m}c_{1j/m}\delta_{i-j}^{0},\\
f_{20}^{(m)}(i,j,l,7) &= \frac{a_{10}^{(m)}}{N}b_{1i/m}b_{2j/m}\delta_{i-j}^{0},\\
f_{20}^{(m)}(i,j,l,8) &= \frac{a_{10}^{(m)}}{N}c_{1i/m}c_{2j/m}\delta_{i-j}^{0},\\
f_{20}^{(m)}(i,j,l,9) &= \frac{1}{4}b_{1i/m}b_{1j/m}b_{2l/m}(\delta_{i-j-l}^{0} + \delta_{i-j+l}^{0} + \delta_{i+j-l}^{0}),\\
f_{20}^{(m)}(i,j,l,10) &= \frac{1}{4}b_{2i/m}c_{1j/m}c_{1l/m}(\delta_{i+j-l}^{0} + \delta_{i-j+l}^{0} - \delta_{i-j-l}^{0}),\\
f_{20}^{(m)}(i,j,l,11) &= \frac{1}{2}c_{2i/m}c_{1j/m}b_{1l/m}(\delta_{i-j-l}^{0} + \delta_{i-j+l}^{0} - \delta_{i+j-l}^{0});
\end{aligned}
\tag{5.17}
$$

and

$$
\begin{aligned}
f_{2k/m}^{(c)}(i,j,l,1) &= \frac{3a_{20}^{(m)}}{2N}b_{2i/m}b_{2j/m}(\delta_{|i-j|}^{k} + \delta_{i+j}^{k}),\\
f_{2k/m}^{(c)}(i,j,l,2) &= \frac{3a_{20}^{(m)}}{2N}c_{2i/m}c_{2j/m}(\delta_{|i-j|}^{k} - \delta_{i+j}^{k}),\\
f_{2k/m}^{(c)}(i,j,l,3) &= \frac{1}{4}b_{2i/m}b_{2j/m}b_{2l/m}(\delta_{|i-j-l|}^{k} + \delta_{|i-j+l|}^{k} + \delta_{|i+j-l|}^{k} + \delta_{i+j+l}^{k}),\\
f_{2k/m}^{(c)}(i,j,l,4) &= \frac{3}{4}b_{2i/m}c_{2j/m}c_{2l/m}(\delta_{|i-j+l|}^{k} + \delta_{|i+j-l|}^{k} - \delta_{|i-j-l|}^{k} - \delta_{i+j+l}^{k}),\\
f_{2k/m}^{(c)}(i,j,l,5) &= \frac{a_{20}^{(m)}}{2N}b_{1i/m}b_{1j/m}(\delta_{|i-j|}^{k} + \delta_{i+j}^{k}),\\
f_{2k/m}^{(c)}(i,j,l,6) &= \frac{a_{20}^{(m)}}{2N}c_{1i/m}c_{1j/m}(\delta_{|i-j|}^{k} - \delta_{i+j}^{k}),\\
f_{2k/m}^{(c)}(i,j,l,7) &= \frac{a_{10}^{(m)}}{N}b_{1i/m}b_{2j/m}(\delta_{|i-j|}^{k} + \delta_{i+j}^{k}),\\
f_{2k/m}^{(c)}(i,j,l,8) &= \frac{a_{10}^{(m)}}{N}c_{1i/m}c_{2j/m}(\delta_{|i-j|}^{k} - \delta_{i+j}^{k}),\\
f_{2k/m}^{(c)}(i,j,l,9) &= \frac{1}{4}b_{1i/m}b_{1j/m}b_{2l/m}(\delta_{|i-j-l|}^{k} + \delta_{|i-j+l|}^{k} + \delta_{|i+j-l|}^{k} + \delta_{i+j+l}^{k}),
\end{aligned}
$$

$$
\begin{aligned}
f_{2k/m}^{(c)}(i,j,l,10) &= \frac{1}{4} b_{2i/m} c_{1j/m} c_{1l/m} (\delta_{|i-j+l|}^{k} + \delta_{|i+j-l|}^{k} - \delta_{|i-j-l|}^{k} - \delta_{i+j+l}^{k}),\\
f_{2k/m}^{(c)}(i,j,l,11) &= \frac{1}{2} b_{1i/m} c_{1j/m} c_{2l/m} (\delta_{|i-j+l|}^{k} + \delta_{|i+j-l|}^{k} - \delta_{|i-j-l|}^{k} - \delta_{i+j+l}^{k});
\end{aligned} \tag{5.18}
$$

and

$$
\begin{aligned}
f_{2k/m}^{(s)}(i,j,l,1) &= \frac{3a_{20}^{(m)}}{N} b_{2i/m} c_{2j/m} [\delta_{i+j}^{k} - \mathrm{sgn}(i-j)\delta_{|i-j|}^{k}],\\
f_{2k/m}^{(s)}(i,j,l,2) &= \frac{1}{4} c_{2i/m} c_{2j/m} c_{2l/m} [\mathrm{sgn}(i-j+l)\delta_{|i-j+l|}^{k} + \mathrm{sgn}(i+j-l)\delta_{|i+j-l|}^{k}\\
&\quad - \mathrm{sgn}(i-j-l)\delta_{|i-j-l|}^{k} - \delta_{i+j+l}^{k}],\\
f_{2k/m}^{(s)}(i,j,l,3) &= \frac{3}{4} b_{2i/m} b_{2j/m} c_{2l/m} [\mathrm{sgn}(i-j+l)\delta_{|i-j+l|}^{k} + \delta_{i+j+l}^{k}\\
&\quad - \mathrm{sgn}(i+j-l)\delta_{|i+j-l|}^{k} - \mathrm{sgn}(i-j-l)\delta_{|i-j-l|}^{k}],\\
f_{2k/m}^{(s)}(i,j,l,4) &= \frac{a_{10}^{(m)}}{N} b_{1i/m} c_{2j/m} [\delta_{i+j}^{k} - \mathrm{sgn}(i-j)\delta_{|i-j|}^{k}],\\
f_{2k/m}^{(s)}(i,j,l,5) &= \frac{a_{10}^{(m)}}{N} c_{1i/m} b_{2j/m} [\delta_{i+j}^{k} + \mathrm{sgn}(i-j)\delta_{|i-j|}^{k}],\\
f_{2k/m}^{(s)}(i,j,l,6) &= \frac{a_{20}^{(m)}}{N} b_{1i/m} c_{1j/m} [\delta_{i+j}^{k} - \mathrm{sgn}(i-j)\delta_{|i-j|}^{k}],\\
f_{2k/m}^{(s)}(i,j,l,7) &= \frac{1}{4} c_{1i/m} c_{1j/m} c_{2l/m} [\mathrm{sgn}(i-j+l)\delta_{|i-j+l|}^{k} + \mathrm{sgn}(i+j-l)\delta_{|i+j-l|}^{k}\\
&\quad - \mathrm{sgn}(i-j-l)\delta_{|i-j-l|}^{k} - \delta_{i+j+l}^{k}],\\
f_{2k/m}^{(s)}(i,j,l,8) &= \frac{1}{2} b_{1i/m} b_{2j/m} c_{1l/m} [\mathrm{sgn}(i-j+l)\delta_{|i-j+l|}^{k} + \delta_{i+j+l}^{k}\\
&\quad - \mathrm{sgn}(i+j-l)\delta_{|i+j-l|}^{k} - \mathrm{sgn}(i-j-l)\delta_{|i-j-l|}^{k}],\\
f_{2k/m}^{(s)}(i,j,l,9) &= \frac{1}{4} b_{1i/m} b_{1j/m} c_{2l/m} [\mathrm{sgn}(i-j+l)\delta_{|i-j+l|}^{k} + \delta_{i+j+l}^{k}\\
&\quad - \mathrm{sgn}(i+j-l)\delta_{|i+j-l|}^{k} - \mathrm{sgn}(i-j-l)\delta_{|i-j-l|}^{k}].
\end{aligned} \tag{5.19}
$$

Define

$$
\begin{aligned}
\mathbf{z}^{(m)} &\triangleq (\mathbf{a}_0^{(m)}, \mathbf{b}^{(m)}, \mathbf{c}^{(m)})^{\mathrm{T}}\\
&= (a_{10}^{(m)}, b_{11/m}, \ldots, b_{1N/m}, c_{11/m}, \ldots, c_{1N/m}\\
&\qquad a_{20}^{(m)}, b_{21/m}, \ldots, b_{2N/m}, c_{21/m}, \ldots, c_{2N/m})^{\mathrm{T}}\\
&\equiv (z_1^{(m)}, z_2^{(m)}, \ldots, z_{2N+1}^{(m)}; z_{2N+2}^{(m)}, z_{2N+3}^{(m)}, \ldots, z_{4N+2}^{(m)})^{\mathrm{T}},
\end{aligned}
$$

$$\begin{aligned}\mathbf{z}_1^{(m)} \triangleq \dot{\mathbf{z}}^{(m)} &= (\dot{a}_0^{(m)}, \dot{\mathbf{b}}^{(m)}, \dot{\mathbf{c}}^{(m)})^{\mathrm{T}} \\ &= (\dot{a}_{10}^{(m)}, \dot{b}_{11/m}, \ldots, \dot{b}_{1N/m}, \dot{c}_{11/m}, \ldots, \dot{c}_{1N/m}; \\ &\quad \dot{a}_{20}^{(m)}, \dot{b}_{21/m}, \ldots, \dot{b}_{2N/m}, \dot{c}_{21/m}, \ldots, \dot{c}_{2N/m})^{\mathrm{T}} \\ &\equiv (\dot{z}_1^{(m)}, \dot{z}_2^{(m)}, \ldots, \dot{z}_{2N+1}^{(m)}; \dot{z}_{2N+2}^{(m)}, \dot{z}_{2N+3}^{(m)}, \ldots, \dot{z}_{4N+2}^{(m)})^{\mathrm{T}}.\end{aligned} \tag{5.20}$$

Equations (5.8) and (5.9) are rewritten as

$$\dot{\mathbf{z}}^{(m)} = \mathbf{z}_1^{(m)} \text{ and } \dot{\mathbf{z}}_1^{(m)} = \mathbf{g}^{(m)}(\mathbf{z}^{(m)}, \mathbf{z}_1^{(m)}), \tag{5.21}$$

where

$$\mathbf{g}^{(m)}(\mathbf{z}^{(m)}, \mathbf{z}_1^{(m)}) = \begin{pmatrix} F_{10}^{(m)}(\mathbf{z}^{(m)}, \mathbf{z}_1^{(m)}) \\ \mathbf{F}_{1/m}^{(c)}(\mathbf{z}^{(m)}, \mathbf{z}_1^{(m)}) - 2\mathbf{k}_1 \frac{\Omega}{m} \dot{\mathbf{c}}_1^{(m)} + \mathbf{k}_2 \left(\frac{\Omega}{m}\right)^2 \mathbf{b}_1^{(m)} \\ \mathbf{F}_{1/m}^{(s)}(\mathbf{z}^{(m)}, \mathbf{z}_1^{(m)}) + 2\mathbf{k}_1 \frac{\Omega}{m} \dot{\mathbf{b}}_1^{(m)} + \mathbf{k}_2 \left(\frac{\Omega}{m}\right)^2 \mathbf{c}_1^{(m)} \\ F_{20}^{(m)}(\mathbf{z}^{(m)}, \mathbf{z}_1^{(m)}) \\ \mathbf{F}_{2/m}^{(c)}(\mathbf{z}^{(m)}, \mathbf{z}_1^{(m)}) - 2\mathbf{k}_1 \frac{\Omega}{m} \dot{\mathbf{c}}_2^{(m)} + \mathbf{k}_2 \left(\frac{\Omega}{m}\right)^2 \mathbf{b}_2^{(m)} \\ \mathbf{F}_{2/m}^{(s)}(\mathbf{z}^{(m)}, \mathbf{z}_1^{(m)}) + 2\mathbf{k}_1 \frac{\Omega}{m} \dot{\mathbf{b}}_2^{(m)} + \mathbf{k}_2 \left(\frac{\Omega}{m}\right)^2 \mathbf{c}_2^{(m)} \end{pmatrix} \tag{5.22}$$

and

$$\begin{aligned}&\mathbf{k}_1 = diag(1, 2, \ldots, N), \\ &\mathbf{k}_2 = diag(1, 2^2, \ldots, N^2), \\ &\mathbf{F}_{1/m}^{(c)} = (F_{11/m}^{(c)}, F_{12/m}^{(c)}, \ldots, F_{1N/m}^{(s)})^{\mathrm{T}}, \\ &\mathbf{F}_{1/m}^{(s)} = (F_{11/m}^{(s)}, F_{12/m}^{(s)}, \ldots, F_{1N/m}^{(s)})^{\mathrm{T}}, \\ &\mathbf{F}_{2/m}^{(c)} = (F_{21/m}^{(c)}, F_{22/m}^{(c)}, \ldots, F_{2N/m}^{(c)})^{\mathrm{T}}, \\ &\mathbf{F}_{2/m}^{(s)} = (F_{21/m}^{(s)}, F_{22/m}^{(s)}, \ldots, F_{2N/m}^{(s)})^{\mathrm{T}} \\ &\text{for } N = 1, 2, \ldots, \infty.\end{aligned} \tag{5.23}$$

Setting

$$\mathbf{y}^{(m)} \equiv (\mathbf{z}^{(m)}, \mathbf{z}_1^{(m)}) \text{ and } \mathbf{f}^{(m)} = (\mathbf{z}_1^{(m)}, \mathbf{g}^{(m)})^{\mathrm{T}}, \tag{5.24}$$

Equation (5.21) becomes

$$\dot{\mathbf{y}}^{(m)} = \mathbf{f}^{(m)}(\mathbf{y}^{(m)}). \tag{5.25}$$

The steady-state solutions for periodic motion in Equation (5.1) can be obtained by setting $\dot{\mathbf{y}}^{(m)} = \mathbf{0}$, that is,

$$\begin{aligned}&F_{10}^{(m)}(\mathbf{z}^{(m)}, \mathbf{z}_1^{(m)}) = 0, \\ &\mathbf{F}_{1/m}^{(c)}(\mathbf{z}^{(m)}, \mathbf{z}_1^{(m)}) + \mathbf{k}_2 \left(\frac{\Omega}{m}\right)^2 \mathbf{b}_1^{(m)} = 0,\end{aligned}$$

$$\mathbf{F}^{(s)}_{1/m}(\mathbf{z}^{(m)}, \mathbf{z}_1^{(m)}) + \mathbf{k}_2\left(\frac{\Omega}{m}\right)^2 \mathbf{c}_1^{(m)} = 0;$$
$$F^{(m)}_{20}\left(\mathbf{z}^{(m)}, \mathbf{z}_1^{(m)}\right) = 0,$$
$$\mathbf{F}^{(c)}_{2/m}(\mathbf{z}^{(m)}, \mathbf{z}_1^{(m)}) + \mathbf{k}_2\left(\frac{\Omega}{m}\right)^2 \mathbf{b}_2^{(m)} = 0,$$
$$\mathbf{F}^{(s)}_{2/m}(\mathbf{z}^{(m)}, \mathbf{z}_1^{(m)}) + \mathbf{k}_2\left(\frac{\Omega}{m}\right)^2 \mathbf{c}_2^{(m)} = 0. \tag{5.26}$$

The $(4N+2)$ nonlinear equations in Equation (5.26) are solved by the Newton–Raphson method. As in Luo (2012a), the linearized equation at equilibrium $\mathbf{y}^{(m)*} = (\mathbf{z}^{(m)*}, \mathbf{0})^{\mathrm{T}}$ is given by

$$\Delta\dot{\mathbf{y}}^{(m)} = D\mathbf{f}(\mathbf{y}^{(m)^*})\Delta\mathbf{y}^{(m)} \tag{5.27}$$

where

$$D\mathbf{f}(\mathbf{y}^{(m)^*}) = \partial\mathbf{f}(\mathbf{y}^{(m)})/\partial\mathbf{y}^{(m)}|_{\mathbf{y}^{(m)*}}. \tag{5.28}$$

The corresponding eigenvalues are determined by

$$|D\mathbf{f}(\mathbf{y}^{(m)^*}) - \lambda\mathbf{I}_{4(2N+1)\times 4(2N+1)}| = 0 \tag{5.29}$$

where

$$D\mathbf{f}(\mathbf{y}^{(m)^*}) = \begin{bmatrix} \mathbf{0}_{2(2N+1)\times 2(2N+1)} & \mathbf{I}_{2(2N+1)\times 2(2N+1)} \\ \mathbf{G}_{2(2N+1)\times 2(2N+1)} & \mathbf{H}_{2(2N+1)\times 2(2N+1)} \end{bmatrix} \tag{5.30}$$

and

$$\mathbf{G} = \frac{\partial \mathbf{g}^{(m)}}{\partial \mathbf{z}^{(m)}} = (\mathbf{G}^{(10)}, \mathbf{G}^{(1c)}, \mathbf{G}^{(1s)}, \mathbf{G}^{(20)}, \mathbf{G}^{(2c)}, \mathbf{G}^{(2s)})^{\mathrm{T}} \tag{5.31}$$

$$\mathbf{G}^{(i0)} = (G_0^{(i0)}, G_1^{(i0)}, \ldots, G_{4N+1}^{(i0)}),$$
$$\mathbf{G}^{(ic)} = (\mathbf{G}_1^{(ic)}, \mathbf{G}_2^{(ic)}, \ldots, \mathbf{G}_N^{(ic)})^{\mathrm{T}},$$
$$\mathbf{G}^{(is)} = (\mathbf{G}_1^{(1s)}, \mathbf{G}_2^{(is)}, \ldots, \mathbf{G}_N^{(is)})^{\mathrm{T}} \tag{5.32}$$

for $i = 1, 2$; and $N = 1, 2, \ldots \infty$ with

$$\mathbf{G}_k^{(ic)} = (G_{k0}^{(ic)}, G_{k1}^{(ic)}, \ldots, G_{k(4N+1)}^{(ic)}),$$
$$\mathbf{G}_k^{(is)} = (G_{k0}^{(is)}, G_{k1}^{(is)}, \ldots, G_{k(4N+1)}^{(is)}) \tag{5.33}$$

for $k = 1, 2, \ldots, N$. The corresponding components are

$$G_r^{(10)} = -(1+\gamma)(\alpha\delta_0^r + \beta g_r^{(10)}),$$
$$G_{kr}^{(1c)} = \left(\frac{k\Omega}{m}\right)^2 \delta_k^r - \delta\frac{k\Omega}{m}\delta_{k+N}^r - (1+\gamma)(\alpha\delta_k^r + \beta g_{kr}^{(1c)}),$$
$$G_{kr}^{(1s)} = \left(\frac{k\Omega}{m}\right)^2 \delta_{k+N}^r + \delta\frac{k\Omega}{m}\delta_k^r - (1+\gamma)(\alpha\delta_{k+N}^r + \beta g_{kr}^{(1s)}),$$
$$G_r^{(10)} = -(1+\gamma)(\alpha\delta_0^r + \beta g_r^{(10)}),$$

$$
\begin{aligned}
G_{kr}^{(1c)} &= \left(\frac{k\Omega}{m}\right)^2 \delta_k^r - \delta\frac{k\Omega}{m}\delta_{k+N}^r - (1+\gamma)(\alpha\delta_k^r + \beta g_{kr}^{(1c)}),\\
G_{kr}^{(1s)} &= \left(\frac{k\Omega}{m}\right)^2 \delta_{k+N}^r + \delta\frac{k\Omega}{m}\delta_k^r - (1+\gamma)(\alpha\delta_{k+N}^r + \beta g_{kr}^{(1s)}),\\
G_r^{(20)} &= -\alpha\delta_{2N+1}^r - \beta g_r^{(20)},\\
G_{kr}^{(2c)} &= \left(\frac{k\Omega}{m}\right)^2 \delta_{k+2N+1}^r - \delta\frac{k\Omega}{m}\delta_{k+3N+1}^r - \alpha\delta_{k+2N+1}^r - \beta g_{kr}^{(2c)},\\
G_{kr}^{(2s)} &= \left(\frac{k\Omega}{m}\right)^2 \delta_{k+3N+1}^r + \delta\frac{k\Omega}{m}\delta_{k+2N+1}^r - \alpha\delta_{k+3N+1}^r - \beta g_{kr}^{(2s)}
\end{aligned} \tag{5.34}
$$

for $r = 0, 1, 2, \ldots 4N+1$; where

$$
\begin{aligned}
g_r^{(10)} &= 3(a_{10}^{(m)})^2\delta_0^r + (a_{20}^{(m)})^2\delta_0^r + 2a_{10}^{(m)}a_{20}^{(m)}\delta_{2N+1}^r\\
&\quad + \sum_{q=1}^{11}\sum_{l=1}^{N}\sum_{j=1}^{N}\sum_{i=1}^{N} g_r^{(10)}(i,j,l,q)
\end{aligned} \tag{5.35}
$$

with

$$
\begin{aligned}
g_r^{(10)}(i,j,l,1) &= \frac{3}{2N}(b_{1i/m}b_{1j/m}\delta_0^r + 2a_{10}^{(m)}b_{1i/m}\delta_j^r)\delta_{i-j}^0,\\
g_r^{(10)}(i,j,l,2) &= \frac{3}{2N}(c_{1i/m}c_{1j/m}\delta_0^r + 2a_{10}^{(m)}c_{1i/m}\delta_{j+N}^r)\delta_{i-j}^0,\\
g_r^{(10)}(i,j,l,3) &= \frac{3}{4}b_{1i/m}b_{1j/m}\delta_l^r(\delta_{i-j-l}^0 + \delta_{i-j+l}^0 + \delta_{i+j-l}^0),\\
g_r^{(10)}(i,j,l,4) &= \frac{3}{4}(c_{1j/m}c_{1l/m}\delta_i^r + 2b_{1i/m}c_{1j/m}\delta_{l+N}^r)(\delta_{i+j-l}^0 + \delta_{i-j+l}^0 - \delta_{i-j-l}^0),\\
g_r^{(10)}(i,j,l,5) &= \frac{1}{2N}(b_{2i/m}b_{2j/m}\delta_0^r + 2a_{10}^{(m)}b_{2i/m}\delta_{j+2N+1}^r)\delta_{i-j}^0,\\
g_r^{(10)}(i,j,l,6) &= \frac{1}{2N}(c_{2i/m}c_{2j/m}\delta_0^r + 2a_{10}^{(m)}c_{2i/m}\delta_{j+3N+1}^r)\delta_{i-j}^0,\\
g_r^{(10)}(i,j,l,7) &= \frac{1}{N}(b_{1i/m}b_{2j/m}\delta_{2N+1}^r + a_{20}^{(m)}b_{2j/m}\delta_i^r + a_{20}^{(m)}b_{1i/m}\delta_{j+2N+1}^r)\delta_{i-j}^0,\\
g_r^{(10)}(i,j,l,8) &= \frac{1}{N}(c_{1i/m}c_{2j/m}\delta_{2N+1}^r + a_{20}^{(m)}c_{2j/m}\delta_{i+N}^r + a_{20}^{(m)}c_{1i/m}\delta_{j+3N+1}^r)\delta_{i-j}^0,\\
g_r^{(10)}(i,j,l,9) &= \frac{1}{4}(b_{2j/m}b_{2l/m}\delta_i^r + 2b_{1i/m}b_{2j/m}\delta_{l+2N+1}^r)(\delta_{i-j-l}^0 + \delta_{i-j+l}^0 + \delta_{i+j-l}^0),\\
g_r^{(10)}(i,j,l,10) &= \frac{1}{4}(c_{2j/m}c_{2l/m}\delta_i^r + 2b_{1i/m}c_{2j/m}\delta_{l+3N+1}^r)(\delta_{i+j-l}^0 + \delta_{i-j+l}^0 - \delta_{i-j-l}^0),\\
g_r^{(10)}(i,j,l,11) &= \frac{1}{2}(c_{2j/m}b_{2l/m}\delta_{i+N}^r + c_{1i/m}b_{2l/m}\delta_{j+3N+1}^r + c_{1i/m}c_{2j/m}\delta_{l+2N+1}^r)\\
&\quad \times(\delta_{i-j-l}^0 + \delta_{i-j+l}^0 - \delta_{i+j-l}^0)];
\end{aligned} \tag{5.36}
$$

and

$$
\begin{aligned}
g_{kr}^{(1c)} &= 3(a_{10}^{(m)})^2\delta_k^r + 6a_{10}^{(m)}b_{1k/m}\delta_0^r + 2a_{20}^{(m)}b_{2k/m}\delta_0^r + 2a_{10}^{(m)}b_{2k/m}\delta_{2N+1}^r \\
&\quad + 2a_{10}^{(m)}a_{20}^{(m)}\delta_{k+2N+1}^r + 2a_{20}^{(m)}b_{1k/m}\delta_{2N+1}^r + (a_{20}^{(m)})^2\delta_k^r \\
&\quad + \sum_{q=1}^{11}\sum_{l=1}^{N}\sum_{j=1}^{N}\sum_{i=1}^{N} g_{kr}^{(1c)}(i,j,l,q)
\end{aligned} \tag{5.37}
$$

with

$$
\begin{aligned}
g_{kr}^{(1c)}(i,j,l,1) &= \frac{3}{2N}(b_{1i/m}b_{1j/m}\delta_0^r + 2a_{01}^{(m)}b_{1i/m}\delta_j^r)(\delta_{|i-j|}^k + \delta_{i+j}^k), \\
g_{kr}^{(1c)}(i,j,l,2) &= \frac{3}{2N}(c_{1i/m}c_{1j/m}\delta_0^r + 2a_{10}^{(m)}c_{1i/m}\delta_{j+N}^r)(\delta_{|i-j|}^k - \delta_{i+j}^k), \\
g_{kr}^{(1c)}(i,j,l,3) &= \frac{3}{4}b_{1i/m}b_{1j/m}\delta_l^r(\delta_{|i-j-l|}^k + \delta_{|i-j+l|}^k + \delta_{|i+j-l|}^k + \delta_{|i+j+l|}^k), \\
g_{kr}^{(1c)}(i,j,l,4) &= \frac{3}{4}(c_{1j/m}c_{1l/m}\delta_i^r + 2b_{1i/m}c_{1j/m}\delta_{l+N}^r) \\
&\quad \times (\delta_{|i-j+l|}^k + \delta_{|i+j-l|}^k - \delta_{|i-j-l|}^k - \delta_{i+j+l}^k), \\
g_{kr}^{(1c)}(i,j,l,5) &= \frac{1}{2N}(b_{2i/m}b_{2j/m}\delta_0^r + 2a_{10}^{(m)}b_{2i/m}\delta_{j+2N+1}^r)(\delta_{|i-j|}^k + \delta_{i+j}^k), \\
g_{kr}^{(1c)}(i,j,l,6) &= \frac{1}{2N}(c_{2i/m}c_{2j/m}\delta_0^r + a_{10}^{(m)}c_{2i/m}\delta_{j+3N+1}^r)(\delta_{|i-j|}^k - \delta_{i+j}^k), \\
g_{kr}^{(1c)}(i,j,l,7) &= \frac{1}{N}(b_{1i/m}b_{2j/m}\delta_{2N+1}^r + a_{20}^{(m)}b_{2j/m}\delta_i^r \\
&\quad + a_{20}^{(m)}b_{1i/m}\delta_{j+2N+1}^r)(\delta_{|i-j|}^k + \delta_{i+j}^k), \\
g_{kr}^{(1c)}(i,j,l,8) &= \frac{1}{N}(c_{1i/m}c_{2j/m}\delta_{2N+1}^r + a_{20}^{(m)}c_{2j/m}\delta_{i+N}^r \\
&\quad + a_{20}^{(m)}c_{1i/m}\delta_{j+3N+1}^r)(\delta_{|i-j|}^k - \delta_{i+j}^k), \\
g_{kr}^{(1c)}(i,j,l,9) &= \frac{1}{4}(b_{2j/m}b_{2l/m}\delta_i^r + 2b_{1i/m}b_{2j/m}\delta_{l+2N+1}^r) \\
&\quad \times (\delta_{|i-j-l|}^k + \delta_{|i-j+l|}^k + \delta_{|i+j-l|}^k + \delta_{|i+j+l|}^k), \\
g_{kr}^{(1c)}(i,j,l,10) &= \frac{1}{4}(c_{2j/m}c_{2l/m}\delta_i^r + 2b_{1i/m}c_{2j/m}\delta_{l+3N+1}^r) \\
&\quad \times (\delta_{|i-j+l|}^k + \delta_{|i+j-l|}^k - \delta_{|i-j-l|}^k - \delta_{i+j+l}^k), \\
g_{kr}^{(1c)}(i,j,l,11) &= \frac{1}{2}(c_{2j/m}c_{1l/m}\delta_{i+2N+1}^r + b_{2i/m}c_{1l/m}\delta_{j+3N+1}^r \\
&\quad + b_{2i/m}c_{2j/m}\delta_{l+N}^r)(\delta_{|i-j+l|}^k + \delta_{|i+j-l|}^k - \delta_{|i-j-l|}^k - \delta_{i+j+l}^k);
\end{aligned} \tag{5.38}
$$

and

$$\begin{aligned}g_{kr}^{(1s)} &= 3(a_{10}^{(m)})^2\delta_{k+N}^r + 6a_{10}^{(m)}c_{1k/m}\delta_0^r + 2a_{20}^{(m)}c_{2k/m}\delta_0^r + 2a_{10}^{(m)}c_{2k/m}\delta_{2N+1}^r\\ &\quad + 2a_{10}^{(m)}a_{20}^{(m)}\delta_{k+3N+1}^r + 2a_{20}^{(m)}c_{1k/m}\delta_{2N+1}^r + (a_{20}^{(m)})^2\delta_{k+N}^r\\ &\quad + \sum_{q=1}^{9}\sum_{l=1}^{N}\sum_{j=1}^{N}\sum_{i=1}^{N} g_{kr}^{(1s)}(i,j,l,q)\end{aligned} \tag{5.39}$$

with

$$\begin{aligned}g_{kr}^{(1s)}(i,j,l,1) &= \frac{3}{N}(a_{10}^{(m)}c_{1j/m}\delta_i^r + a_{10}^{(m)}b_{1i/m}\delta_{j+N}^r + b_{1i/m}c_{1j/m}\delta_0^r)\\ &\quad \times[\delta_{i+j}^k - \mathrm{sgn}(i-j)\delta_{|i-j|}^k],\\ g_{kr}^{(1s)}(i,j,l,2) &= \frac{3}{4}c_{1i/m}c_{1j/m}\delta_{l+N}^r[\mathrm{sgn}(i-j+l)\delta_{|i-j+l|}^k + \mathrm{sgn}(i+j-l)\delta_{|i+j-l|}^k\\ &\quad - \mathrm{sgn}(i-j-l)\delta_{|i-j-l|}^k - \delta_{i+j+l}^k],\\ g_{kr}^{(1s)}(i,j,l,3) &= \frac{3}{4}(b_{1i/m}b_{1j/m}\delta_{l+N}^r + 2b_{1i/m}c_{1l/m}\delta_j^r)[\mathrm{sgn}(i-j+l)\delta_{|i-j+l|}^k\\ &\quad + \delta_{i+j+l}^k - \mathrm{sgn}(i+j-l)\delta_{|i+j-l|}^k - \mathrm{sgn}(i-j-l)\delta_{|i-j-l|}^k],\\ g_{kr}^{(1s)}(i,j,l,4) &= \frac{1}{N}(a_{20}^{(m)}c_{2j/m}\delta_i^r + a_{20}^{(m)}b_{1i/m}\delta_{j+3N+1}^r + b_{1i/m}c_{2j/m}\delta_{2N+1}^r)\\ &\quad \times[\delta_{i+j}^k - \mathrm{sgn}(i-j)\delta_{|i-j|}^k],\\ g_{kr}^{(1s)}(i,j,l,5) &= \frac{1}{N}(a_{20}^{(m)}b_{2j/m}\delta_{i+N}^r + a_{20}^{(m)}c_{1i/m}\delta_{j+2N+1}^r + c_{1i/m}b_{2j/m}\delta_{2N+1}^r)\\ &\quad \times[\delta_{i+j}^k + \mathrm{sgn}(i-j)\delta_{|i-j|}^k],\\ g_{kr}^{(1s)}(i,j,l,6) &= \frac{1}{N}(a_{10}^{(m)}c_{2j/m}\delta_{i+2N+1}^r + a_{10}^{(m)}b_{2i/m}\delta_{j+3N+1}^r + b_{2i/m}c_{yj/m}\delta_0^r)\\ &\quad \times[\delta_{i+j}^k - \mathrm{sgn}(i-j)\delta_{|i-j|}^k],\\ g_{kr}^{(1s)}(i,j,l,7) &= \frac{1}{4}(c_{2j/m}c_{2l/m}\delta_{i+N}^r + 2c_{1i/m}c_{2j/m}\delta_{l+3N+1}^r)[\mathrm{sgn}(i-j+l)\delta_{|i-j+l|}^k\\ &\quad + \mathrm{sgn}(i+j-l)\delta_{|i+j-l|}^k - \mathrm{sgn}(i-j-l)\delta_{|i-j-l|}^k - \delta_{i+j+l}^k],\\ g_{kr}^{(1s)}(i,j,l,8) &= \frac{1}{2}(b_{2j/m}c_{2l/m}\delta_i^r + b_{1i/m}c_{2l/m}\delta_{j+2N+1}^r + b_{1i/m}b_{2j/m}\delta_{l+3N+1}^r)\\ &\quad \times[\mathrm{sgn}(i-j+l)\delta_{|i-j+l|}^k + \delta_{i+j+l}^k - \mathrm{sgn}(i+j-l)\delta_{|i+j-l|}^k\\ &\quad - \mathrm{sgn}(i-j-l)\delta_{|i-j-l|}^k],\\ g_{kr}^{(1s)}(i,j,l,9) &= \frac{1}{4}(2b_{2j/m}c_{1l/m}\delta_{i+2N+1}^r + b_{2i/m}b_{2j/m}\delta_{l+N}^r)[\mathrm{sgn}(i-j+l)\delta_{|i-j+l|}^k\\ &\quad + \delta_{i+j+l}^k - \mathrm{sgn}(i+j-l)\delta_{|i+j-l|}^k - \mathrm{sgn}(i-j-l)\delta_{|i-j-l|}^k];\end{aligned} \tag{5.40}$$

and

$$g_r^{(20)} = 3(a_{20}^{(m)})^2\delta_{2N+1}^r + (a_{10}^{(m)})^2\delta_{2N+1}^r + 2a_{10}^{(m)}a_{20}^{(m)}\delta_0^r + \sum_{q=1}^{11}\sum_{l=1}^{N}\sum_{j=1}^{N}\sum_{i=1}^{N} g_r^{(20)}(i,j,l,q) \tag{5.41}$$

with

$$\begin{aligned}
g_r^{(20)}(i,j,l,1) &= \frac{3}{2N}(b_{2i/m}b_{2j/m}\delta_{2N+1}^r + 2a_{20}^{(m)}b_{2i/m}\delta_{j+2N+1}^r)\delta_{i-j}^0,\\
g_r^{(20)}(i,j,l,2) &= \frac{3}{2N}(c_{2i/m}c_{2j/m}\delta_{2N+1}^r + 2a_{20}^{(m)}c_{2i/m}\delta_{j+3N+1}^r)\delta_{i-j}^0,\\
g_r^{(20)}(i,j,l,3) &= \frac{3}{4}b_{2i/m}b_{2j/m}\delta_{l+2N+1}^r(\delta_{i-j-l}^0 + \delta_{i-j+l}^0 + \delta_{i+j-l}^0),\\
g_r^{(20)}(i,j,l,4) &= \frac{3}{4}(c_{2j/m}c_{2l/m}\delta_{i+2N+1}^r + 2b_{2i/m}c_{2j/m}\delta_{l+3N+1}^r)\\
&\quad \times(\delta_{i+j-l}^0 + \delta_{i-j+l}^0 - \delta_{i-j-l}^0),\\
g_r^{(20)}(i,j,l,5) &= \frac{1}{2N}(b_{1i/m}b_{1j/m}\delta_{2N+1}^r + 2a_{20}^{(m)}b_{1i/m}\delta_j^r)\delta_{i-j}^0,\\
g_r^{(20)}(i,j,l,6) &= \frac{1}{2N}(c_{1i/m}c_{1j/m}\delta_{2N+1}^r + 2a_{20}^{(m)}c_{1i/m}\delta_{j+N}^r)\delta_{i-j}^0,\\
g_r^{(20)}(i,j,l,7) &= \frac{1}{N}(b_{1i/m}b_{2j/m}\delta_r^0 + a_{10}^{(m)}b_{1i/m}\delta_{j+2N+1}^r + a_{01}^{(m)}b_{yj/m}\delta_i^r)\delta_{i-j}^0,\\
g_r^{(20)}(i,j,l,8) &= \frac{1}{N}(c_{1i/m}c_{2j/m}\delta_r^0 + a_{10}^{(m)}c_{1i/m}\delta_{j+3N+1}^r + a_{10}^{(m)}c_{2j/m}\delta_{i+N}^r)\delta_{i-j}^0,\\
g_r^{(20)}(i,j,l,9) &= \frac{1}{4}(2b_{1j/m}b_{2l/m}\delta_i^r + b_{1i/m}b_{1j/m}\delta_{l+2N+1}^r)\\
&\quad \times(\delta_{i-j-l}^0 + \delta_{i-j+l}^0 + \delta_{i+j-l}^0),\\
g_r^{(20)}(i,j,l,10) &= \frac{1}{4}(c_{1j/m}c_{1l/m}\delta_{i+2N+1}^r + 2b_{2i/m}c_{1j/m}\delta_{l+N}^r)\\
&\quad \times(\delta_{i+j-l}^0 + \delta_{i-j+l}^0 - \delta_{i-j-l}^0),\\
g_r^{(20)}(i,j,l,11) &= \frac{1}{2}(c_{1j/m}b_{1l/m}\delta_{i+3N+1}^r + c_{2i/m}b_{1l/m}\delta_{j+N}^r + c_{2i/m}c_{1j/m}\delta_l^r)\\
&\quad \times(\delta_{i-j-l}^0 + \delta_{i-j+l}^0 - \delta_{i+j-l}^0);
\end{aligned} \tag{5.42}$$

and

$$\begin{aligned}
g_{kr}^{(2c)} &= 3(a_{20}^{(m)})^2\delta_{k+2N+1}^r + 6a_{20}^{(m)}b_{2k/m}\delta_{2N+1}^r + 2a_{20}^{(m)}b_{1k/m}\delta_0^r\\
&\quad + 2a_{01}^{(m)}b_{1k/m}\delta_{2N+1}^r + 2a_{10}^{(m)}a_{20}^{(m)}\delta_k^r + 2a_{10}^{(m)}b_{2k/m}\delta_0^r + (a_{10}^{(m)})^2\delta_{k+2N+1}^r\\
&\quad + \sum_{q=1}^{11}\sum_{l=1}^{N}\sum_{j=1}^{N}\sum_{i=1}^{N} g_{kr}^{(2c)}(i,j,l,q)
\end{aligned} \tag{5.43}$$

with

$$
\begin{aligned}
g_{kr}^{(2c)}(i,j,l,1) &= \frac{3}{2N}(b_{2i/m}b_{2j/m}\delta_{2N+1}^{r} + 2a_{20}^{(m)}b_{2i/m}\delta_{j+2N+1}^{r})(\delta_{|i-j|}^{k} + \delta_{i+j}^{k}),\\
g_{kr}^{(2c)}(i,j,l,2) &= \frac{3}{2N}(c_{2i/m}c_{2j/m}\delta_{2N+1}^{r} + 2a_{20}^{(m)}c_{2i/m}\delta_{j+3N+1}^{r})(\delta_{|i-j|}^{k} - \delta_{i+j}^{k}),\\
g_{kr}^{(2c)}(i,j,l,3) &= \frac{3}{4}b_{2i/m}b_{2j/m}\delta_{l+2N+1}^{r}(\delta_{|i-j-l|}^{k} + \delta_{|i-j+l|}^{k} + \delta_{|i+j-l|}^{k} + \delta_{|i+j+l|}^{k}),\\
g_{kr}^{(2c)}(i,j,l,4) &= \frac{3}{4}(c_{2j/m}c_{2l/m}\delta_{i+2N+1}^{r} + 2b_{2i/m}c_{2j/m}\delta_{l+3N+1}^{r})\\
&\quad\times(\delta_{|i-j+l|}^{k} + \delta_{|i+j-l|}^{k} - \delta_{|i-j-l|}^{k} - \delta_{i+j+l}^{k}),\\
g_{kr}^{(2c)}(i,j,l,5) &= \frac{1}{2N}(b_{1i/m}b_{1j/m}\delta_{2N+1}^{r} + 2a_{20}^{(m)}b_{1i/m}\delta_{j}^{r})(\delta_{|i-j|}^{k} + \delta_{i+j}^{k}),\\
g_{kr}^{(2c)}(i,j,l,6) &= \frac{1}{2N}(c_{1i/m}c_{1j/m}\delta_{2N+1}^{r} + 2a_{20}^{(m)}c_{1i/m}\delta_{j+N}^{r})(\delta_{|i-j|}^{k} - \delta_{i+j}^{k}),\\
g_{kr}^{(2c)}(i,j,l,7) &= \frac{1}{N}(b_{1i/m}b_{2j/m}\delta_{0}^{r} + a_{10}^{(m)}b_{2j/m}\delta_{i}^{r} + a_{10}^{(m)}b_{1i/m}\delta_{j+2N+1}^{r})\\
&\quad\times(\delta_{|i-j|}^{k} + \delta_{i+j}^{k}),\\
g_{kr}^{(2c)}(i,j,l,8) &= \frac{1}{N}(c_{1i/m}c_{2j/m}\delta_{0}^{r} + a_{10}^{(m)}c_{2j/m}\delta_{i+N}^{r} + a_{01}^{(m)}c_{1i/m}\delta_{j+3N+1}^{r})\\
&\quad\times(\delta_{|i-j|}^{k} - \delta_{i+j}^{k}),\\
g_{kr}^{(2c)}(i,j,l,9) &= \frac{1}{4}(2b_{1j/m}b_{2l/m}\delta_{i}^{r} + b_{1i/m}b_{1j/m}\delta_{l+2N+1}^{r})\\
&\quad\times(\delta_{|i-j-l|}^{k} + \delta_{|i-j+l|}^{k} + \delta_{|i+j-l|}^{k} + \delta_{|i+j+l|}^{k}),\\
g_{kr}^{(2c)}(i,j,l,10) &= \frac{1}{4}(c_{1j/m}c_{1l/m}\delta_{i+2N+1}^{r} + 2b_{2i/m}c_{1j/m}\delta_{l+N}^{r})\\
&\quad\times(\delta_{|i-j+l|}^{k} + \delta_{|i+j-l|}^{k} - \delta_{|i-j-l|}^{k} - \delta_{i+j+l}^{k}),\\
g_{kr}^{(2c)}(i,j,l,11) &= \frac{1}{2}(c_{1j/m}c_{2l/m}\delta_{i}^{r} + b_{1i/m}c_{2l/m}\delta_{j+N}^{r} + b_{1i/m}c_{1j/m}\delta_{l+3N+1}^{r})\\
&\quad\times(\delta_{|i-j+l|}^{k} + \delta_{|i+j-l|}^{k} - \delta_{|i-j-l|}^{k} - \delta_{i+j+l}^{k});
\end{aligned}
\tag{5.44}
$$

and

$$
\begin{aligned}
g_{kr}^{(2s)} &= 3(a_{20}^{(m)})^{2}\delta_{k+3N+1}^{r} + 6a_{20}^{(m)}c_{yk/m}\delta_{2N+1}^{r} + 2a_{20}^{(m)}c_{1k/m}\delta_{r}^{0}\\
&\quad + 2a_{10}^{(m)}c_{1k/m}\delta_{2N+1}^{r} + 2a_{10}^{(m)}a_{20}^{(m)}\delta_{k+N}^{r} + 2a_{10}^{(m)}c_{2k/m}\delta_{0}^{r} + (a_{10}^{(m)})^{2}\delta_{k+3N+1}^{r}\\
&\quad + \sum_{q=1}^{9}\sum_{l=1}^{N}\sum_{j=1}^{N}\sum_{i=1}^{N}g_{kr}^{(2s)}(i,j,l,q)
\end{aligned}
\tag{5.45}
$$

with

$$
\begin{aligned}
g_{kr}^{(2s)}(i,j,l,1) &= \frac{3}{N}(a_{20}^{(m)}c_{2j/m}\delta_{i+2N+1}^{r} + a_{20}^{(m)}b_{2i/m}\delta_{j+3N+1}^{r} + b_{2i/m}c_{2j/m}\delta_{2N+1}^{r}) \\
&\quad \times [\delta_{i+j}^{k} - \operatorname{sgn}(i-j)\delta_{|i-j|}^{k}], \\
g_{kr}^{(2s)}(i,j,l,2) &= \frac{3}{4}c_{2i/m}c_{2j/m}\delta_{l+3N+1}^{r}[\operatorname{sgn}(i-j+l)\delta_{|i-j+l|}^{k} \\
&\quad + \operatorname{sgn}(i+j-l)\delta_{|i+j-l|}^{k} - \operatorname{sgn}(i-j-l)\delta_{|i-j-l|}^{k} - \delta_{i+j+l}^{k}], \\
g_{kr}^{(2s)}(i,j,l,3) &= \frac{3}{4}(b_{2i/m}b_{2j/m}\delta_{l+3N+1}^{r} + 2b_{2i/m}c_{2l/m}\delta_{j+2N+1}^{r}) \\
&\quad \times [\operatorname{sgn}(i-j+l)\delta_{|i-j+l|}^{k} + \delta_{i+j+l}^{k} \\
&\quad - \operatorname{sgn}(i+j-l)\delta_{|i+j-l|}^{k} - \operatorname{sgn}(i-j-l)\delta_{|i-j-l|}^{k}], \\
g_{kr}^{(2s)}(i,j,l,4) &= \frac{1}{N}(a_{10}^{(m)}c_{2j/m}\delta_{i}^{r} + a_{10}^{(m)}b_{1i/m}\delta_{j+3N+1}^{r} + b_{1i/m}c_{2j/m}\delta_{0}^{r}) \\
&\quad \times [\delta_{i+j}^{k} - \operatorname{sgn}(i-j)\delta_{|i-j|}^{k}], \\
g_{kr}^{(2s)}(i,j,l,5) &= \frac{1}{N}(a_{10}^{(m)}b_{2j/m}\delta_{i+N}^{r} + a_{10}^{(m)}c_{1i/m}\delta_{j+2N+1}^{r} + c_{1i/m}b_{2j/m}\delta_{0}^{r}) \\
&\quad \times [\delta_{i+j}^{k} + \operatorname{sgn}(i-j)\delta_{|i-j|}^{k}], \\
g_{kr}^{(2s)}(i,j,l,6) &= \frac{1}{N}(a_{20}^{(m)}c_{1j/m}\delta_{i}^{r} + a_{20}^{(m)}b_{1i/m}\delta_{j+N}^{r} + b_{1i/m}c_{1j/m}\delta_{2N+1}^{r}) \\
&\quad \times [\delta_{i+j}^{k} - \operatorname{sgn}(i-j)\delta_{|i-j|}^{k}], \\
g_{kr}^{(2s)}(i,j,l,7) &= \frac{1}{4}(2c_{1j/m}c_{2l/m}\delta_{i+N}^{r} + c_{1i/m}c_{1j/m}\delta_{l+3N+1}^{r})[\operatorname{sgn}(i-j+l)\delta_{|i-j+l|}^{k} \\
&\quad + \operatorname{sgn}(i+j-l)\delta_{|i+j-l|}^{k} - \operatorname{sgn}(i-j-l)\delta_{|i-j-l|}^{k} - \delta_{i+j+l}^{k}], \\
g_{kr}^{(2s)}(i,j,l,8) &= \frac{1}{2}(b_{2j/m}c_{1l/m}\delta_{i}^{r} + b_{1i/m}c_{1l/m}\delta_{j+2N+1}^{r} + b_{1i/m}b_{2j/m}\delta_{l+N}^{r}) \\
&\quad \times [\operatorname{sgn}(i-j+l)\delta_{|i-j+l|}^{k} + \delta_{i+j+l}^{k} - \operatorname{sgn}(i+j-l)\delta_{|i+j-l|}^{k} \\
&\quad - \operatorname{sgn}(i-j-l)\delta_{|i-j-l|}^{k}], \\
g_{kr}^{(2s)}(i,j,l,9) &= \frac{1}{4}(2b_{1j/m}c_{2l/m}\delta_{i}^{r} + b_{1i/m}b_{1j/m}\delta_{l+3N+1}^{r})[\operatorname{sgn}(i-j+l)\delta_{|i-j+l|}^{k} \\
&\quad + \delta_{i+j+l}^{k} - \operatorname{sgn}(i+j-l)\delta_{|i+j-l|}^{k} - \operatorname{sgn}(i-j-l)\delta_{|i-j-l|}^{k}];
\end{aligned}
\tag{5.46}
$$

and

$$
\mathbf{H} = \frac{\partial \mathbf{g}^{(m)}}{\partial \mathbf{z}_{1}^{(m)}} = (\mathbf{H}^{(10)}, \mathbf{H}^{(1c)}, \mathbf{H}^{(1s)}, \mathbf{H}^{(20)}, \mathbf{H}^{(2c)}, \mathbf{H}^{(2s)})^{\mathrm{T}} \tag{5.47}
$$

where

$$\mathbf{H}^{(i0)} = (H_0^{(i0)}, H_1^{(i0)}, \ldots, H_{4N+1}^{(i0)}),$$
$$\mathbf{H}^{(ic)} = (\mathbf{H}_1^{(ic)}, \mathbf{H}_2^{(ic)}, \ldots, \mathbf{H}_N^{(ic)})^{\mathrm{T}},$$
$$\mathbf{H}^{(is)} = (\mathbf{H}_1^{(is)}, \mathbf{H}_2^{(is)}, \ldots, \mathbf{H}_N^{(is)})^{\mathrm{T}} \tag{5.48}$$

for $i = 1, 2$ and $N = 1, 2, \ldots \infty$, with

$$\mathbf{H}_k^{(ic)} = (H_{k0}^{(ic)}, H_{k1}^{(ic)}, \ldots, H_{k(4N+1)}^{(ic)}),$$
$$\mathbf{H}_k^{(is)} = (H_{k0}^{(is)}, H_{k1}^{(is)}, \ldots, H_{k(4N+1)}^{(is)}) \tag{5.49}$$

for $k = 1, 2, \ldots N$. The corresponding components are

$$H_r^{(10)} = -\delta\delta_0^r,$$
$$H_{kr}^{(1c)} = -2\frac{k\Omega}{m}\delta_{k+N}^r - \delta\delta_k^r,$$
$$H_{kr}^{(1s)} = 2\frac{k\Omega}{m}\delta_k^r - \delta\delta_{k+N}^r,$$
$$H_r^{(20)} = -\delta\delta_{2N+1}^r,$$
$$H_{kr}^{(2c)} = -2\frac{k\Omega}{m}\delta_{k+3N+1}^r - \delta\delta_{k+2N+1}^r,$$
$$H_{kr}^{(2s)} = 2\frac{k\Omega}{m}\delta_{k+2N+1}^r - \delta\delta_{k+3N+1}^r \tag{5.50}$$

for $r = 0, 1, \ldots, 4N + 1$. From Luo (2012a), the eigenvalues of $D\mathbf{f}(\mathbf{y}^{(m)*})$ are classified as

$$(n_1, n_2, n_3 | n_4, n_5, n_6). \tag{5.51}$$

The corresponding boundary between the stable and unstable solutions is given by the saddle-node bifurcation and Hopf bifurcation.

5.2 Frequency-Amplitude Characteristics

As in Huang and Luo (2014), the equilibrium solution of Equations (5.7) and (5.8) can be obtained from Equation (5.26) by using the Newton–Raphson method, and the stability analysis will be discussed. The amplitude varying with rotation frequency Ω are illustrated. The harmonic amplitude and phase are defined by

$$A_{(i)k/m} \equiv A_{ik/m} = \sqrt{b_{ik/m}^2 + c_{ik/m}^2} \text{ and } \varphi_{(i)k/m} = \arctan\frac{c_{ik/m}}{b_{ik/m}} \tag{5.52}$$

where $i = 1, 2$. The corresponding solution in Equation (5.1) becomes

$$x^*(t) = a_{10}^{(m)} + \sum_{k=1}^{N} A_{(1)k/m} \cos\left(\frac{k}{m}\Omega t - \varphi_{(1)k/m}\right),$$

$$y^*(t) = a_{20}^{(m)} + \sum_{k=1}^{N} A_{(2)k/m} \cos\left(\frac{k}{m}\Omega t - \varphi_{(2)k/m}\right). \tag{5.53}$$

As in Huang and Luo (2014b), consider system parameters as

$$\delta = 0.02, \alpha = 0.68, \beta = 10, \gamma = 1.0, e = 1.5 \tag{5.54}$$

The acronyms "USN" and "SN" are used to represent the saddle-unstable node and saddle-node bifurcations, respectively. The acronyms "UHB" and "HB" are used to represent the unstable Hopf bifurcation (subcritical) and stable Hopf bifurcation (supercritical), respectively. Solid and dashed curves represent stable and unstable period-m motions, respectively.

5.2.1 Period-1 Motions

From the above parameters, the frequency-amplitude curves of period-1 motions in the x-direction and y-direction of the rotor are presented in Figures 5.1 and 5.2 that are based on 13 harmonic terms.

In Figure 5.1, the period-1 motion of the nonlinear rotor in the x-direction is presented. In Figure 5.1(i), the constant a_{10} versus rotation speed Ω is presented. For the symmetric period-1 motion, $a_{10} = 0$ is observed. For the asymmetric period-1 motion, the rotation speed lies in the approximate range of $\Omega \in (0.883, 2.017)$. From the symmetric to asymmetric period-1 motion, the two saddle-node bifurcations occurs at $\Omega \approx 1.592, 2.017$. The unstable Hopf bifurcations (UHBs) of the asymmetric period-1 motion are located at $\Omega \approx 1.687, 1.94865, 2.01$. The stable Hopf bifurcations of the asymmetric period-1 motion occur at $\Omega \approx 0.895$, $1.28, 1.58722, 1.709$. The other saddle-node bifurcations of the asymmetric period-1 motion are at $\Omega \approx 0.883, 1.714, 1.813$. The asymmetric period-1 motion possesses the unstable saddle-node bifurcation, (stable) Hopf bifurcation, and saddle-node bifurcation. The range of the constant is $a_{10} < 0.18$. The stable asymmetric period-1 motion has six segments and the unstable asymmetric period-1 motion possesses five segments. One of six segments is very short with $\Omega \approx (0.883, 0.895)$ and $a_{10} \sim 4 \times 10^{-3}$, which is zoomed. Only the positive constant of $a_{10} > 0$ is presented, from which the center of the period-1 motion is located on the positive x-axis. The constant of $a_{10} < 0$ for asymmetric period-1 motion will not be presented. Such a constant has the same magnitude and the center of periodic motion is located on the negative x-axis. In Figure 5.1(ii), the harmonic amplitude A_1 varying with rotation speed Ω is presented. In addition to the stable and unstable Hopf bifurcations and saddle-node bifurcations for the asymmetric period-1 motion, the Hopf bifurcation and saddle-node bifurcations for symmetric period-1 motion can be determined. The saddle-node bifurcations for symmetric period-1 motion are at $\Omega \approx 0.426, 0.472$. The Hopf bifurcations of the symmetric period-1 motion occur at $\Omega \approx 0.572, 0.611, 2.859$. From such Hopf bifurcations, the quasi-periodic motions are observed. The harmonic amplitudes for the symmetric and asymmetric period-1 motions are presented with $A_{1(1)} < 1.0$. From the traditional analysis, one may obtain the approximate frequency-amplitude curves for

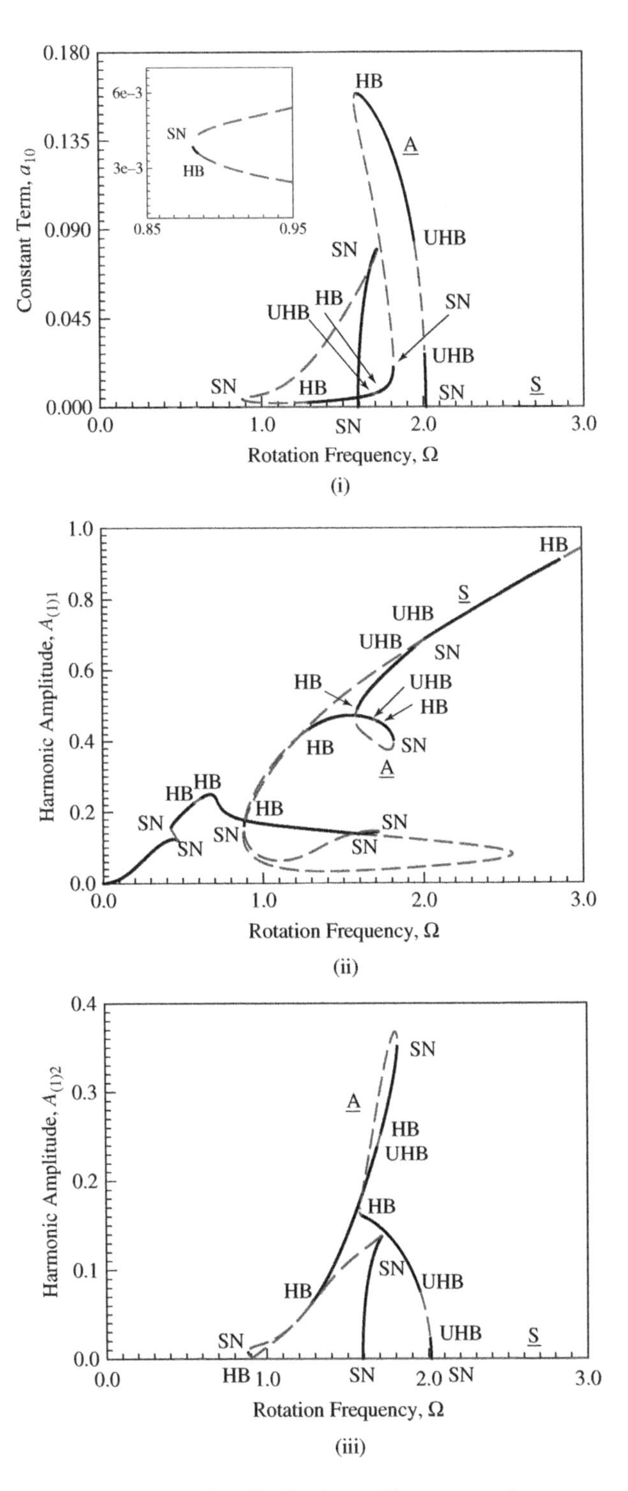

Figure 5.1 Period-1 motion of the x-direction in the nonlinear rotor: frequency-amplitude curves of harmonic terms based on 13 harmonic terms: (i) a_{10}, (ii)–(vi) $A_{(1)k}$ ($k = 1, 2, 3, 12, 13$), ($\delta = 0.02$, $\alpha = 0.68$, $\beta = 10$, $\gamma = 1.0$, $e = 1.5$)

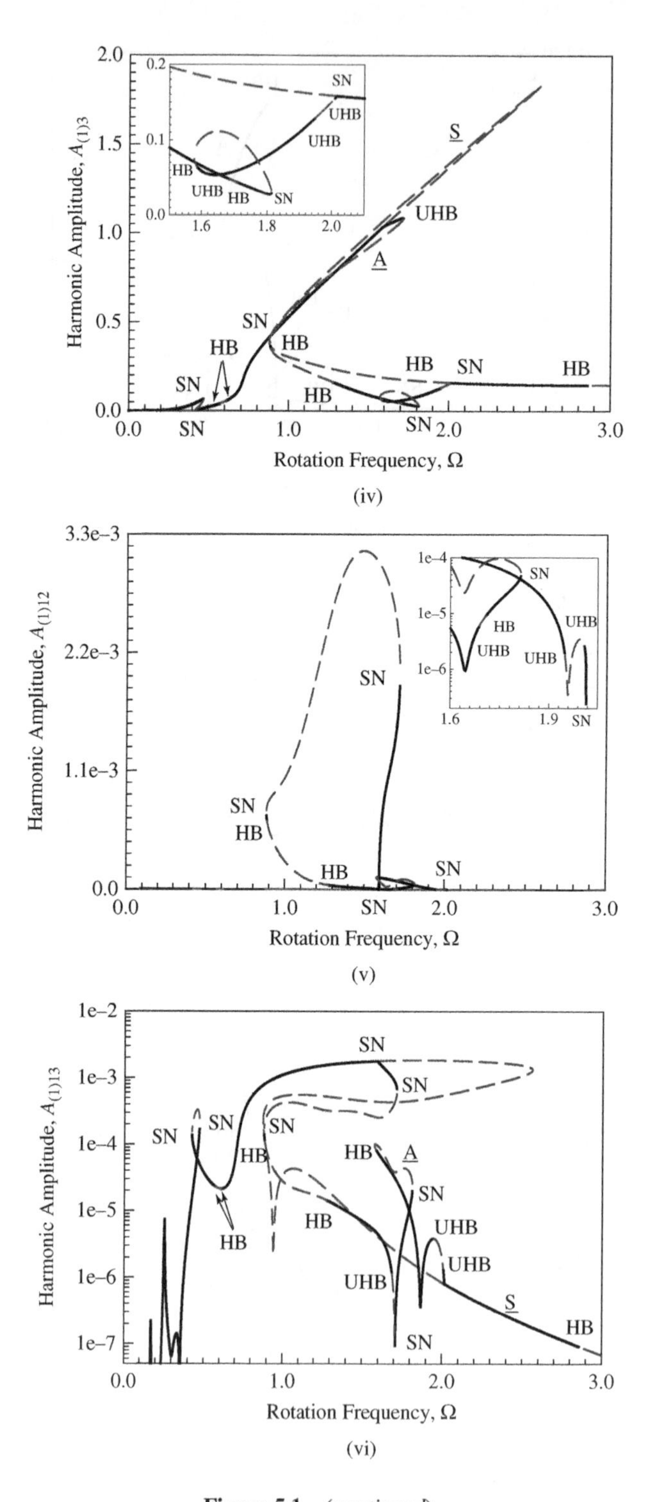

Figure 5.1 *(continued)*

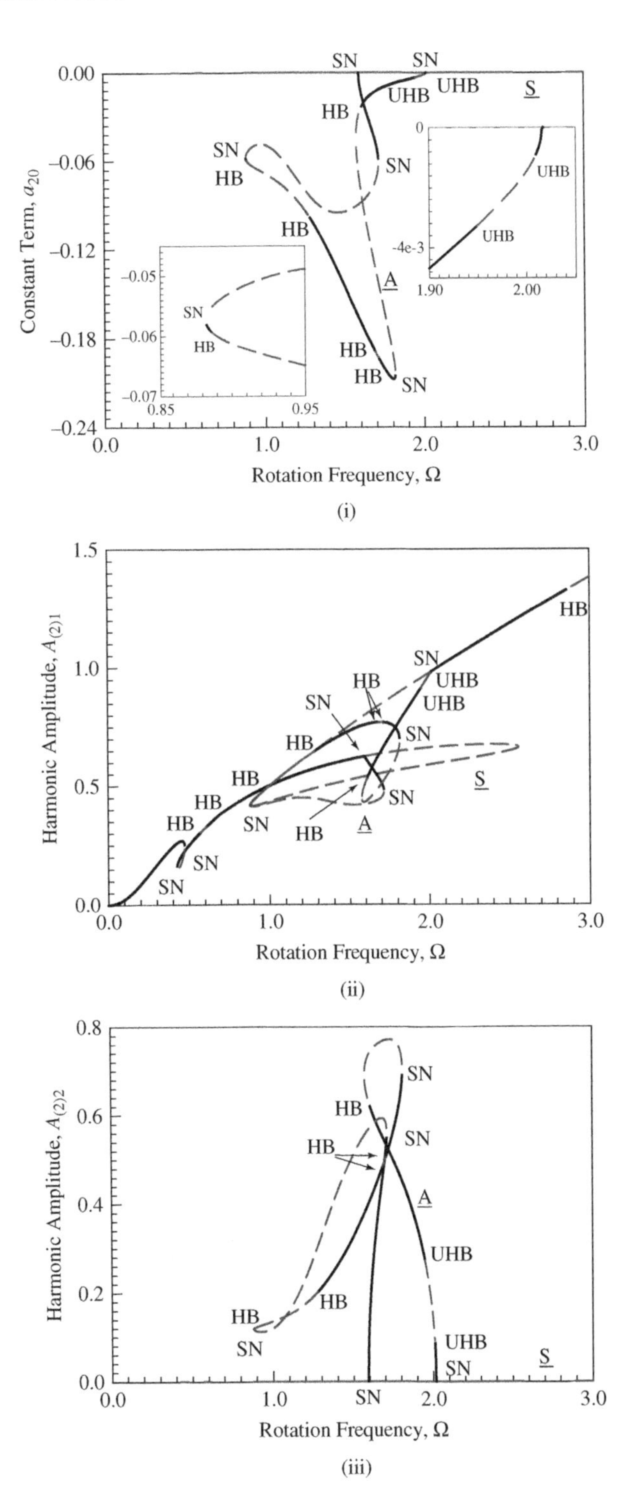

Figure 5.2 Period-1 motion of the y-direction in the nonlinear rotor: frequency-amplitude curves of harmonic terms based on 13 harmonic terms: (i) a_{20}, (ii)–(vi) $A_{(2)k}$ ($k = 1, 2, 3, 12, 13$), ($\delta = 0.02$, $\alpha = 0.68$, $\beta = 10$, $\gamma = 1.0$, $e = 1.5$)

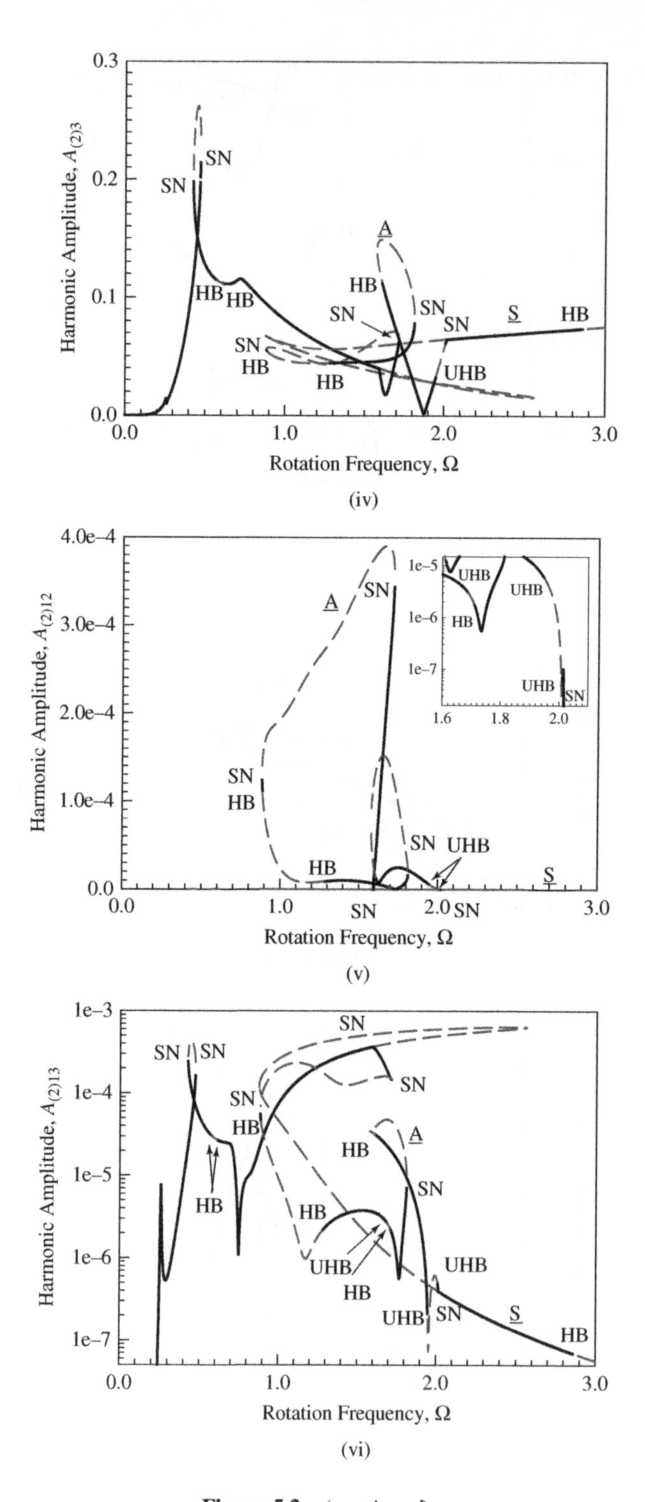

Figure 5.2 *(continued)*

symmetric motion. In Figure 5.1(iii), the harmonic amplitude $A_{(1)2}$ versus rotation frequency Ω are presented. Only asymmetric period-1 motion with $A_{(1)2} < 0.4$ exists as for constant a_{10}. For symmetric period-1 motion, $A_{(1)2} = 0$ is observed. In Figure 5.1(iv), the harmonic amplitude $A_{(1)3}$ varying with rotation speed Ω is presented. For asymmetric motion, its local view is zoomed for the stability detail. The harmonic amplitude $A_{(1)3} < 2.0$ possesses largest values compared to the primary harmonic amplitude $A_{(1)1} < 1.0$. To avoid abundant illustrations, the harmonic amplitudes $A_{(1)12}$ and $A_{(1)13}$ are presented in Figure 5.1(v) and (vi), respectively. The quantity levels of $A_{(1)12}$ and $A_{(1)13}$ are close to 10^{-3}. As usual, the more harmonic terms should be included. The local view of asymmetric period-1 motion of $A_{(1)12}$ is zoomed, and the logarithm scale is used to present the harmonic amplitude of $A_{(1)13}$.

For this nonlinear rotor system, the period-1 motion of the nonlinear rotor in the y-direction is presented in Figure 5.2. In Figure 5.2(i), the constant term a_{20} versus rotation speed Ω is presented. Only the asymmetric period-1 motion exists as in the x-direction, but the values of a_{20} lie in the range of $a_{20} \in (-0.24, 0)$. Two local areas are zoomed to view the details. The positive a_{20} can be obtained with mirror symmetry as for negative a_{10}. The harmonic amplitude $A_{(2)1}$ varying with rotation speed is presented in Figure 5.2(ii), and the primary harmonic amplitude in the y-direction is in $A_{(2)1} \in (0, 1.5)$ different from $A_{(1)1}$ in the x-direction. The symmetric and asymmetric period-1 motions are presented. In Figure 5.2(iii), the harmonic amplitude $A_{(2)2}$ versus rotation speed Ω is presented only for the asymmetric period-2 motion with $A_{(2)2} \sim 1$ because the symmetric period-1 motion has $A_{(2)2} = 0$. The harmonic amplitude $A_{(2)3}$ versus rotation speed is presented in Figure 5.2(iv). The quantity level of $A_{(2)3}$ is about $A_{(2)3} \sim 0.3$ much less than $A_{(1)3} \sim 2$. To avoid abundant illustrations, the harmonic amplitudes $A_{(2)12}$ and $A_{(2)13}$ are presented in Figure 5.2(v) and (vi), respectively. Their quantity levels are $A_{(2)12} \sim 4 \times 10^{-4}$ and $A_{(2)13} \sim 10^{-3}$.

5.2.2 *Analytical Bifurcation Trees*

From the Hopf bifurcations of symmetric period-1 motion, the quasi-periodic motion or other periodic motions may exist. However, for asymmetric period-1 motion, its Hopf bifurcation may be the onset of the period-2 motions. For the stable Hopf bifurcation, the stable period-2 motion will be obtained. For the unstable Hopf bifurcation, the unstable period-2 motion will be achieved. The analytical solutions of period-2 motions are based on the 26 harmonic terms (HB26) in the Fourier serious solutions. For more accurate solutions, more harmonic terms should be considered. However, the stability and bifurcation cannot be changed too much from the approximate period-2 solutions with 26 harmonic terms.

In Figure 5.3, the first branch of bifurcation tree of the period-1 motion to period-2 motion of the nonlinear rotor in the x-direction is presented. The constant term $a_{10}^{(m)}$ $(m = 1, 2)$ versus rotation speed is presented in Figure 5.3(i). Two local areas are zoomed to show the bifurcation characteristics. The saddle-node bifurcations of period-2 motions occur at $\Omega \approx 1.548, 2.01022$ and the Hopf bifurcation of period-2 motion occurs at $\Omega \approx 1.5870$. The unstable Hopf bifurcation of unstable asymmetric period-1 motion occurs at $\Omega \approx 1.5874$. The unstable period-2 motions appear from the unstable Hopf bifurcations. The positive constant $a_{10}^{(m)}$ lies in the range of $a_{10}^{(m)} \in (0, 0.18)$. The bifurcation relation of period-1 motion to period-2 motion is clearly illustrated. For period-2 motion, harmonic amplitude $A_{(1)1/2}$ is presented in Figure 5.3(ii). The appearances of period-2 motions take place at the Hopf bifurcation of period-1 motions, which is the saddle-node bifurcation of period-2 motion for appearance at $\Omega \approx 1.58722$. The unstable Hopf bifurcations of period-1 motion give the unstable saddle-node bifurcation (USN) of

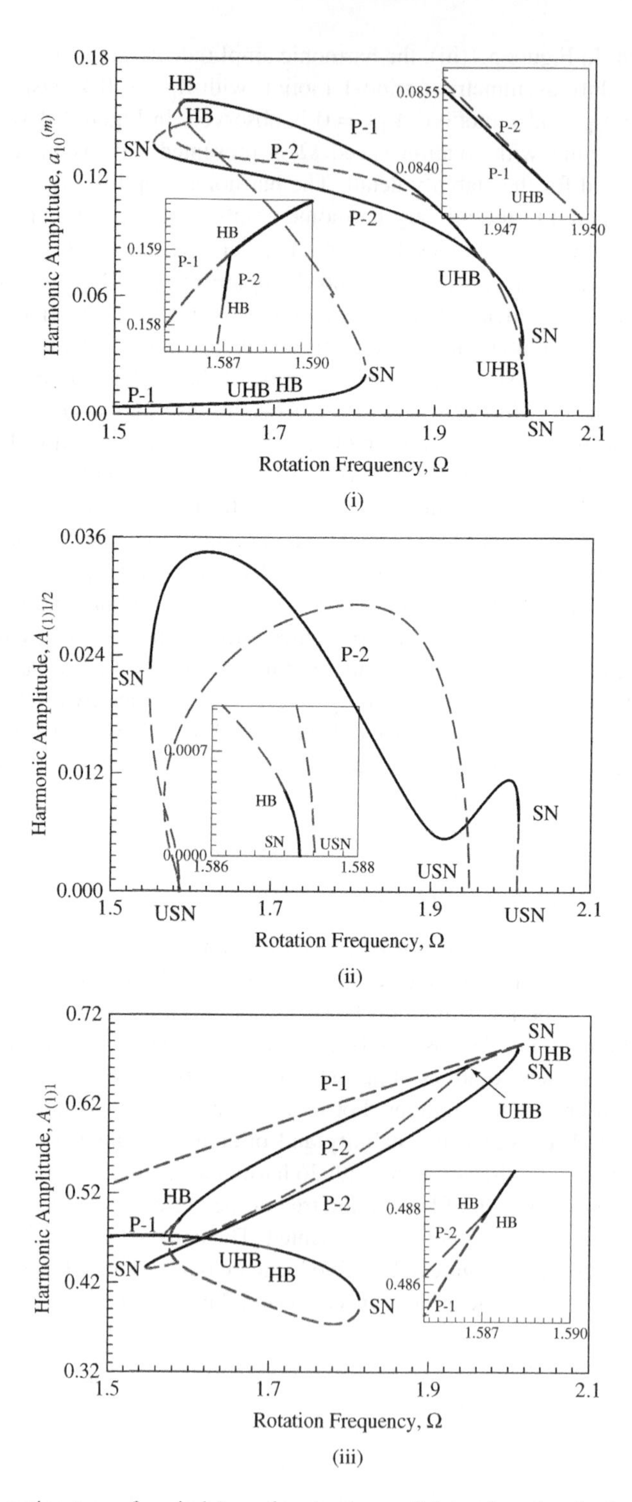

Figure 5.3 Bifurcation tree of period-1 motion to chaos of the x-direction in the nonlinear rotor: frequency-amplitude curves of harmonic terms based on 26 harmonic terms: (i) $a_{10}^{(m)}$, (ii)–(vi) $A_{(1)k/m}$ ($k = 1, 2, 3, 4, 26, m = 2$), ($\delta = 0.02, \alpha = 0.68, \beta = 10, \gamma = 1.0, e = 1.5$)

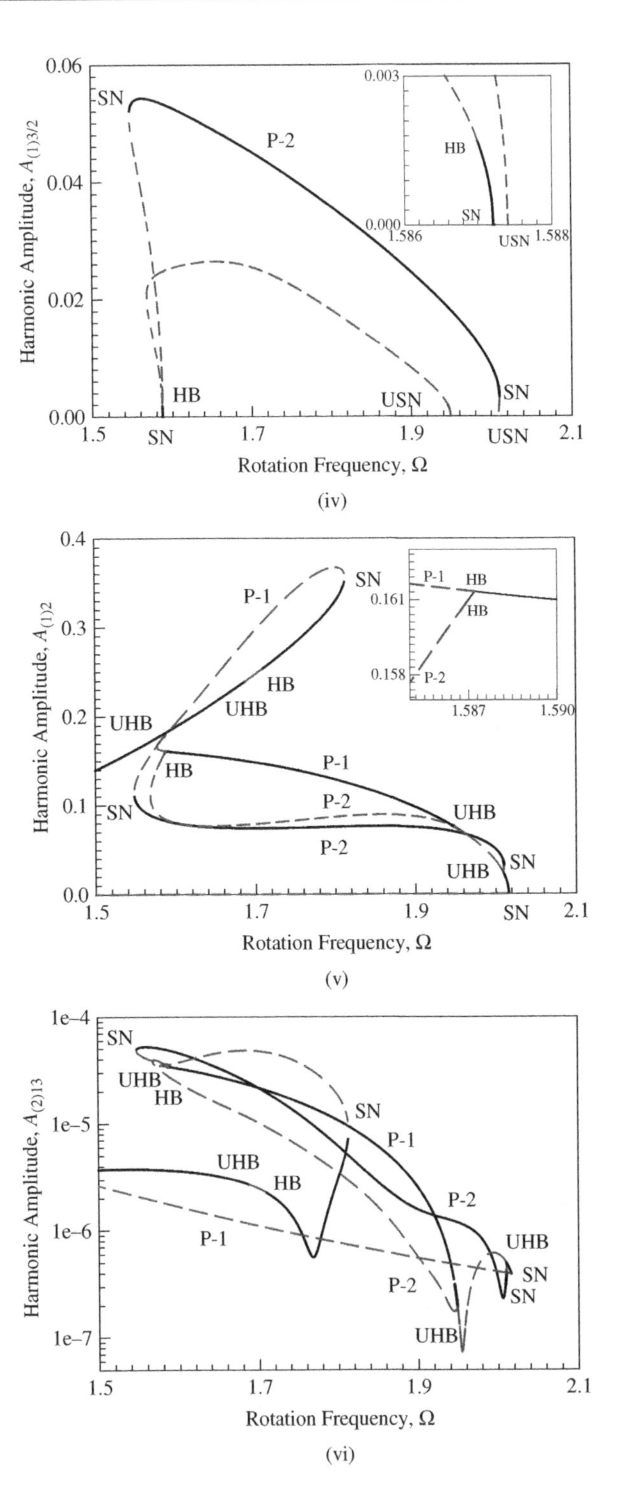

Figure 5.3 (*continued*)

period-2 motions for the onset of period-2 motion. The unstable saddle-node bifurcations of period-2 motions are at $\Omega \approx 1.5874, 1.94865, 2.01$. The quantity level of $A_{(1)1/2} \sim 4 \times 10^{-2}$ is observed and the period-2 motion for this branch is in the range of $\Omega \in (1.5, 2.1)$. The harmonic amplitude $A_{(1)1}$ versus rotation speed is presented in Figure 5.3(iii) for the bifurcation tree of period-1 motion to period-2 motion via the Hopf bifurcations. The quantity level of $A_{(1)1}$ for period-2 motion is in the range of $A_{(1)1} \sim 0.7$ for $\Omega \in (1.5, 2.1)$. Compared to $A_{(1)1/2}$, the harmonic amplitude $A_{(1)3/2}$ versus rotation speed is presented in Figure 5.3(iv) with $A_{(1)3/2} \sim 6 \times 10^{-2}$ in $\Omega \in (1.5, 2.1)$. Compared to the period-1 motion, the harmonic amplitude $A_{(1)2}$ versus rotation speed is illustrated in Figure 5.3(v). The quantity level is $A_{(1)2} \sim 0.4$. No symmetric period-1 motion exists, and both asymmetric period-1 motion and period-2 motion exist to show the bifurcation tree much more clearly. To reduce abundant illustrations, the harmonic amplitude $A_{(1)13}$ versus rotation speed is illustrated in Figure 5.3(vi), respectively. The quantity level for the harmonic amplitude for period-2 motions is $A_{(1)13} \sim 1 \times 10^{-4}$.

As in Figure 5.3, the first branch of the bifurcation tree of the period-1 motion to period-2 motion of the nonlinear rotor in the y-direction is presented in Figure 5.4. In Figure 5.4(i), the constant term $a_{20}^{(m)}$ $(m = 1, 2)$ versus rotation speed is presented. A local area is zoomed to show the bifurcation characteristics. The bifurcation points are the same as discussed in Figure 5.3. The negative constant $a_{20}^{(m)}$ lies in the range of $a_{20}^{(m)} \in (-0.24, 0)$. The bifurcation relation of period-1 to period-2 motion is clearly illustrated. For period-2 motion, the harmonic amplitude $A_{(2)1/2}$ in the y-direction is different in the x-direction, as shown in Figure 5.4(ii). The quantity level of $A_{(2)1/2} \sim 5 \times 10^{-2}$ is observed in the range of $\Omega \in (1.5, 2.1)$. The harmonic amplitude $A_{(2)1}$ varying with rotation speed is presented in Figure 5.4(iii) for the bifurcation tree of period-1 to period-2 motion. The quantity level of $A_{(2)1} \sim 1$ for period-2 motion is observed for $\Omega \in (1.5, 2.1)$. The harmonic amplitude $A_{(2)3/2} \sim 5 \times 10^{-2}$ versus rotation speed is presented in Figure 5.4(iv), and the harmonic amplitude $A_{(2)2} \sim 1$ versus rotation speed are illustrated in Figure 5.4(v). To reduce abundant illustrations, the harmonic amplitude $A_{(1)13}$ versus rotation speed is illustrated in Figure 5.4(vi), and the quantity level of the harmonic amplitude is $A_{(2)13} \sim 10^{-4}$ for period-2 motions.

In Figure 5.5, the second branch of bifurcation tree of the period-1 motion to period-2 motion of the nonlinear rotor in the x-direction is presented in $\Omega \in (1.30, 1.80)$. Only the asymmetric period-1 motion relative to period-2 motion is presented herein. The constant term $a_{10}^{(m)}$ $(m = 1, 2)$ versus rotation speed is presented in Figure 5.5(i). The Hopf bifurcations of period-1 motions occur at $\Omega \approx 1.70872$ and the Hopf bifurcation of period-2 motion occurs at $\Omega \approx 1.5875$. The unstable Hopf bifurcation of unstable asymmetric period-1 motion occurs at $\Omega \approx 1.689$. The positive constant $a_{10}^{(m)}$ lies in the range of $a_{10}^{(m)} \in (0, 0.015)$. The bifurcation relation of period-1 to period-2 motion is clearly illustrated. For period-2 motion, harmonic amplitude $A_{(1)1/2}$ is presented in Figure 5.5(ii). The onset of period-2 motion is at the Hopf bifurcation of period-1 motions, which is the saddle-node bifurcation of period-2 motion at $\Omega \approx 1.70872$. The unstable saddle-node bifurcations of period-2 motions are at $\Omega \approx 1.689$. The quantity level of $A_{(1)1/2} \sim 9 \times 10^{-2}$ is observed in $\Omega \in (1.4, 1.75)$. The harmonic amplitude $A_{(1)1}$ versus rotation speed is presented in Figure 5.5(iii). The quantity level of $A_{(1)1}$ for period-2 motion is in the range of $A_{(1)1} \sim 0.5$ for $\Omega \in (1.4, 1.75)$. The harmonic amplitude $A_{(1)3/2}$ versus rotation speed is presented in Figure 5.5(iv) with $A_{(1)3/2} \sim 10^{-1}$ in $\Omega \in (1.4, 1.75)$. The harmonic amplitude $A_{(1)2} \sim 0.3$ versus rotation speed is presented in Figure 5.5(v). To reduce abundant illustrations, the harmonic amplitude $A_{(1)13} \sim 2 \times 10^{-4}$ versus rotation speed are illustrated in Figure 5.5(vi).

The second branch of bifurcation tree of the period-1 motion to period-2 motion of the nonlinear rotor in the y-direction is presented in Figure 5.6. In Figure 5.6(i), the constant

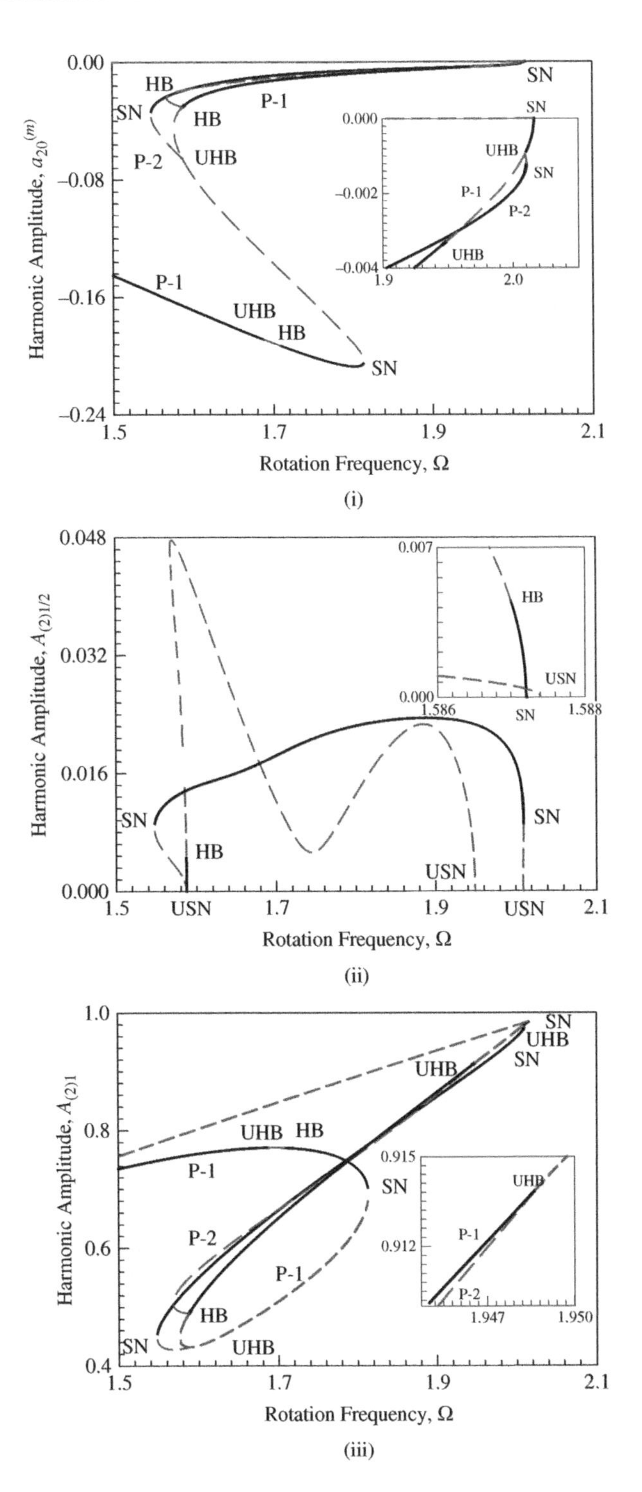

Figure 5.4 Bifurcation tree of period-1 motion to chaos of the y-direction in the nonlinear rotor: frequency-amplitude curves of harmonic terms based on 26 harmonic terms: (i) $a_{20}^{(m)}$, (ii)–(vi) $A_{(2)k/m}$ ($k = 1, 2, 3, 4, 26, m = 2$), ($\delta = 0.02, \alpha = 0.68, \beta = 10, \gamma = 1.0, e = 1.5$)

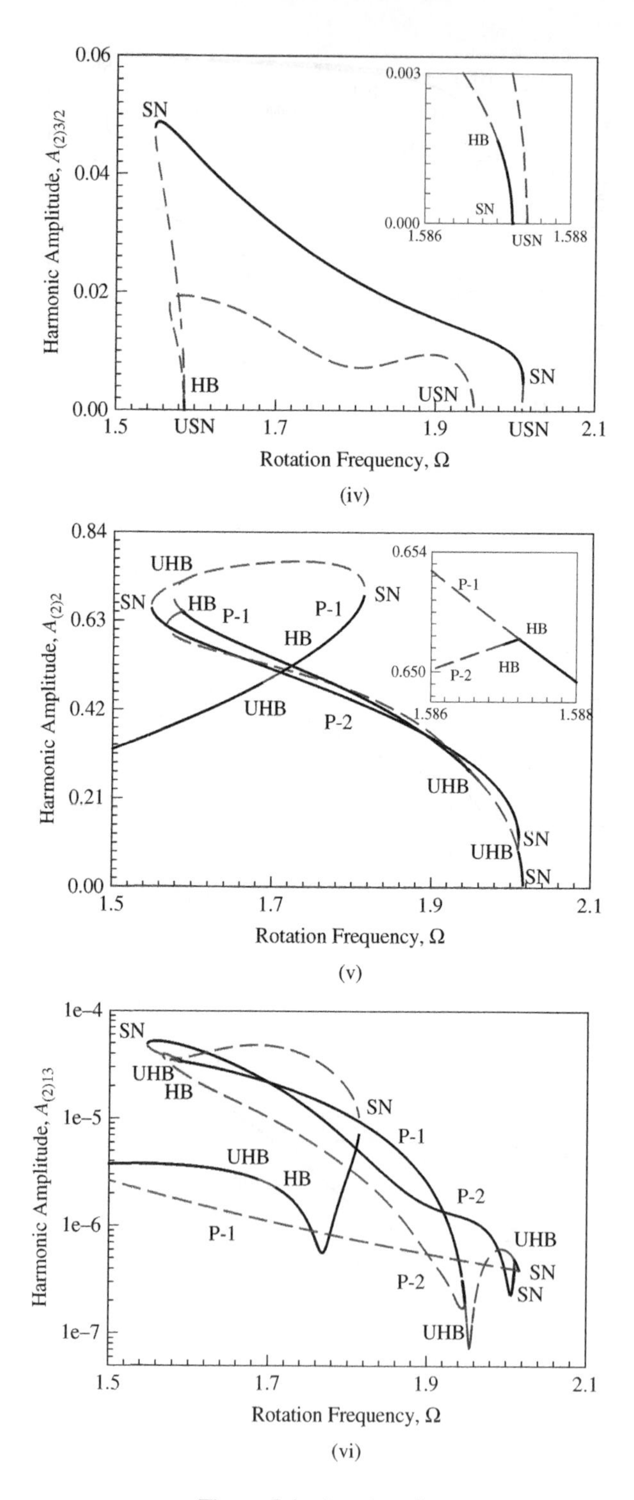

Figure 5.4 (*continued*)

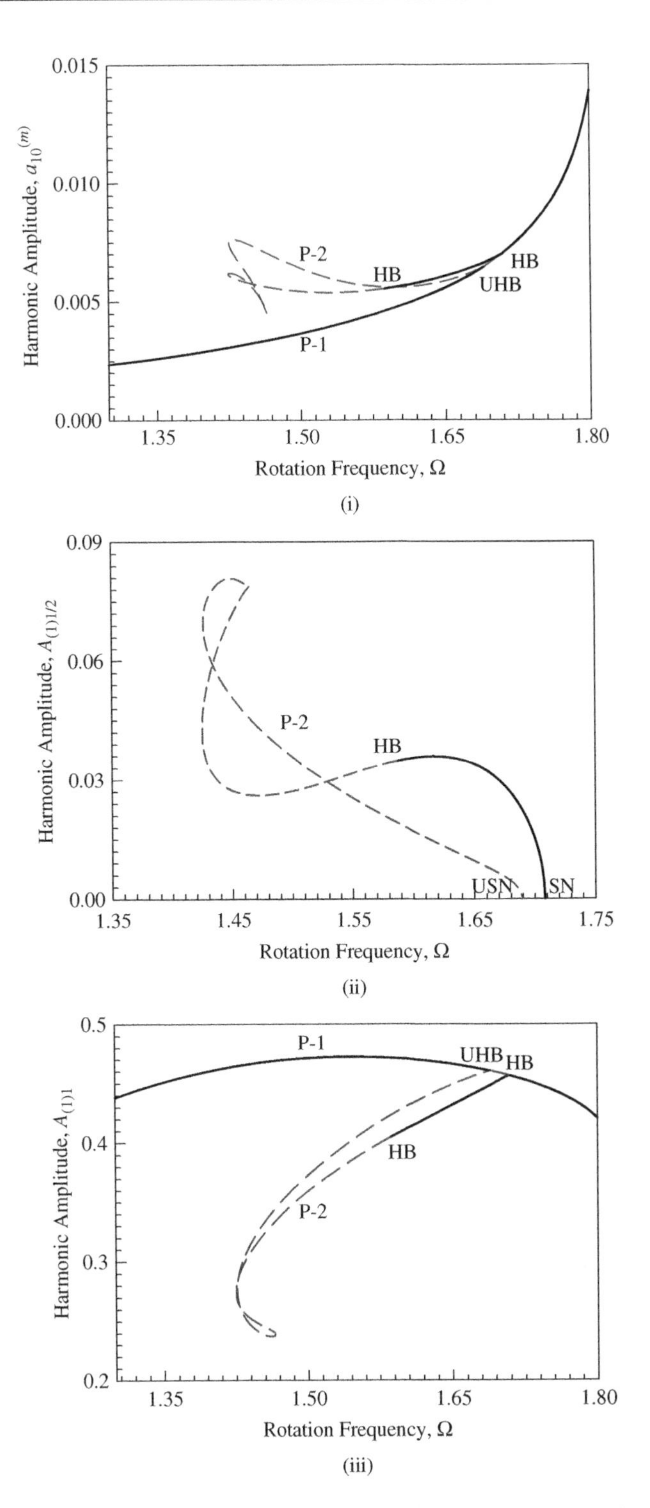

Figure 5.5 The second bifurcation tree of period-1 motion to chaos of the x-direction in the nonlinear rotor: frequency-amplitude curves of harmonic terms based on 26 harmonic terms: (i) $a_{20}^{(m)}$, (ii)–(vi) $A_{(2)k/m}$ ($k = 1, 2, 3, 4, 26, m = 2$), ($\delta = 0.02, \alpha = 0.68, \beta = 10, \gamma = 1.0, e = 1.5$)

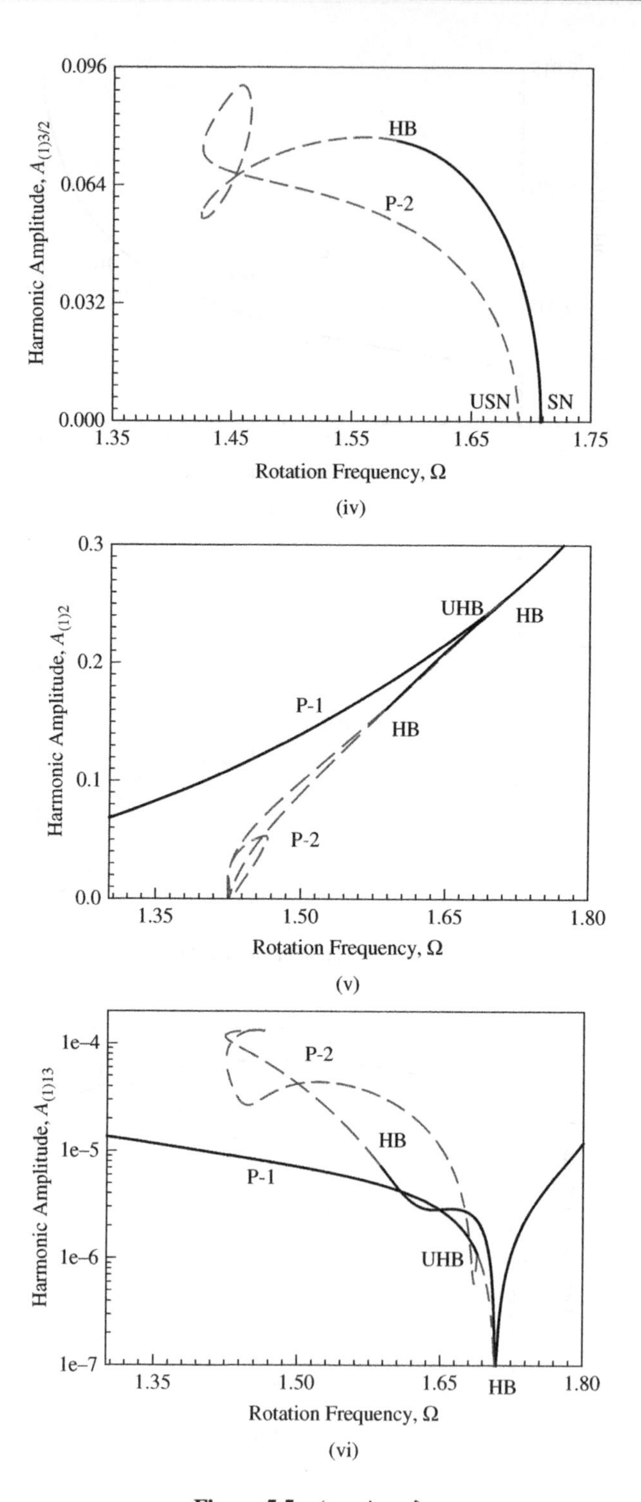

Figure 5.5 *(continued)*

term $a_{20}^{(m)}$ $(m = 1, 2)$ versus rotation speed is presented. The positive constant $a_{20}^{(m)}$ lies in the range of $a_{20}^{(m)} \in (-0.2, -0.1)$. The bifurcation relation of period-1 to period-2 motion is clearly depicted. For period-2 motion, harmonic amplitude $A_{(2)1/2}$ in the y-direction is presented in Figure 5.6(ii). The quantity level of $A_{(2)1/2} \sim 7 \times 10^{-2}$ is observed in $\Omega \in (1.4, 1.75)$. The harmonic amplitude $A_{(2)1}$ varying with rotation speed is presented in Figure 5.6(iii) for the bifurcation tree of period-1 motion to period-2 motion. The quantity level of $A_{(2)1} \sim 0.8$ for period-2 motion is observed for $\Omega \in (1.5, 2.1)$. Thee harmonic amplitude $A_{(2)3/2} \sim 0.12$ versus rotation speed is presented in Figure 5.6(iv). The harmonic amplitudes $A_{(2)2} \sim 1$ varying with rotation speed is presented in Figure 5.6(v). To reduce abundant illustrations, the harmonic amplitude $A_{(2)13} \sim 10^{-4}$ versus rotation speed are illustrated in Figure 5.6(vi).

5.2.3 Independent Period-5 Motion

In the previous section, the period-1 motion to period-2 motions is presented via the bifurcation tree. Herein, an independent period-5 motion based on 55 harmonic terms is considered, and the frequency-amplitude curve is shown in Figures 5.7 and 5.8. Since the independent period-5 motion is symmetric, $a_{10}^{(5)} = 0$ and $A_{(1)k/m} = 0$ $(k = 2l,\ l = 1, 2, \ldots,$ and $m = 5)$ is obtained. Thus, only $A_{(1)k/m}$ $(k = 2l + 1,\ l = 1, 2, \ldots,$ and $m = 5)$ are presented herein.

In Figure 5.7, the frequency-amplitude characteristics of period-5 motion for the x-direction of the nonlinear rotor are presented. In Figure 5.7(i), the harmonic amplitude $A_{(1)1/5}$ versus rotation speed Ω is presented. The two saddle-node bifurcations occur at $\Omega = 2.485, 2.995$. The frequency-amplitude curve forms a closed loop for the stable and unstable period-5 motions. The quantity levels of stable and unstable period-5 motions are $A_{(1)1/5} \sim 0.08$ and $A_{(1)1/5} \sim 0.24$, respectively. The harmonic amplitude $A_{(1)3/5}$ varying with rotation speed Ω is presented in Figure 5.7(ii). The quantity levels of $A_{(1)1/5}$ and $A_{(1)3/5}$ are quite similar. The primary harmonic amplitude $A_{(1)1} \sim 1$ versus rotation speed is arranged in Figure 5.7(iii). To reduce abundant illustrations, the harmonic amplitudes $A_{(1)3}$ and $A_{(1)5}$ are presented in Figure 5.7(iv) and (v), respectively. The corresponding quantity levels of the harmonic amplitudes are $A_{(1)3} \sim 0.2$ and $A_{(1)5} \sim 0.04$, respectively. The 55th order harmonic amplitude varying with rotation speed is presented in Figure 5.7(vi), and the quantity levels of harmonic amplitudes for the stable and unstable period-5 motions are $A_{(1)11} \sim 10^{-4}$ and $A_{(1)11} \sim 10^{-3}$, respectively.

In Figure 5.8, the frequency-amplitude of period-5 motion for the y-direction of the nonlinear rotor is also presented. In Figure 5.8(i), the harmonic amplitude $A_{(2)1/5}$ versus rotation speed Ω is presented. The quantity levels of stable and unstable period-5 motions are $A_{(2)1/5} \sim 0.2$ and $A_{(2)1/5} \sim 0.3$, respectively. The harmonic amplitude $A_{(1)3/5}$ varying with rotation speed Ω is presented in Figure 5.8(ii). The quantity level of $A_{(1)3/5}$ reduces to $A_{(2)3/5} \sim 0.15$. The primary harmonic amplitude $A_{(2)1}$ versus rotation speed is arranged in Figure 5.8(iii). The quantity level of $A_{(2)1}$ is very large with $A_{(2)1} \sim 1.5$. To reduce abundant illustrations, the harmonic amplitudes $A_{(2)3}$ and $A_{(2)5}$ are presented in Figure 5.8(iv) and (v), respectively. The corresponding quantity levels of the harmonic amplitudes are $A_{(2)3} \sim 0.08$ and $A_{(2)5} \sim 0.02$, respectively. The 55th order harmonic amplitude varying with rotation speed is presented in Figure 5.8(vi). The quantity levels of harmonic amplitudes for the stable and unstable period-5 motions are $A_{(2)11} \sim 2 \times 10^{-5}$ and $A_{(2)11} \sim 2.4 \times 10^{-4}$, respectively. The asymmetrical period-5 motions can be similarly obtained analytically.

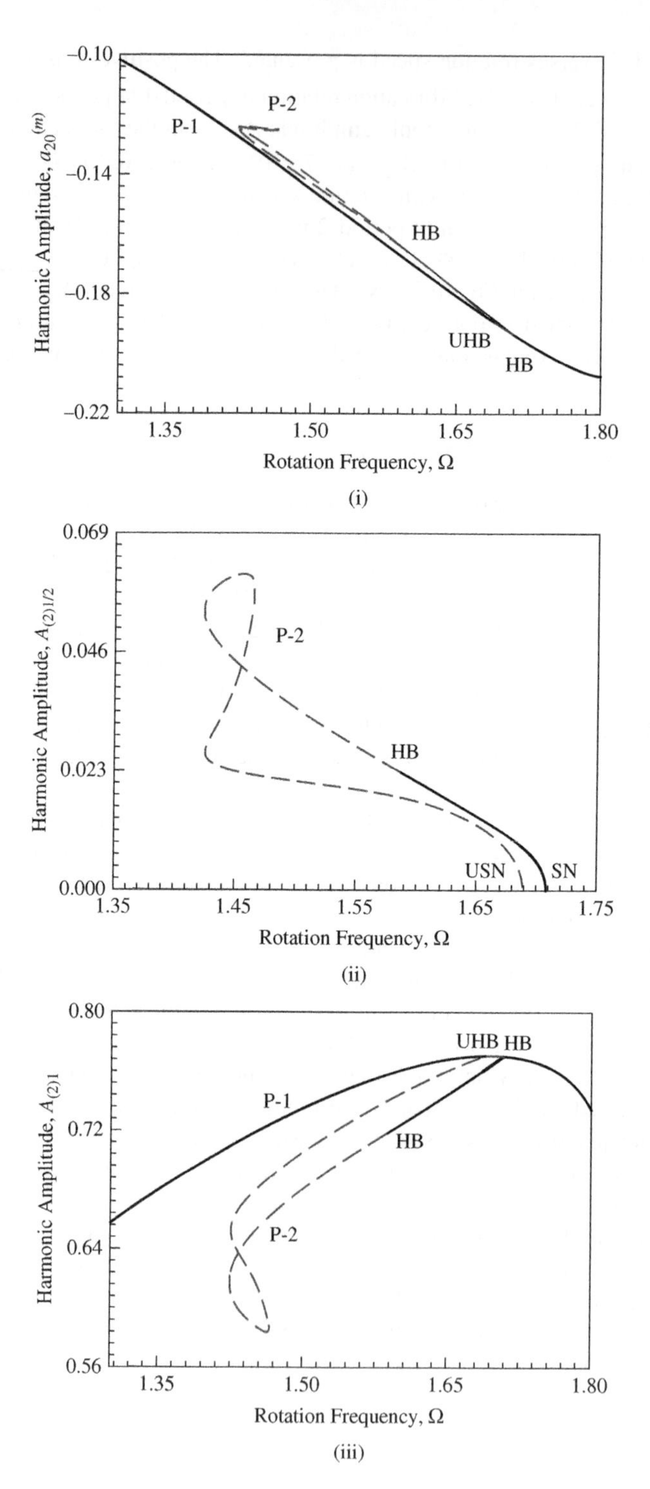

Figure 5.6 The second bifurcation tree of period-1 motion to chaos of the y-direction in the nonlinear rotor: frequency-amplitude curves of harmonic terms based on 26 harmonic terms: (i) $a_{20}^{(m)}$, (ii)–(vi) $A_{(2)k/m}$ $(k = 1, 2, 3, 4, 26, m = 2)$, $(\delta = 0.02, \alpha = 0.68, \beta = 10, \gamma = 1.0, e = 1.5)$

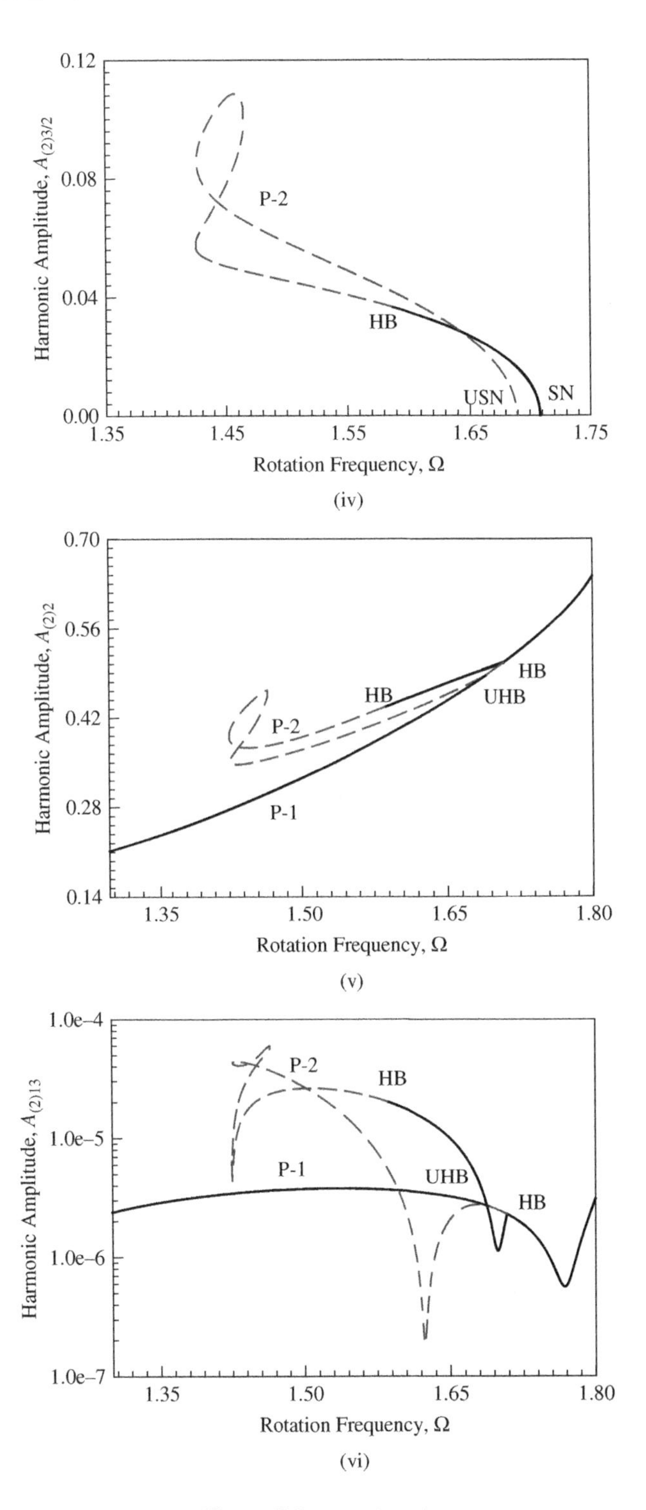

Figure 5.6 *(continued)*

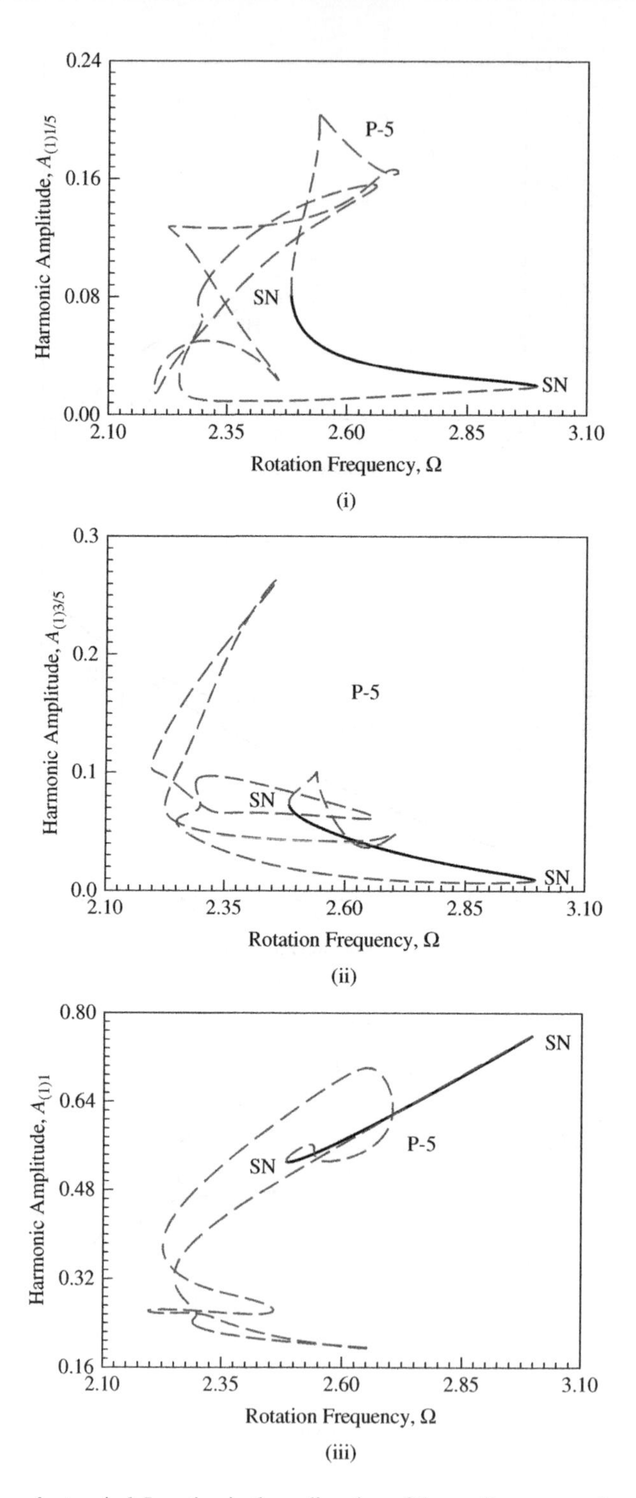

Figure 5.7 Independent period-5 motion in the x-direction of the nonlinear rotor: frequency-amplitude curves of harmonic terms based on 55 harmonic terms (HB55): (i)–(vi) $A_{(1)k/m}$ ($k = 1, 3, 5, 10, 15, 55$, $m = 2$), ($\delta = 0.02, \alpha = 0.68, \beta = 10, \gamma = 1.0, e = 1.5$)

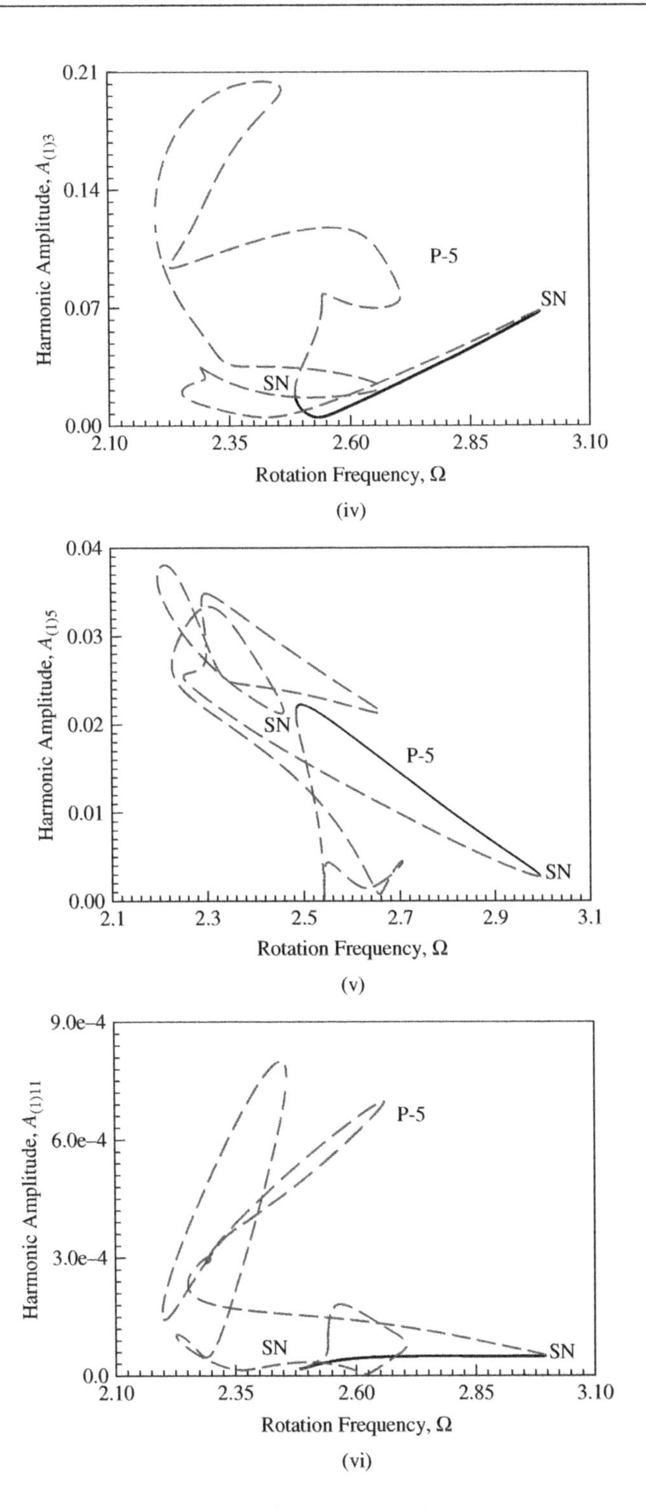

Figure 5.7 (*continued*)

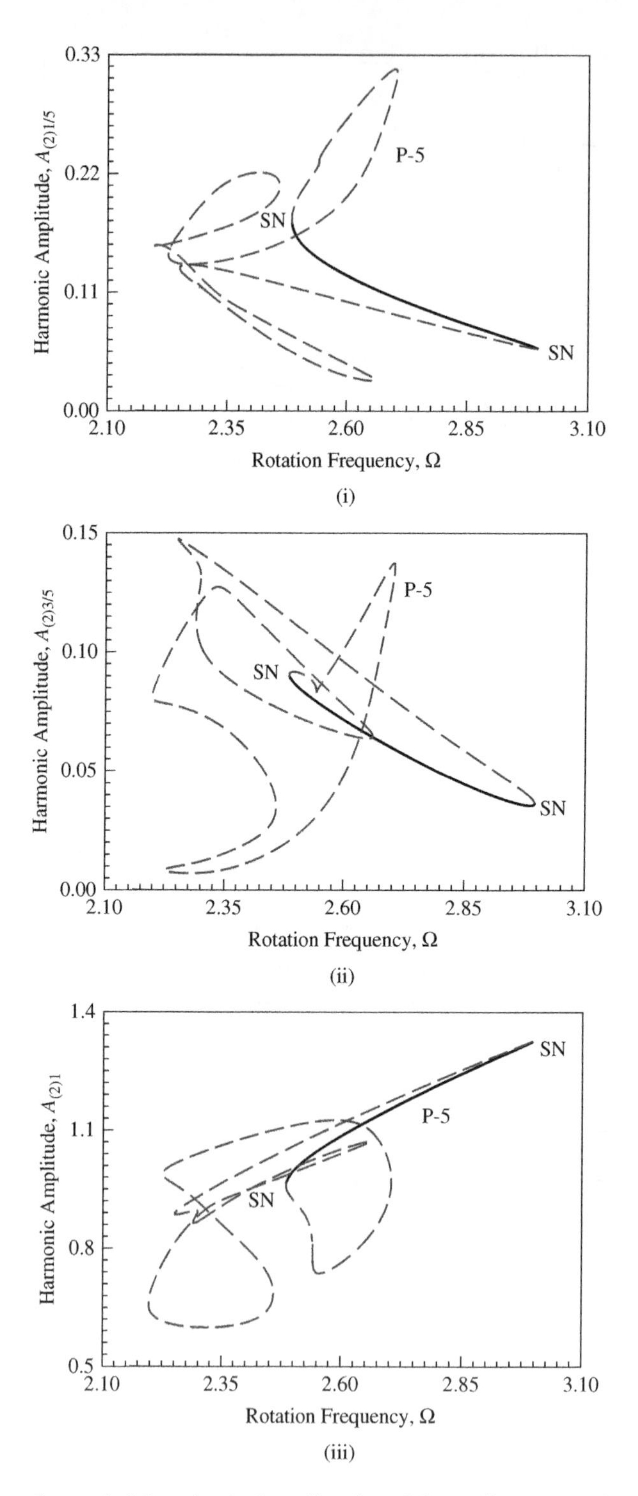

Figure 5.8 Independent period-5 motion in the y-direction of the nonlinear rotor: frequency-amplitude curves of harmonic terms based on 55 harmonic terms (HB55): (i)–(vi) $A_{(1)k/m}$ ($k = 1, 3, 5, 10, 15, 55$, $m = 2$), ($\delta = 0.02$, $\alpha = 0.68$, $\beta = 10$, $\gamma = 1.0$, $e = 1.5$)

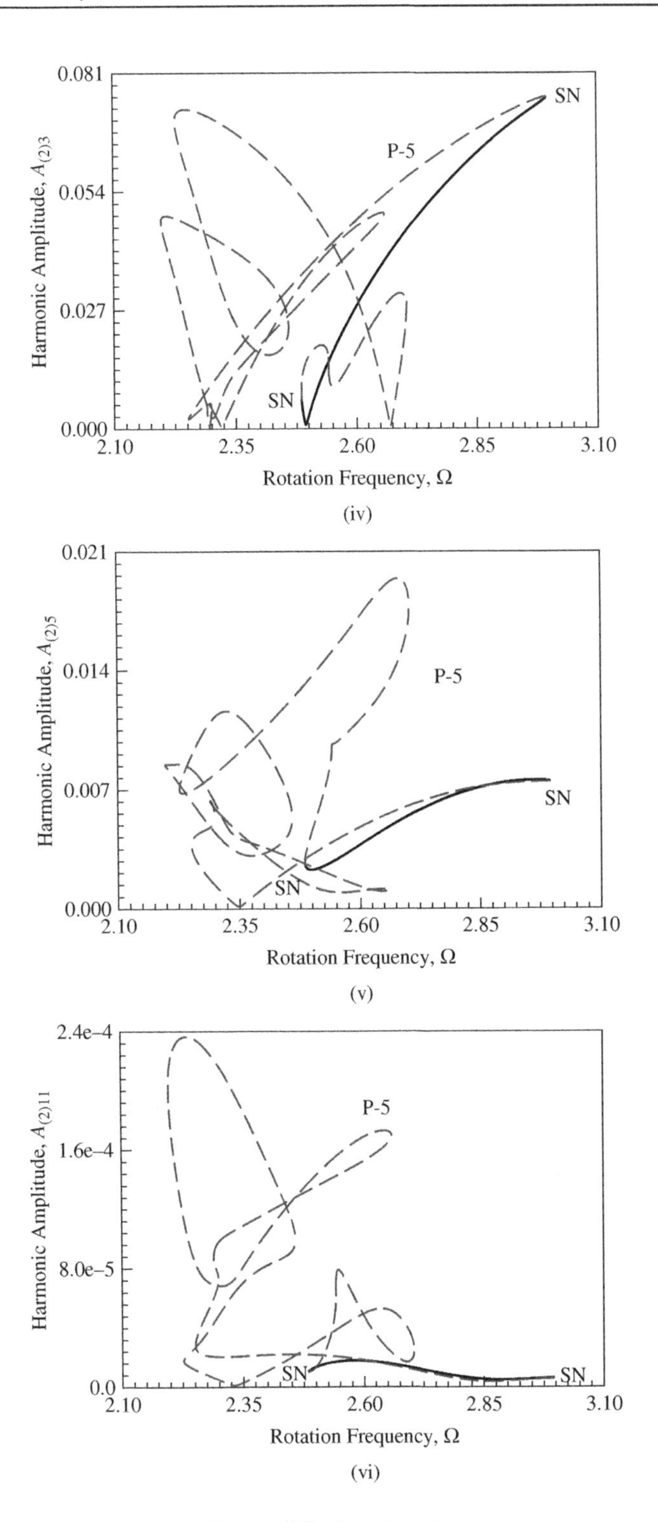

Figure 5.8 (*continued*)

5.3 Numerical Simulations

To illustrate period-m motions in the nonlinear rotor, numerical simulations and analytical solutions will be presented. The initial conditions for numerical simulations are computed from approximate analytical solutions of periodic solutions. In all plots, circular symbols gives approximate solutions, and solid curves give numerical simulation results. The acronym "I.C." with a large circular symbol represents initial condition for all plots. The numerical solutions of periodic motions are generated via the symplectic scheme.

In Figure 5.9, period-1 motion based on 13 harmonic terms (HB13) are presented for $\Omega = 2.01$ with other parameters in Equation (5.54). The displacement and velocity responses in the x-direction of the nonlinear rotor are presented in Figure 5.9(i) and (ii), respectively. One period (T) for the period-1 motion response is labeled in the two plots. Similarly, the displacement and velocity responses in the y-direction of the nonlinear rotor are also presented in Figure 5.9(iii) and (iv), respectively. The analytical and numerical solutions match very well. The two trajectories for x and y-directions are presented for over 40 periods in Figure 5.9(v) and (vi), respectively. Both of them are different because of the interaction. The initial conditions are marked by large circular symbols and labeled by I.C. In engineering, one is interested in displacement orbit. The displacement orbit of rotor in x and y-directions is presented in Figure 5.9(vii). For better understanding of harmonic contributions, the harmonic amplitude spectrums of rotor in x and y-directions are presented in Figure 5.9(viii) and (ix). The harmonic amplitude spectrums are computed from analytical solutions. The main harmonic amplitudes of rotor in the x-direction are $a_{10} \approx 0.026626$, $A_{(1)1} \approx 0.685920$, $A_{(1)2} \approx 0.022759$, $A_{(1)3} \approx 0.155060$, and $A_{(1)4} \approx 0.014091$. The other harmonic amplitudes in the x-direction are $A_{(1)5} \sim 5 \times 10^{-3}$, $A_{(1)k} \sim 10^{-3}$ $(k = 6, 7)$, $A_{(1)8} \sim 5 \times 10^{-5}$, $A_{(1)9} \sim 10^{-4}$, $A_{(1)10} \sim 3 \times 10^{-5}$, $A_{(1)k} \sim 10^{-6}$ $(k = 11, 12, 13)$. However, the main harmonic amplitudes of rotor in the y-direction are $a_{20} \sim -10^{-3}$, $A_{(2)1} \approx 0.977544$, $A_{(2)2} \approx 0.085401$, $A_{(2)3} \approx 0.061716$, and $A_{(2)4} \approx 0.010812$. The other harmonic amplitudes in the y-direction are $A_{(2)5} \sim 2.5 \times 10^{-3}$, $A_{(1)k} \sim 10^{-4}$ $(k = 6, 8)$, $A_{(2)7} \sim 6 \times 10^{-4}$, $A_{(2)9} \sim 3 \times 10^{-5}$, $A_{(2)10} \sim 10^{-5}$, $A_{(2)11} \sim 3 \times 10^{-6}$, $A_{(2)12} \sim 5 \times 10^{-8}$, and $A_{(2)12} \sim 5 \times 10^{-7}$.

In Figure 5.10, a period-2 motion based on 26 harmonic terms (HB26) are presented for $\Omega = 2.01$ with other parameters in Equation (5.54). The time-histories of displacement and velocity in the x-direction of the nonlinear rotor are presented in Figure 5.10(i) and (ii), respectively. Compared to the coexisting period-1 motion, the displacement and velocity of period-2 motion cannot keep the period-1 motion patterns. Two periods ($2T$) for the period-2 motion is labeled in the two plots. The time-histories of displacement and velocity in the y-direction of the nonlinear rotor are also presented in Figure 5.10(iii) and (iv), respectively. The two trajectories for x and y-directions are presented for over 40 periods in Figure 5.10(v) and (vi), respectively. Compared to period-1 motion, the period-doubling responses are clearly observed. The initial conditions are marked by large circular symbols and also labeled by "I.C." The displacement orbit of rotor in the x and y-directions is presented in Figure 5.10(vii). To show harmonic contributions on the period-2 motion, the harmonic amplitude spectrums of the stable period-2 motion of the nonlinear rotor in x and y-directions are presented in Figure 5.10(viii) and (ix). The harmonic amplitude spectrums of the stable period-2 motion are given by analytical solutions. The main harmonic amplitudes in the x-direction are $a_{10} \approx 0.038564$, $A_{(1)1} \approx 0.680973$, $A_{(1)2} \approx 0.034070$, $A_{(1)3} \approx 0.149972$, and $A_{(1)4} \approx 0.021612$. However, $A_{(1)1/2} \approx 8.34 \times 10^{-3}$, $A_{(1)3/2} \approx 4.04 \times 10^{-3}$, $A_{(1)5/2} \approx 0.067610$, $A_{(1)7/2} \approx 9.36 \times 10^{-4}$. The other harmonic amplitudes in the x-direction are $A_{(1)9/2} \sim 2 \times 10^{-4}$, $A_{(1)5} \sim 4 \times 10^{-3}$, $A_{(1)11/2} \sim 1.2 \times 10^{-4}$,

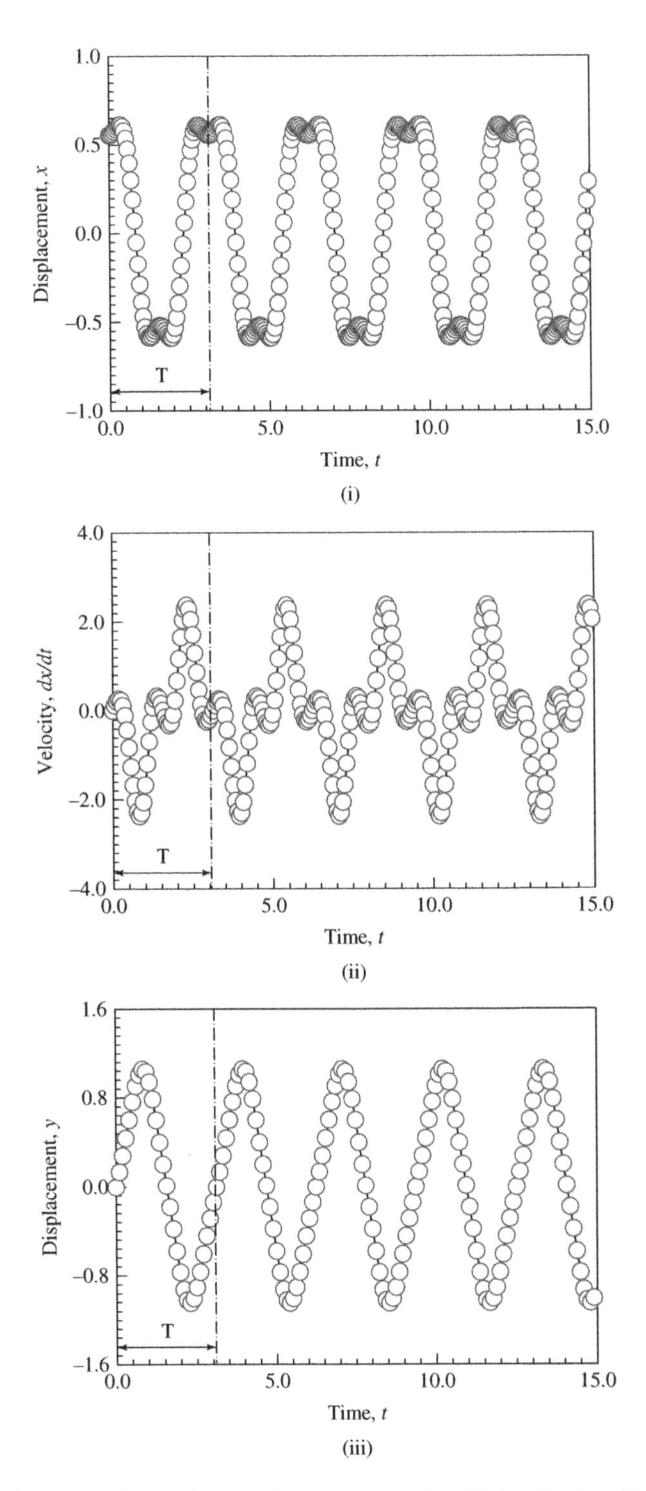

Figure 5.9 Period-1 motion of a nonlinear rotor ($\Omega = 2.01$, HB13): (i) x-displacement, (ii) x-velocity, (iii) y-displacement, (iv) y-velocity, (v) x-trajectory, (vi) y-trajectory, (vii) displacement orbit, (viii) x-harmonic amplitude, and (ix) y-harmonic amplitude. Initial condition $(x_0, \dot{x}_0) = (0.552759, 2.309360\text{E-}3)$ and $(y_0, \dot{y}_0) = (-7.387180\text{E-}3, 1.315090)$. ($\delta = 0.02$, $\alpha = 0.68$, $\beta = 10$, $\gamma = 1.0$, $e = 1.5$)

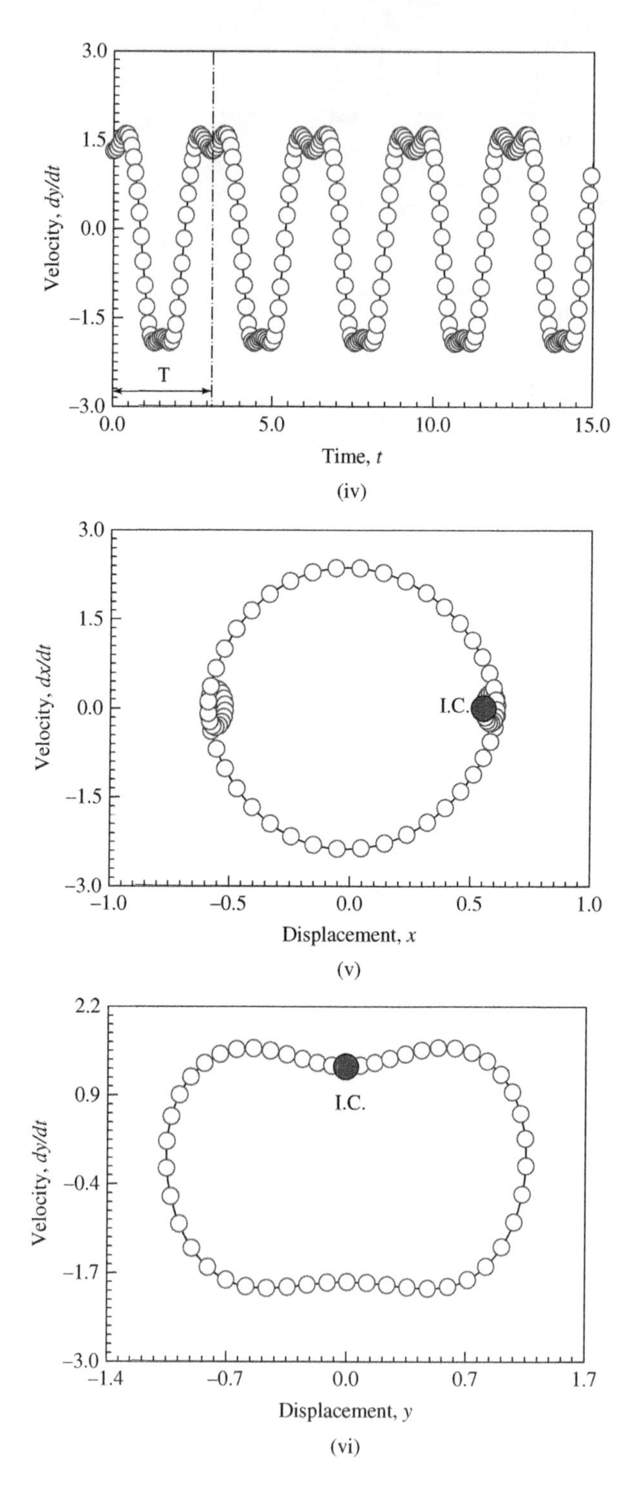

Figure 5.9 (*continued*)

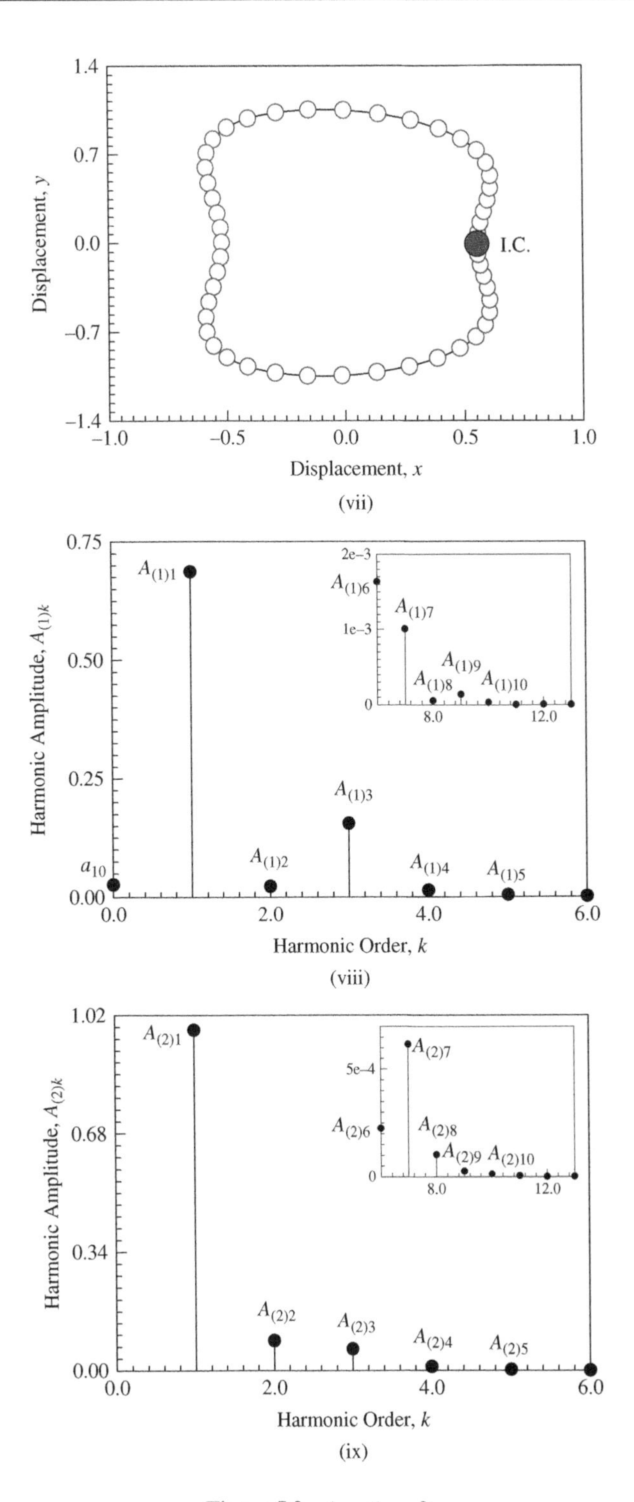

Figure 5.9 (*continued*)

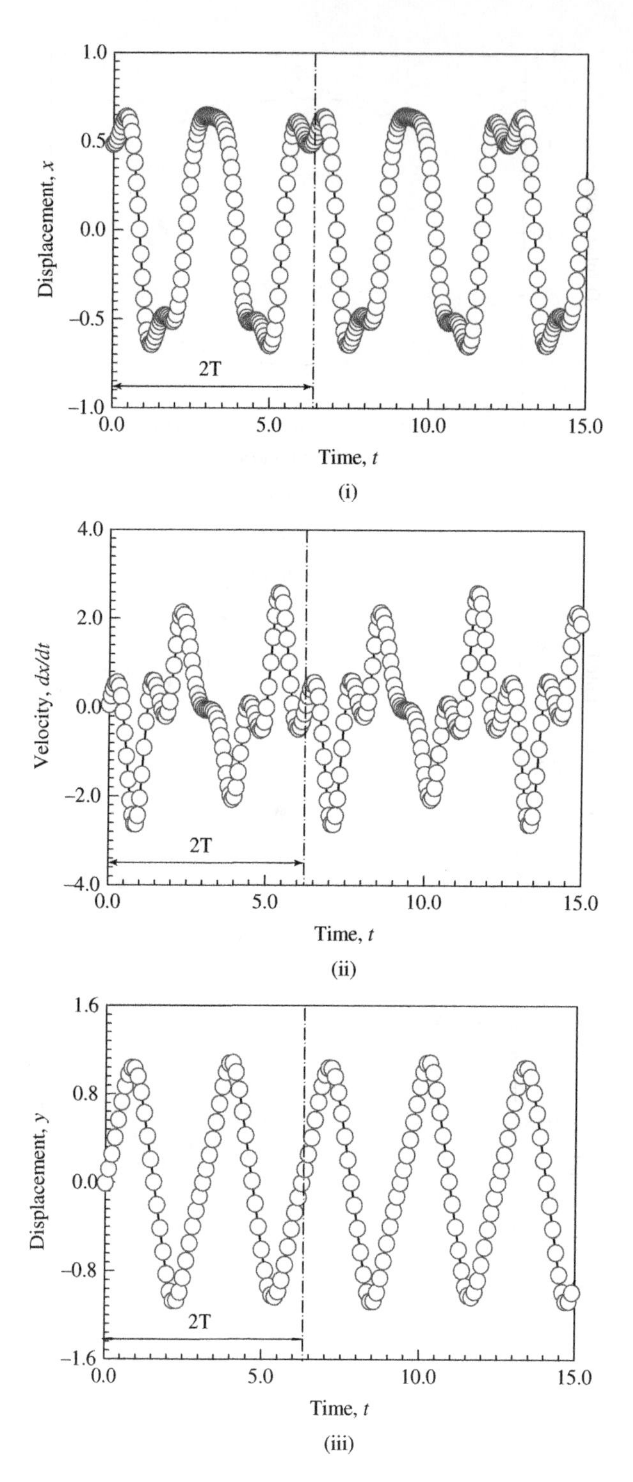

Figure 5.10 Stable period-2 motion of a nonlinear rotor ($\Omega = 2.01$, HB26): (i) x-displacement, (ii) x-velocity, (iii) y-displacement, (iv) y-velocity, (v) x-trajectory, (vi) y-trajectory, (vii) displacement orbit, (viii) x-harmonic amplitude, and (ix) y-harmonic amplitude. Initial conditions $(x_0, \dot{x}_0) = (0.482128, 0.059237)$ and $(y_0, \dot{y}_0) = (-0.010820, 1.208560)$. ($\delta = 0.02$, $\alpha = 0.68$, $\beta = 10$, $\gamma = 1.0$, $e = 1.5$)

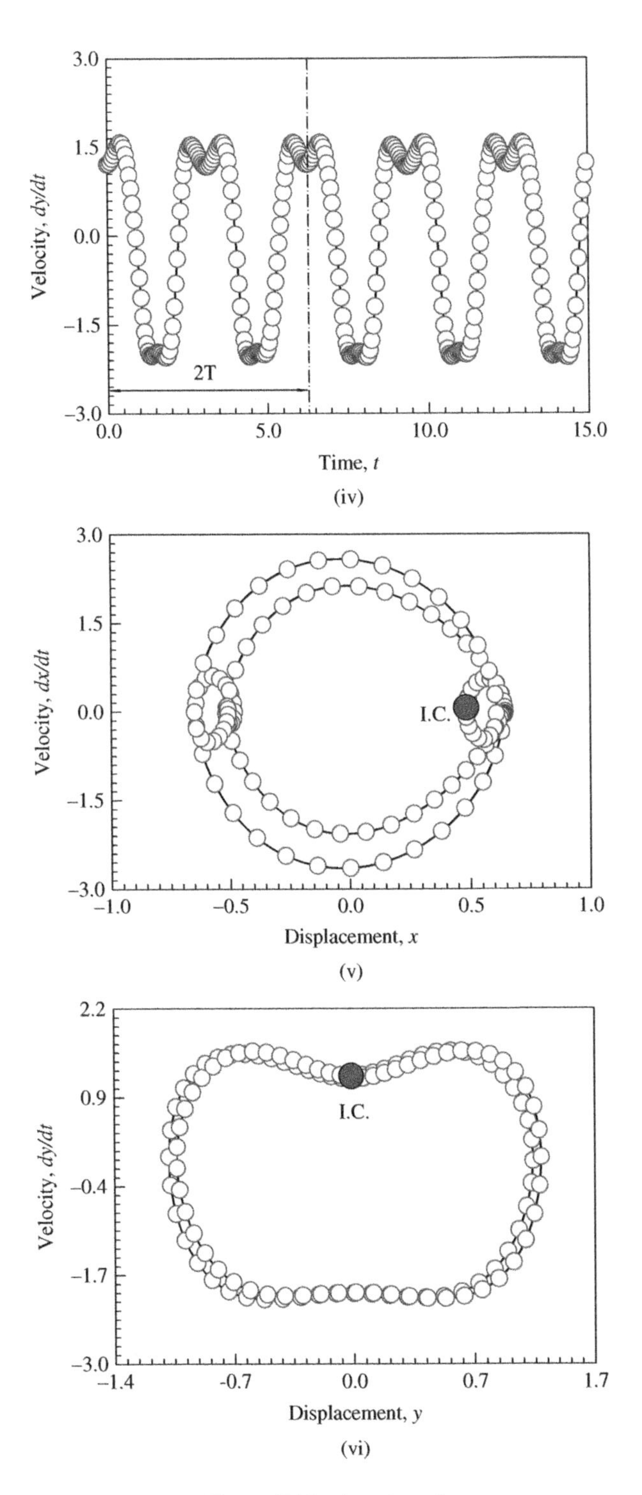

Figure 5.10 *(continued)*

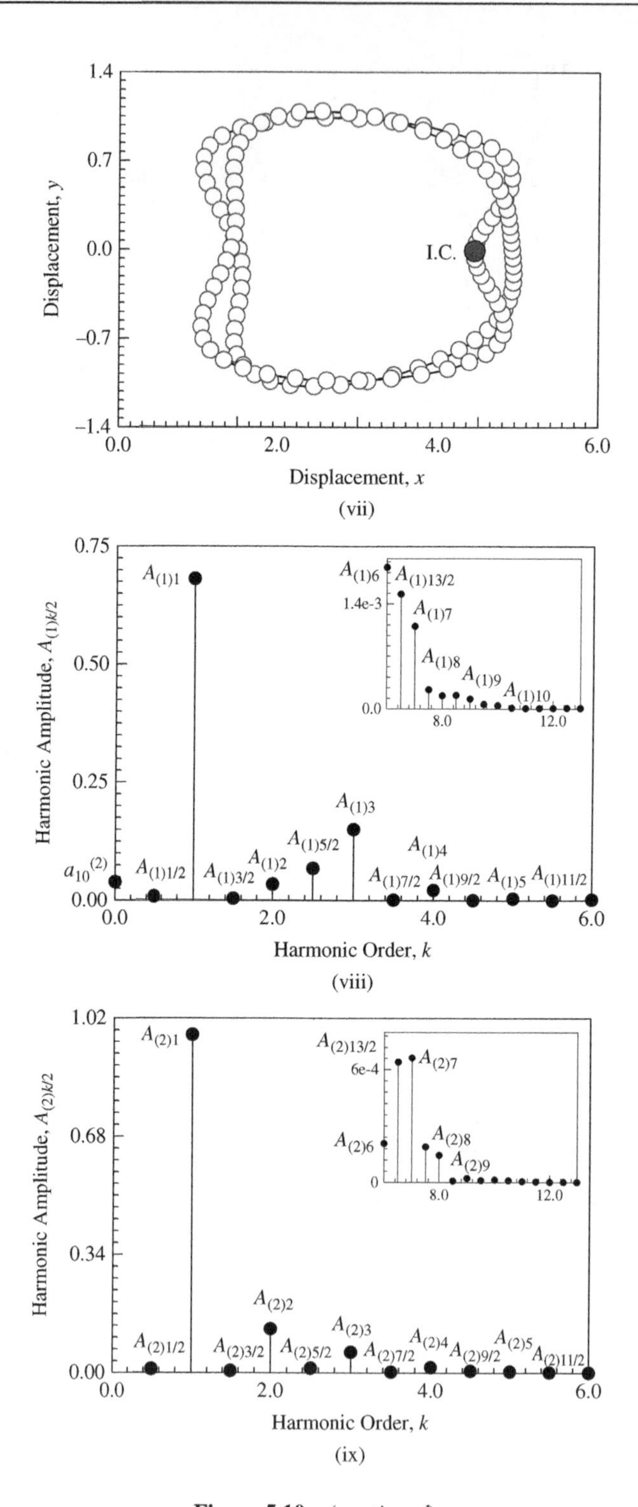

Figure 5.10 *(continued)*

$A_{(1)6} \sim 2\times10^{-4}, A_{(1)13/2} \sim 1.5\times10^{-3}, A_{(1)7} \sim 10^{-3}, A_{k/2} \in (10^{-6}, 10^{-4})$ $(k = 16, 17, \ldots 26)$. However, the main harmonic amplitudes in the y-direction are $a_{20} \approx -1.3\times10^{-3}$, $A_{(2)1} \approx 0.972574$, $A_{(2)2} \approx 0.126123$, $A_{(2)3} \approx 0.058972$, and $A_{(2)4} \approx 0.015588$. However, $A_{(2)1/2} \approx 0.010661$, $A_{(2)3/2} \approx 5.50\times10^{-3}$, $A_{(2)5/2} \approx 0.012854$, $A_{(2)7/2} \approx 2.56\times10^{-4}$. The other harmonic amplitudes in the y-direction are $A_{(2)9/2} \sim 5.39\times10^{-3}$, $A_{(2)5} \sim 3.26\times10^{-3}$, $A_{(2)11/2} \sim 9.65\times10^{-4}$, $A_{(2)6} \sim 2.06\times10^{-4}$, $A_{(2)13/2} \sim 6.4\times10^{-4}$, $A_{(2)7} \sim 6.6\times10^{-4}$, $A_{(2)k/2} \in (10^{-7}, 10^{-4})$ $(k = 16, 17, \ldots 26)$.

For the Hopf bifurcation of symmetric period-1 motion at $\Omega = 2.859$, the period-1 motion jumps to the period-5 motion. The symmetric period-5 motion is observed. In Figure 5.11, a period-5 motion based on 55 harmonic terms (HB55) are presented for $\Omega = 2.7$. The time-histories of displacement and velocity in the x-direction of the nonlinear rotor are presented in Figure 5.11(i) and (ii), respectively. Five periods ($5T$) for the period-5 motion is labeled in the two plots. The time-histories of displacement and velocity in the y-direction of the nonlinear rotor are also presented in Figure 5.11(iii) and (iv), respectively. The two trajectories for x and y-directions are presented for over 40 periods in Figure 5.11(v) and (vi), respectively. The initial conditions are marked by large circular symbols and also labeled by I.C. The displacement orbit of rotor in the x and y-directions is presented in Figure 5.11(vii). To show harmonic contributions on the period-5 motion, the harmonic amplitude spectrums of the stable period-5 motion of the nonlinear rotor in x and y-directions are presented in Figure 5.11(viii) and (ix). The harmonic amplitude spectrums of the stable period-5 motion are given by analytical solutions. The main harmonic amplitudes in the x-direction are $A_{(1)1/5} \approx 0.031127$, $A_{(1)3/5} \approx 0.033035$, $A_{(1)1} \approx 0.614421$, $A_{(1)7/5} \approx 0.083210$, $A_{(1)9/5} \approx 0.092028$, $A_{(1)11/5} \approx 0.550803$, $A_{(1)13/5} \approx 0.029438$, $A_{(1)3} \approx 0.025041$, $A_{(1)17/5} \approx 0.066635$, $A_{(1)19/5} \approx 0.047846$, $A_{(1)21/5} \approx 0.012867$, $A_{(1)23/5} \approx 2.76\times10^{-3}$, $A_{(1)5} \approx 0.014469$, and $A_{(1)27/5} \approx 0.014365$. The other harmonic amplitudes in the x-direction are $A_{(1)(2l+1)/5} \in (10^{-5}, 10^{-3})$ $(l = 14, 15, \ldots 26)$. Due to symmetric period-5 motion, $a_{10}^{(5)} \approx 0$ and $A_{(1)2l/5} = 0$ $(l = 1, 2, \ldots)$. The harmonic amplitude spectrums of the stable period-5 motion are given by analytical solutions. The main harmonic amplitudes in the y-direction are $A_{(1)1/5} \approx 0.105302$, $A_{(2)3/5} \approx 0.059754$, $A_{(2)1} \approx 1.152670$, $A_{(2)7/5} \approx 0.073500$, $A_{(2)9/5} \approx 0.416582$, $A_{(2)11/5} \approx 0.045259$, $A_{(2)13/5} \approx 0.050286$, $A_{(2)3} \approx 0.044222$, $A_{(2)17/5} \approx 0.045093$, $A_{(2)19/5} \approx 0.043544$, $A_{(2)21/5} \approx 6.83\times10^{-3}$, $A_{(2)23/5} \approx 0.012223$, $A_{(2)5} \approx 5.42\times10^{-3}$, and $A_{(2)27/5} \approx 0.013007$. The other harmonic amplitudes in the x-direction are $A_{(2)(2l+1)/5} \in (10^{-5}, 10^{-3})$ $(l = 14, 15, \ldots 26)$. Due to symmetric period-5 motion, $a_{10}^{(5)} \approx 0$ and $A_{(1)2l/5} = 0$ $(l = 1, 2, \ldots)$.

From the bifurcation analysis, after some Hopf bifurcations, the period-1 and period-2 motions become quasi-periodic motions. From the unstable period-1 motions, the quasi-periodic motions are presented in Figure 5.12. The input data for numerical simulation is listed in Table 5.1. The analytical solutions of unstable period-1 motions are based on 13 harmonic terms (HB13). The quasi-periodic motion is presented after 1000 periods ($1000T$) from the analytical unstable period-1 motion. The unstable period-1 motion is the central curves of the quasi-periodic motion, which is depicted by the circular symbol.

In Figure 5.12(i)–(iii), displacement orbit with x and y-harmonic spectrum for quasi-periodic motion relative to unstable period-1 motion are presented for $\Omega = 0.573$. The displacement orbit of the quasi-periodic motion is illustrated in Figure 5.12(i). In Figure 5.12(ii), the main harmonic amplitudes of the unstable period-1 motion in the x-direction are $A_{(1)1} \approx 0.225903$,

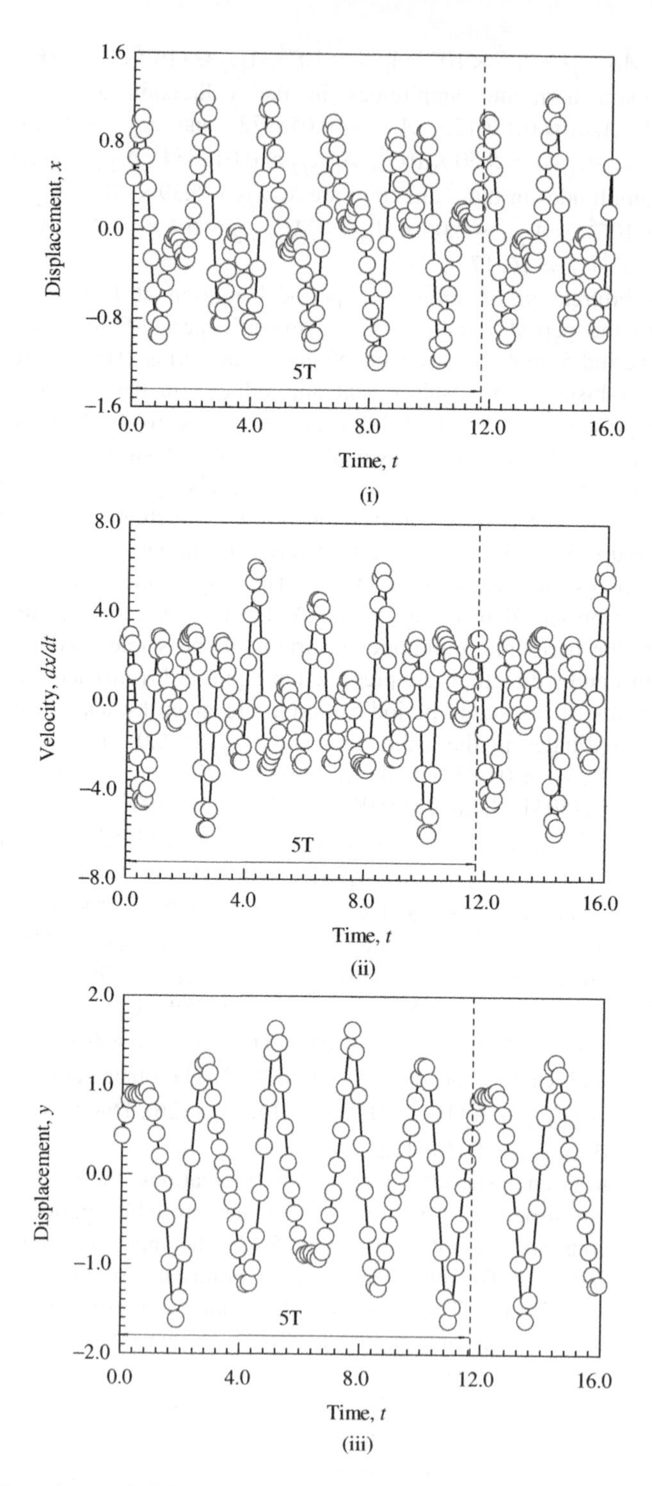

Figure 5.11 Independent period-5 motion for a nonlinear rotor ($\Omega = 2.7$, HB55): (i) x-displacement, (ii) x-velocity; (iii) y-displacement, (iv) y-velocity; (v) x-trajectory and (vi) y-trajectory; (vii) displacement orbit, (viii) x-harmonic amplitude, and (ix) y-harmonic amplitude. Initial condition $(x_0, \dot{x}_0) = (0.463286, 2.668610)$ and $(y_0, \dot{y}_0) = (0.430493, 2.100350)$. ($\delta = 0.02$, $\alpha = 0.68$, $\beta = 10$, $\gamma = 1.0$, $e = 1.5$)

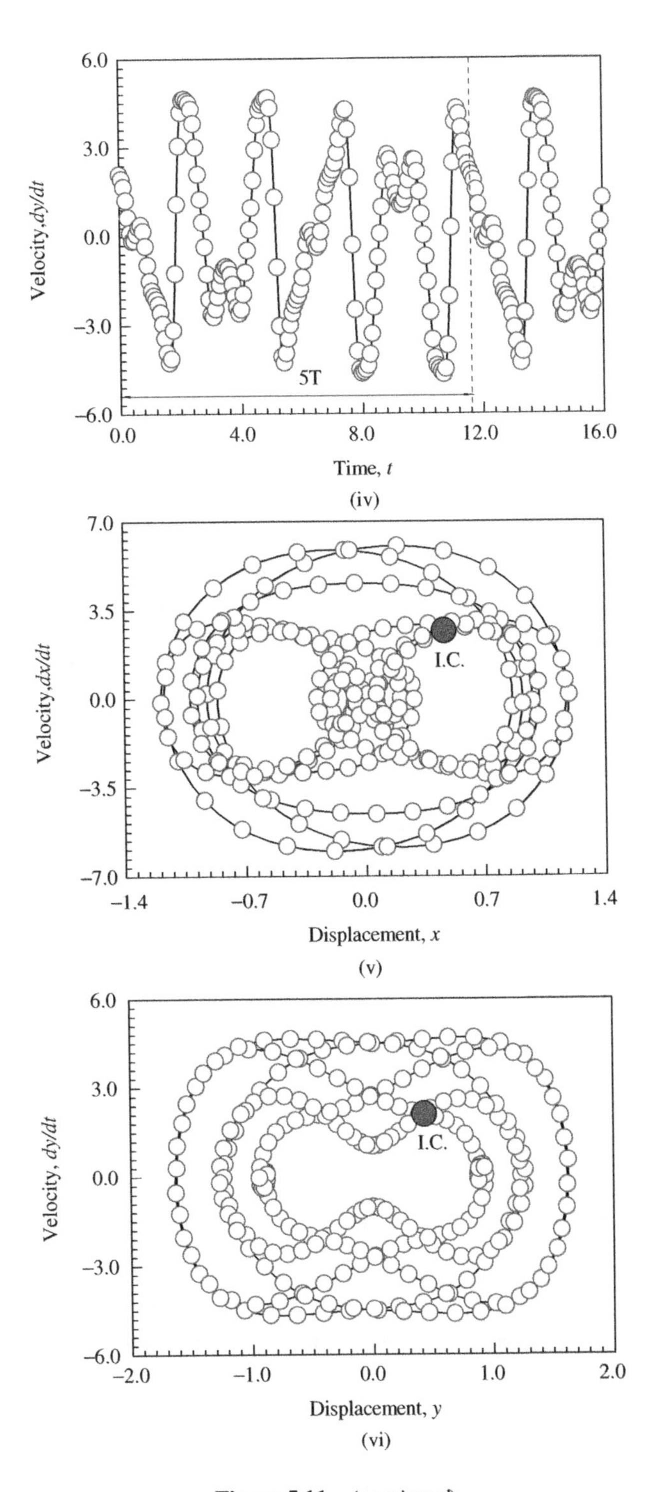

Figure 5.11 *(continued)*

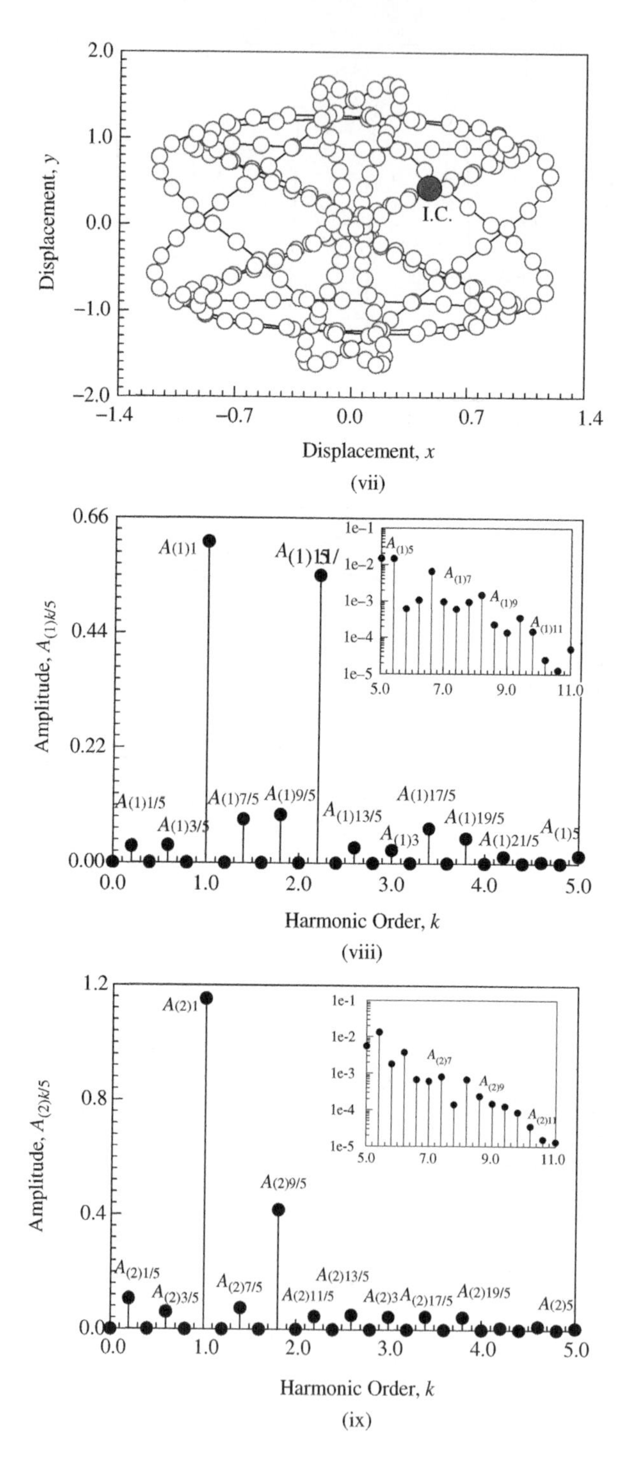

Figure 5.11 (*continued*)

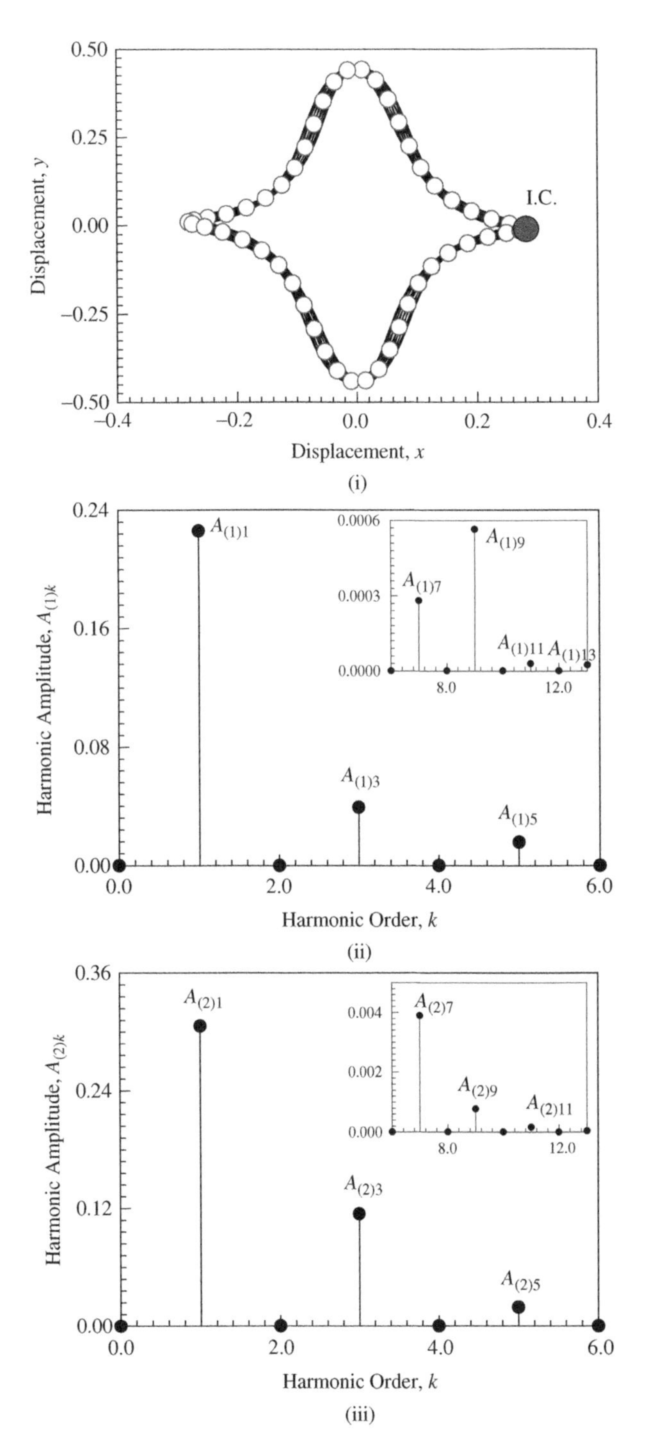

Figure 5.12 Displacement orbit, x-harmonic amplitude, and y-harmonic amplitude of quasi-periodic motions relative to period-1 motion for a nonlinear rotor: (i)–(iii) $\Omega = 0.573$ (HB13); (iv)–(vi) $\Omega = 0.6$ (HB13); (vii)–(ix) $\Omega = 0.896$ (HB13). ($\delta = 0.02$, $\alpha = 0.68$, $\beta = 10$, $\gamma = 1.0$, $e = 1.5$)

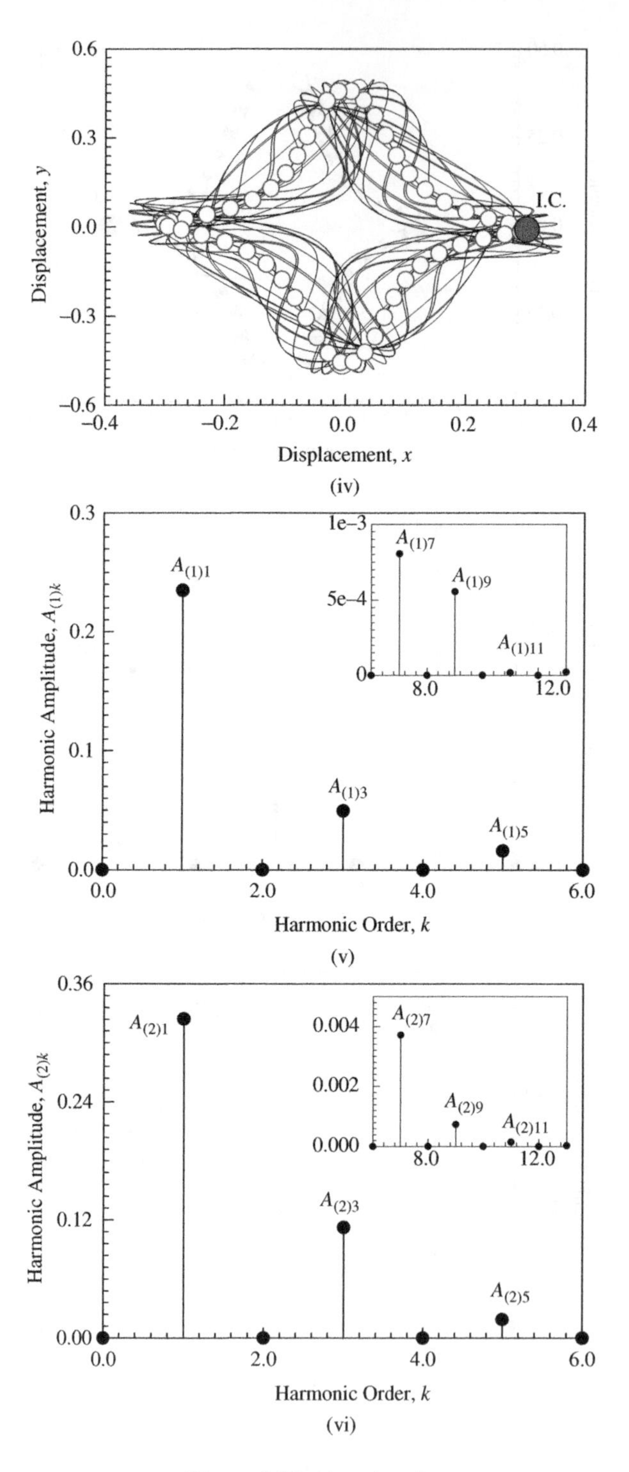

Figure 5.12 (*continued*)

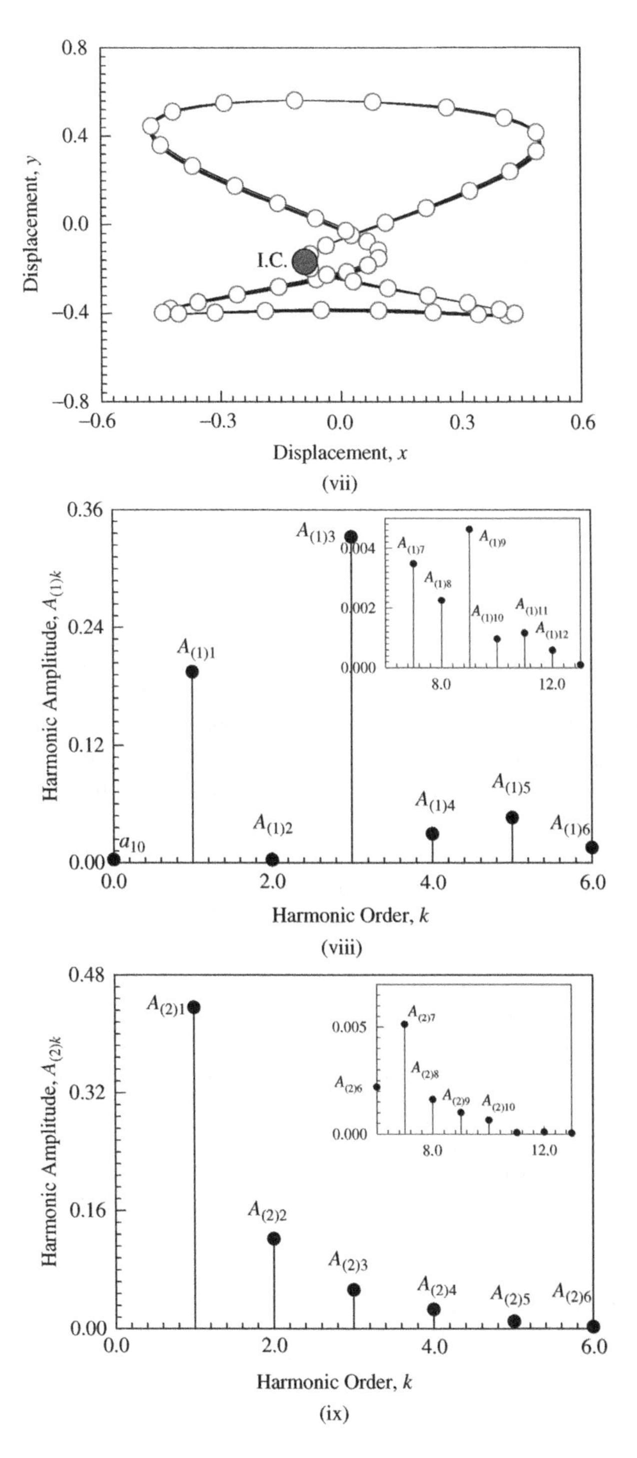

Figure 5.12 *(continued)*

Table 5.1 Input data for numerical simulations of quasi-periodic motion ($\delta = 0.02$, $\alpha = 0.68$, $\beta = 10$, $\gamma = 1.0$, $e = 1.5$).

Ω	x_0	$\dot{x}_0$	y_0	$\dot{y}_0$	Type	Stability
0.573	0.281391	6.698350e-3	−9.653220e-3	0.021055	P-1 (HB13)	Unstable
0.6	0.301636	9.452720e-3	−8.685960e-3	0.036671	P-1 (HB13)	Unstable
0.896	−0.089768	4.633820e-4	−0.167986	0.226290	P-1 (HB13)	Unstable

$A_{(1)2} \approx 0.181789$, and $A_{(1)4} \approx 0.056185$. The other harmonic amplitudes of the unstable period-1 motion in the x-direction are $A_{(1)2l+1} \in (10^{-6}, 10^{-3})$ $(l = 3, 4, \ldots, 6)$. Since the unstable period-1 motion is symmetric, $a_{10} \approx 0$, and $A_{(1)2l} = 0$ $(l = 1, 2, \ldots)$ are obtained. However, in Figure 5.12(iii), the main harmonic amplitudes of the stable period-1 motion in the y-direction are $A_{(2)1} \approx 0.306048$, $A_{(2)3} \approx 0.114292$, and $A_{(2)5} \approx 0.019007$. The other harmonic amplitudes in the y-direction are $A_{(2)2l+1} \in (10^{-6}, 10^{-3})$ $(l = 3, 4, \ldots, 6)$. Since the unstable period-1 motion is symmetric, $a_{20} \approx 0$, and $A_{(2)2l} = 0$ $(l = 1, 2, \ldots)$ are obtained.

The displacement orbit with x and y-harmonic spectrum for quasi-periodic motion relative to the unstable period-1 motion are presented in Figure 5.12(iv)–(vi) for $\Omega = 0.6$. The displacement orbit of the quasi-periodic motion is shown in Figure 5.12(iv). The analytical solution of the unstable period-1 motion is also presented by circular symbols, which is the central curve of the quasi-periodic motion. Such a quasi-periodic motion experiences a large amplitude, compared to the previous quasi-periodic motion for $\Omega = 0.573$. In Figure 5.12(v), the main harmonic amplitudes of the unstable period-1 motion in the x-direction are $A_{(1)1} \approx 0.235004$, $A_{(1)2} \approx 0.049452$, and $A_{(1)4} \approx 0.016000$. The other harmonic amplitudes of the unstable period-1 motion in the x-direction are $A_{(1)2l+1} \in (10^{-5}, 10^{-4})$ $(l = 3, 4, \ldots, 6)$. Since the unstable period-1 motion is also symmetric, $a_{10} \approx 0$, and $A_{(1)2l} = 0$ $(l = 1, 2, \ldots)$ are obtained. In Figure 5.12(vi), the main harmonic amplitudes of the stable period-1 motion in the y-direction are $A_{(2)1} \approx 0.324243$, $A_{(2)3} \approx 0.112356$, and $A_{(2)5} \approx 0.018902$. The other harmonic amplitudes in the y-direction are $A_{(2)2l+1} \in (10^{-5}, 10^{-3})$ $(l = 3, 4, \ldots, 6)$. Since the unstable period-1 motion is symmetric, $a_{20} \approx 0$, and $A_{(2)2l} = 0$ $(l = 1, 2, \ldots)$ are obtained.

The displacement orbit with x and y-harmonic spectrum for a quasi-periodic motion relative to the unstable, asymmetric period-1 motion are presented in Figure 5.12(vii)–(ix) for $\Omega = 0.896$. The displacement orbit of the quasi-periodic motion is presented in Figure 5.12(vii), and the unstable asymmetric period-1 motion is also presented via the circular symbols, which is still the central curve of the quasi-periodic motion. The x and $y-$harmonic spectrums of the unstable period-1 motion are presented in Figure 5.12(viii) and (ix). The main harmonic amplitudes of the second stable period-1 motions of the nonlinear rotor in the x-direction are $a_{10} \approx 3.24 \times 10^{-3}$, $A_{(1)1} \approx 0.194586$, $A_{(1)2} \approx 3.00 \times 10^{-3}$, $A_{(1)3} \approx 0.332163$, $A_{(1)4} \approx 0.029424$, $A_{(1)5} \approx 0.045866$, and $A_{(1)6} \approx 0.015542$. The other harmonic amplitudes in the x-direction are $A_{(1)k} \in (10^{-6}, 10^{-3})$ $(k = 7, 8, \ldots, 13)$. However, the main harmonic amplitudes of the unstable period-1 motion in the y-direction are $a_{20} \approx -0.060800$, $A_{(2)1} \approx 0.435707$, $A_{(2)2} \approx 0.121917$, $A_{(2)3} \approx 0.052810$, $A_{(2)4} \approx 0.025832$, $A_{(2)5} \approx 9.56 \times 10^{-3}$, and $A_{(2)6} \approx 2.81 \times 10^{-3}$. The other harmonic amplitudes in the y-direction are $A_{(2)k} \in (10^{-5}, 10^{-3})$ $(k = 7, 8, \ldots, 13)$. The quasi-periodic motion should be further investigated from the theory of analytical quasi-periodic motions in nonlinear dynamical systems in Luo (2014).

References

Begg, I.G. (1974) Friction induced rotor whirl-A study in stability. *ASME Journal of Engineering for Industry*, **96**(2), 450–453.

Birkhoff, G.D. (1913) Proof of Poincare's geometric theorem. *Transactions on American Mathematical Society*, **14**, 14–22.

Birkhoff, G.D. (1927) *Dynamical Systems*, American Mathematical Society, New York.

Bogoliubov, N. and Mitropolsky, Y. (1961) *Asymptotic Methods in the Theory of Nonlinear Oscillations*, Gordon and Breach, New York.

Buonomo, A. (1998a) On the periodic solution of the van der Pol equation for the small damping parameter. *International Journal of Circuit Theory and Applications*, **26**, 39–52.

Buonomo, A. (1998b) The periodic solution of van der Pol's equation. *SIAM Journal on Applied Mathematics*, **59**, 156–171.

Cartwright, M.L. and Littlewood, J.E. (1945) On nonlinear differential equations of the second order I. The equation $\ddot{y} - k(1 - y^2)\dot{y} + y = b\lambda k \cos(\lambda t + \alpha)$, k large. *Journal of London Mathematical Society*, **20**, 180–189.

Childs, D.W. (1982) Fractional-frequency rotor motion due to nonsymmetric clearance effects. *ASME Journal of Energy and Power*, **104**, 533–541.

Choi, Y.S. and Noah, S.T. (1987) Nonlinear Steady-state responses of a rotor – support system. *ASME Journal of Vibration Acoustics, Stress, and Reliability in Design*, **109**, 255–261.

Choi, S.K. and Noah, S.T. (1994) Mode-locking and chaos in a Jeffcott rotor with bearing clearances. *ASME Journal of Applied Mechanics*, **61**, 131–138.

Chu, F. and Zhang, Z. (1998) Bifurcation and chaos in a rub-impact Jeffcott with bearing effects. *Journal of Sound and Vibration*, **210**, 1–18.

Coppola, V.T. and Rand, R.H. (1990) Averaging using elliptic functions: approximation of limit cycle. *Acta Mechanica*, **81**, 125–142.

Day, W.B. (1987) Asymmetric expansion in nonlinear rotordynamics. *Quarterly of Applied Mathematics*, **44**, 779–792.

Duffing, G. (1918) Erzwunge schweingungen bei veranderlicher eigenfrequenz, F. Viewig u. Sohu, Branunschweig.

Ehrich, F.F. (1988) Higher order subharmonic responses of high speed rotors in bearing clearance. *ASME Journal of Vibration Acoustics, Stress, and Reliability in Design*, **110**, 9–16.

Fatou, P. (1928) Suré le mouvement d'un systeme soumis á des forces a courte periode. *Bulletin de la Société Mathématique*, **56**, 98–139.

Hayashi, C. (1964) *Nonlinear Oscillations in Physical Systems*, McGraw-Hill Book Company, New York.

Analytical Routes to Chaos in Nonlinear Engineering, First Edition. Albert C. J. Luo.

Holmes, P.J. (1979) A nonlinear oscillator with strange attractor. *Philosophical Transactions of the Royal Society A*, **292**, 419–448.

Holmes, P.J. and Rand, D.A. (1976) Bifurcations of Duffing equation; An application of catastrophe theory. *Quarterly Applied Mathematics*, **35**, 495–509.

Hsu, C.S. (1963) On the parametric excitation of a dynamics system having multiple degrees of freedom. *ASME Journal of Applied Mechanics*, **30**, 369–372.

Hsu, C.S. (1965) Further results on parametric excitation of a dynamics system. *ASME Journal of Applied Mechanics*, **32**, 373–377.

Huang, J. and Luo, A.C.J. (2014) Analytical periodic motions and bifurcations in a nonlinear rotor system. *International Journal of Dynamics and Control*, in press.

Jeffcott, H.H. (1919) The lateral vibration loaded shaft in the neighborhood of a whirling speed-the effect of want of balance. *Philosophical Magazine*, **37**(6), 304–314.

Jiang, J. and Ulbrich, H. (2001) Stability analysis of sliding whirl in a nonlinear Jeffcott rotor with cross-coupling stiffness coefficients. *Nonlinear Dynamics*, **24**, 269–283.

Kao, Y.H., Wang, C.S. and Yang, T.H. (1992) Influences of harmonic coupling on bifurcations in Duffing oscillator with bounded potential wells. *Journal of Sound and Vibration*, **159**, 13–21.

Kim, Y.B. and Noah, Y.B. (1990) Bifurcation analysis for a modified Jeffcott rotor with bearing clearances. *Nonlinear Dynamics*, **1**, 221–241.

Kim, Y.B. and Noah, Y.B. (1996) Quasi-periodic response and stability analysis for a nonlinear Jeffcott rotor. *Journal of Sound and Vibration*, **190**, 239–253.

Kovacic, I. and Mickens, R.E. (2012) A generalized van der Pol type oscillator: investigation of the properties of a limit cycle. *Mathematical and Computer Modelling*, **55**, 645–655.

Krylov, N.M. and Bogolyubov, N.N. (1935) *Methodes approchees de la mecanique non-lineaire dans leurs application a l'Aeetude de la perturbation des mouvements periodiques de divers phenomenes de resonance s'y rapportant*, Academie des Sciences d'Ukraine, Kiev, (in French).

Lagrange, J.L. (1788) *Mecanique Analytique*, 2 Vol., Edition Albert Balnchard, Paris.

Leung, A.Y.T. and Guo, Z. (2012) Bifurcation of the periodic motions in nonlinear delayed oscillators. *Journal of Vibration and Control*. Doi: 10.1177/1077546312464988

Levinson, N. (1948) A simple second order differential equation with singular motions. *Proceedings of the National Academy of Sciences of the United States of America*, **34**(1), 13–15.

Levinson, N. (1949) A second order differential equation with singular solutions. *Annals of Mathematics, Second Series*, **50**(1), 127–153.

Luo, A.C.J. (2004) Chaotic motion in the resonant separatrix bands of a Mathieu-Duffing oscillator with twin-well potential. *Journal of Sound and Vibration*, **273**, 653–666.

Luo, A.C.J. (2012a) *Continuous Dynamical Systems*, Higher Education Press/L&H Scientific, Beijing/Glen Carbon.

Luo, A.C.J. (2012b) *Regularity and Complexity in Dynamical Systems*, Springer, New York.

Luo, A.C.J. (2013) Analytical solutions for nonlinear dynamical systems with/without time-delay. *International Journal of Dynamics and Control*, **1**, 330–359.

Luo, A.C.J. (2014) *Toward Analytical Chaos in Nonlinear Systems*, John Wiley & Sons, Inc., Hoboken, NJ.

Luo, A.C.J. and Han, R.P.S. (1997) A quantitative stability and bifurcation analyses of a generalized Duffing oscillator with strong nonlinearity. *Journal of the Franklin Institute*, **334B**, 447–459.

Luo, A.C.J. and Han, R.P.S. (1999) Analytical predictions of chaos in a nonlinear rod. *Journal of Sound and Vibration*, **227**(3), 523–544.

Luo, A.C.J. and Huang, J. (2012a) Approximate solutions of periodic motions in nonlinear systems via a generalized harmonic balance. *Journal of Vibration and Control*, **18**, 1661–1671.

Luo, A.C.J. and Huang, J.Z. (2012b) Analytical dynamics of period-m flows and chaos in nonlinear Systems. *International Journal of Bifurcation and Chaos*, **22**(4), Article No: 1250093 (29 pp).

Luo, A.C.J. and Huang, J.Z. (2012c) Analytical routes of period-1 motions to chaos in a periodically forced Duffing oscillator with a twin-well potential. *Journal of Applied Nonlinear Dynamics*, **1**, 73–108.

Luo, A.C.J. and Huang, J.Z. (2012d) Unstable and stable period-*m* motions in a twin-well potential Duffing oscillator. *Discontinuity, Nonlinearity and Complexity*, **1**, 113–145.

Luo, A.C.J. and Huang, J.Z. (2013a) Analytical solutions for asymmetric periodic motions to chaos in a hardening Duffing oscillator. *Nonlinear Dynamics*, **72**, 417–438.

Luo, A.C.J. and Huang, J.Z. (2013b), Analytical period-3 motions to chaos in a hardening Duffing oscillator. *Nonlinear Dynamics*, **73**, 1905–1932.

Luo, A.C.J. and Huang, J.Z. (2013c) An analytical prediction of period-1 motions to chaos in a softening Duffing oscillator. *International Journal of Bifurcation and Chaos*, **23**(5), Article No: 1350086 (31 pages).

Luo, A.C.J. and Huang, J.Z. (2014a) Period-3 motions to chaos in a softening Duffing oscillator. *International Journal of Bifurcation and Chaos*, **24**(3), Article No: 1430010 (26 pages).

Luo, A.C.J. and Huang, J.Z. (2014b) Analytical periodic motions and bifurcation in a nonlinear rotor system. *International Journal of Dynamics and Control*, Doi: 10.1007/s4043-014-0058-9.

Luo, A.C.J. and Lakeh, A.B. (2013a), Analytical solutions for period-m motions in a periodically forced van der Pol oscillator, *International Journal of Dynamics and Control*, **1**, 99–115.

Luo, A.C.J. and Lakeh, A.B. (2013b) Period-m motions and bifurcation trees in a periodically forced, van der Pol–Duffing oscillator. *International Journal of Dynamics and Control*, Doi: 10.1007/s40435-013-0058-9.

Luo, A.C.J. and O'Connor, D. (2014) On periodic motions in parametric hardening Duffing oscillator. *International Journal of Bifurcation and Chaos*, **24**(1), Article No. 1430004 (17 pages).

Luo, A.C.J. and Yu, B. (2013a) Analytical solutions for stable and unstable period-1 motion in a periodically forced oscillator with quadratic nonlinearity. *ASME Journal of Vibration and Acoustics*, **135**, Article No: 034503 (5 pp).

Luo, A.C.J. and Yu, B. (2013b) Complex period-1 motions in a periodically forced, quadratic nonlinear oscillator. *Journal of Vibration and Control*, in press.

Luo, A.C.J. and Yu, B. (2013c) Period-*m* motions and bifurcation trees in a periodically excited, quadratic nonlinear oscillator. *Discontinuity, Nonlinearity and Complexity*, **3**, 263–288.

Luo, A.C.J. and Yu, B. (2013d) Bifurcation trees of periodic motions to chaos in a parametric, quadratic nonlinear oscillator. *International Journal of Bifurcation and Chaos*, in press.

Mathieu, E. (1868) Memoire sur le mouvement vibratoire d'une membrance deforme elliptique. *Journal of Matheematique*, **2**(13), 137–203.

Mathieu, E. (1873) *Cours de Physique Methematique*, Gauthier-Villars, Paris.

McLachlan, N.W. (1947) *Theory and Applications of Mathieu Equations*, Oxford University Press: London.

Minorsky, N. (1962) *Nonlinear Oscillations*, Van Nostrand: New York.

Mond, M., Cederbaum, G., Khan, P.B. and Zarmi, Y. (1993) Stability analysis of non-linear Mathieu equation. *Journal of Sound and Vibration*, **167**(1), 77–89.

Nayfeh, A.H. (1973) *Perturbation Methods*, John Wiley & Sons, Inc.: New York.

Nayfeh, A.H. and Mook, D.T. (1979) *Nonlinear Oscillation*, John Wiley & Sons, Inc., New York.

Peng, Z.K., Lang, Z.Q., Billings, S.A. and Tomlinson, G.R. (2008) Comparisons between harmonic balance and nonlinear output frequency response function in nonlinear system analysis. *Journal of Sound and Vibration*, **311**, 56–73.

Poincare, H. (1890) Sur les equations de la dynamique et le probleme de trios corps. *Acta Mathematica*, **13**, 1–270.

Poincare, H. (1899) *Les Methods Nouvelles de la Mechanique Celeste*, vol. 3, Gauthier-Villars, Paris.

Sevin, E. (1961) On the parametric excitation of pendulum-type vibration absorber. *ASME Journal of Applied Mechanics*, **28**, 330–334.

Shen, J.H., Lin, K.C., Chen, S.H. and Sze, K.Y. (2008) Bifurcation and route-to-chaos analysis for Mathieu-Duffing oscillator by the incremental harmonic balance method. *Nonlinear Dynamics*, **52**, 403–414.

Stoker, J.J. (1950) *Nonlinear Vibrations*, John Wiley & Sons, Inc.: New York.

Tso, W.K. and Caughey, T.K. (1965) Parametric excitation of a nonlinear system. *ASME Journal of Applied Mechanics*, **32**, 899–902.

Ueda, Y. (1980) Explosion of strange attractors exhibited by the Duffing equations. *Annals of the New York Academy of Sciences*, **357**, 422–434.

van der Pol, B. (1920) A theory of the amplitude of free and forced triode vibrations. *Radio Review*, **1**, 701–710, 754–762.

van der Pol, B. and van der Mark, J. (1927) Frequency demultiplication. *Nature*, **120**, 363–364.

Verhulst, F. (1991) *Nonlinear Differential Equations and Dynamical Systems*, Springer, Berlin.

Wang, C.S., Kao, Y.H., Huang, J.C. and Gou, Y.H. (1992) Potential dependence of the bifurcation structure in generalized Duffing oscillators. *Physical Review A*, **45**, 3471–3485.

Whittaker, E.T. (1913) General solution of Mathieu's equation. *Proceedings of Edinburgh Mathematics Society*, **32**, 75–80.

Whittaker, E.T. and Watson, G.N. (1935) *A Course of Modern Analysis*, Cambridge University Press: London.

Zounes, R.S. and Rand, R.H. (2000) Transition curves for the quasi-periodic Mathieu equations. *SIAM Journal of Applied Mathematics*, **58**(4), 1094–1115.

Index

Analytical Routes to Chaos in Nonlinear Engineering, First Edition. Albert C. J. Luo.

Printed and bound by CPI Group (UK) Ltd, Croydon, CR0 4YY
16/04/2025
14658551-0002